Management and Industrial Engineering

Series Editor

J. Paulo Davim, Department of Mechanical Engineering, University of Aveiro, Aveiro, Portugal

This series fosters information exchange and discussion on management and industrial engineering and related aspects, namely global management, organizational development and change, strategic management, lean production, performance management, production management, quality engineering, maintenance management, productivity improvement, materials management, human resource management, workforce behavior, innovation and change, technological and organizational flexibility, self-directed work teams, knowledge management, organizational learning, learning organizations, entrepreneurship, sustainable management, etc. The series provides discussion and the exchange of information on principles, strategies, models, techniques, methodologies and applications of management and industrial engineering in the field of the different types of organizational activities. It aims to communicate the latest developments and thinking in what concerns the latest research activity relating to new organizational challenges and changes world-wide. Contributions to this book series are welcome on all subjects related with management and industrial engineering. To submit a proposal or request further information, please contact Professor J. Paulo Davim, Book Series Editor, pdavim@ua.pt

Alina M. Waite · Carolina Machado ·
J. Paulo Davim
Editors

Leading in Industry 5.0

Springer

Editors
Alina M. Waite
Department of Management
Information Systems, and Business
Education
Scott College of Business
Indiana State University
Terre Haute, IN, USA

J. Paulo Davim
Department of Mechanical Engineering
University of Aveiro
Aveiro, Portugal

Carolina Machado
Department of Management
School of Economics
Management and Political Science
University of Minho
Braga, Portugal

Interdisciplinary Centre of Social Sciences
(CICS.NOVA.UMinho)
Braga, Portugal

ISSN 2365-0532 ISSN 2365-0540 (electronic)
Management and Industrial Engineering
ISBN 978-3-032-19206-6 ISBN 978-3-032-19207-3 (eBook)
https://doi.org/10.1007/978-3-032-19207-3

This Springer imprint is published by the registered company Springer Nature Switzerland AG
The registered company address is: Gewerbestrasse 11, 6330 Cham, Switzerland

Preface

The Fifth Industrial Revolution, known as Industry 5.0, is emerging. Is your organization's leadership ready for Industry 5.0? Conscious of this timely and critical question, this book, titled Leading in Industry 5.0, presents practical guidance and strategic insights for moving beyond the Fourth Industrial Revolution (Industry 4.0) and effectively navigating the complexities of this new, promising era to unlock its full potential.

While Industry 4.0 is characterized by the digital transformation of industry, namely the employment of digital technologies to automate manufacturing and business processes, it has led to unintended consequences. Examples include strains on technology integration, data security and privacy, human safety and well-being, job security, and natural resource management. Industry 5.0, on the other hand, retains the essential hallmarks of Industry 4.0 but aims to mitigate the threats to people, society, and the planet. Industry 5.0 balances the current and future needs of workers and society with sustainable practices; elevates human intelligence; integrates newer, more sophisticated technologies like cognitive artificial intelligence (AI), collaborative robots (Cobots), large language models (LLMs), human interaction and recognition technologies, and industrial smart wearables; and provides personalized, on-demand goods and services.

As society increasingly relies on organizations to solve global challenges, the expectations of leaders have become even more daunting. Leadership roles and responsibilities are multifaceted and rapidly evolving, requiring leadership to adapt in Industry 5.0. Astute leaders proactively respond to opportunities and threats in their dynamic environments, integrate AI and advanced technologies, attract and retain top talent, engage with progressively diverse stakeholders, promote sustainability, and uphold ethical standards.

Moreover, heightened competition, global crises, and future needs necessitate innovation and radical change, convincing leaders to align their organizational mission, strategy, and culture with Industry 5.0 principles. With a clear vision, leaders are poised to adapt their business models and champion fundamental changes in individual behaviors to create value, not just economically, but also socially and environmentally. Strong leadership is crucial, now more than ever, for mastering Industry

5.0. By leveraging mission, strategy, and culture toward human-centricity, sustainability, and resiliency, effective leadership stands as the cornerstone of organizational success, prepared to help shape the future of society.

To this end, expert scholars and practitioners in different disciplines and from various higher educational institutions around the world were invited to share their fresh perspectives and ideas, research, strategies, models, and frameworks and more by way of this book to enlighten its readers on organizational leadership and its pivotal role in Industry 5.0.

The content of *Leading in Industry 5.0* is organized into three parts:

- Understanding Industry 5.0.
- Preparing to Lead in Industry 5.0.
- Effectively Leading in Industry 5.0: A Holistic Approach.

Understanding Industry 5.0

The first part of this book contains three chapters. This part introduces the concept of Industry 5.0 by comparing its emergence with the challenges of Industry 4.0 and establishes the strategic context for effective leadership by examining the three core principles of Industry 5.0 (human-centricity, sustainability, and resiliency) alongside advanced technologies, such as cognitive AI, Cobots, and LLMs. Advances in data processing and technologies are presented, along with other areas that may present challenges and opportunities as Industry 5.0 unfolds.

Chapter "Industry 5.0 and the Psychological Well-Being of Employees: Centering Humanity in the Future of Work", explores the concept of Industry 5.0, highlighting the importance of psychological well-being in the workplace and the human-centered approach to innovation and collaboration with intelligent machines. Chapter "Leading through Emergent Technologies in Industry 5.0", underscores the transition from the automation-focused Industry 4.0 to a human-centric, ethically grounded vision of Industry 5.0, including shifts from structured to unstructured data and adoption of emergent technologies (e.g., agentic AI, LLMs, additive manufacturing, blockchain, digital twins, and autonomous systems) toward human–machine collaboration. Priority is given to ethical governance and sustainable practices to ensure that technological advancements serve both societal and environmental needs. Chapter "Learning Curves: Educational Design Principles for Ethical GAI Adoption in Industry 5.0", discusses the responsible integration of generative AI (GenAI) in an educational context, thereby serving as a testbed for industry and offering insights into its adoption. Emphasis is placed on ethical leadership, human-centered design principles, and continuous learning to enhance human intelligence and capabilities rather than replace them.

Preparing to Lead in Industry 5.0

The second part of this book includes four chapters. This part explores the evolving roles and responsibilities of Industry 5.0 leaders, from fostering innovation and talent management to promoting ethics and sustainability, while examining a variety of leadership styles, including digital, transformational, collective, and convertible. This part also provides insights into key areas for leadership development, including systems thinking and distributed decision-making.

Chapter "Leaders' Roles and Responsibilities in Industry 5.0", considers a reorientation of traditional leadership roles and responsibilities in the context of Industry 5.0, creating environments that enable systems thinking, distributed decision-making, and continuous learning. In this way, technology augments human capabilities and fosters meaningful, cross-functional collaboration to address impending challenges. As such, the nature of leadership in Industry 5.0 is multifaceted and adaptive to traverse dynamic environments effectively. Chapter "Human-Centered Leadership in Industry 5.0: Fostering Inclusivity, Virtual Collaboration, and Emotional Intelligence", explores digital leadership among other approaches required by Industry 5.0, emphasizing emotional and cultural intelligences, creativity, adaptability, innovation, and ethical awareness. Strategic human resource management (SHRM) is crucial in adapting an organization's culture to prioritize inclusivity, diversity, and employee satisfaction, while addressing challenges in virtual environments and AI integration. Chapter "Collective Leadership in Action: A Multi-Level Model for Adaptive and Collaborative Organizations", moves beyond traditional, hierarchical, and individual-centric leadership models to propose the Multi-Dimensional Collective Leadership (MDCL) model, which integrates various leadership theories to emphasize co-creation, co-adaptation, and co-governance across individual, team, and organizational levels. This novel model addresses the complexities of modern, adaptive organizations through collaborative and distributed leadership practices to enhance resilience and effectiveness in dynamic systems. Chapter "Convertible Leadership in Industry 5.0", discusses the Convertible Leadership framework, requiring leaders to adapt between directive, collaborative, coaching, and visionary styles based on situational needs and stakeholder expectations. Based on seven core capabilities, this approach responds to the dynamic demands of Industry 5.0 while maintaining alignment with ethical principles and organizational values.

Effectively Leading in Industry 5.0: A Holistic Approach

The third part of this book has six chapters. This part examines the interdependent relationships between leadership, mission, strategy, and culture, illustrating how their alignment with Industry 5.0 principles can drive transformational change and sustainable value creation. By integrating case studies of successful leadership initiatives with novel theoretical models, it provides a practical roadmap for adapting business practices to technological integration, thereby achieving economic, social, and environmental goals.

Chapter "Leadership Strategy for Building Industry 5.0: Human-centric Societal Organizations", discusses leadership strategies for operationalizing Industry 5.0 in today's volatile, uncertain, complex, and ambiguous (VUCA) environments, harmonizing business operations with societal needs. This chapter illustrates the integration of the Burke-Litwin Model (BLM) and Complexity Leadership Theory (CLT) through a case study of a North American automotive manufacturing plant that successfully transformed its culture to enhance employee engagement, reduce turnover, and improve organizational resilience. Chapter "Shadow Leadership: An Emerging Organizational Influence, Hiding in Plain Sight", discusses the concept of shadow leadership, highlighting its emergence from misaligned agendas between a prominent leader and leadership peers and progression through five stages: descant, departure, dissonance, discord, and destruction. Recognizing early signs of shadow leadership and addressing them promptly help maintain long-term organizational stability and viability. Chapter "Thriving in Industry 5.0: Leadership and Business Models for Value Creation", emphasizes human-centric leadership and ethical technology integration for fostering sustainable and resilient organizational cultures. Walgreens and Sephora are examples of organizations that have implemented inclusive practices, such as hiring individuals with disabilities, fostering diverse teams, and creating supportive work environments, which have enhanced employee engagement, productivity, and retention. Chapter "AI-Driven Leadership: Strategies and Tactics", explores leadership strategies for effectively integrating analytical AI and genAI within organizations, aligning AI integration with organizational values and strategic goals while fostering organizational readiness and a culture of data excellence, ethics, and collaborative practices. Chapter "Leading Workforce Ecosystems in Industry 5.0: A Dynamic Capabilities Playbook", describes the concept of human-centric, sustainable, and resilient workforce ecosystems and the role of dynamic capabilities in leadership, providing structured routines for sensing, seizing, and reconfiguring, which collectively enable leaders to guide their organizations to become resilient, sustainable, and innovative. Last, Chapter "Human-Centric Leadership in the Age of AI: Redefining HRD for Industry 5.0 Transformation", recognizes the vital role that Human Resource Development (HRD) professionals can play in leading the transformation from Industry 4.0 to Industry 5.0. Leveraging frameworks like nested complexity theory, wicked problem constructs, and systems thinking to bridge the gap between technological advancements and human flourishing, HRD professionals are positioned to harness synergies that address the complexities of Industry 5.0, which neither humans nor machines can manage alone.

Leading in Industry 5.0 can benefit a variety of stakeholders, including executives, managers, engineers, human resources professionals, and aspiring leaders who are interested in emerging trends and tasked with navigating the complexities of Industry 5.0 effectively. This book can also serve leadership educators, researchers, and students (at the undergraduate, master's, and doctoral levels) in various disciplines who seek fresh perspectives on organizational leadership and its pivotal role in Industry 5.0.

We sincerely thank the chapter authors of this book for their valuable contributions, bringing forth novel ideas, empirical research, theories, and new models and frameworks for leading in Industry 5.0. We extend our appreciation for their interest and willingness to collaborate on this project, as well as their genuine enthusiasm for leadership in the context of emerging technologies.

The Editors also express their gratitude to the publishing team at Springer for the opportunity to pursue the book *Leading in Industry 5.0* and for their professional support and assistance throughout this project.

Terre Haute, IN, USA — Alina M. Waite
Braga, Portugal — Carolina Machado
Aveiro, Portugal — J. Paulo Davim

Contents

Editors and Contributors

About the Editors

Alina M. Waite is Full Professor at Indiana State University in Indiana, USA. She received her Ph.D. in Education and Human Resource Studies, specializing in Organizational Performance and Change, from Colorado State University; a M.S. in Biomedical Engineering from the New Jersey Institute of Technology; and a B.S. in Electrical Engineering with a Biomedical Engineering Option from the University of New Mexico. Her teaching and research interests include leadership, creativity, innovation, and STEM careers. She has been Member of the Academy of Human Resource Development since 2005 and the Society for Human Resource Management since 2018. She actively supports her profession by serving different international journal outlets as a board member and peer reviewer. She was Associate Editor of the *Advances in Developing Human Resources* journal from 2017 to 2023. She also completed a Fulbright exchange as Specialist in Business Administration in 2017 and continues serving the program as Peer Reviewer. She held various appointments at ISU including Interim Department Chair in 2023, Interim Director of the Ph.D. Consortium Program in 2022, and Faculty Fellow to the Dean's Office in Technology between 2020 and 2022. Prior to joining ISU in 2012, she was Assistant Professor of the OPC Program at Colorado State University. Before entering academia in 2007, she was Director of Research and Development for an international healthcare company specializing in medical devices and innovative technologies and held various other leadership positions over the span of fifteen years while working in Europe and the USA. e-mail: Alina.Waite@indstate.edu

Carolina Machado received her Ph.D. degree in Management Sciences (Organizational and Policies Management area/Human Resources Management) from the University of Minho in 1999, a Master's degree in Management (Strategic Human Resource Management) from the Technical University of Lisbon in 1994, and a Degree in Business Administration from the University of Minho in 1989. Teaching in Human Resources Management subjects since 1989 at University of Minho, she is since 2004 Associate Professor (with Habilitation since 2022), with experience and research interest areas in the field of Human Resource Management, International Human Resource Management, Human Resource Management in SMEs, Training and Development, Emotional Intelligence, Management Change, Knowledge Management, and Management/HRM in the Digital Age/Business Analytics, Industry 5.0, and Higher Education Sustainability. She is Head of the Human Resources Management Work Group at the School of Economics, Management and Political Science at University of Minho, Coordinator of Advanced Training Courses at the Interdisciplinary Centre of Social Sciences, Member of the Interdisciplinary Centre of Social Sciences (CICS.NOVA.UMinho), University of Minho, as well as Chief Editor of the *International Journal of Applied Management Sciences and Engineering* (IJAMSE), Guest Editor of journals, Books Editor and Book Series Editor, as well as Reviewer in different international prestigious journals. In addition, she has also published as both Editor and Co-editor, as well as Author/Co-author, of several books, book chapters, and articles in journals and conferences. e-mail: carolina@eeg.uminho.pt

J. Paulo Davim is Full Professor at the University of Aveiro, Portugal. He is also distinguished as Honorary Professor in several universities/colleges/institutes in China, India, and Spain. He received his Ph.D. in Mechanical Engineering, M.Sc. in Mechanical Engineering (materials and manufacturing processes) and Mechanical Engineering degree (5 years), from the University of Porto (FEUP), the Aggregate title (Full Habilitation) from the University of Coimbra and the D.Sc. (Higher Doctorate) from London Metropolitan University. He is Fellow (FIET) of IET-London and Fellow (FIMF) of IMF-Birmingham and Eur Ing by Engineers Europe FEANI-Brussels. He is also Engenheiro Conselheiro and Specialist Engineer by the Portuguese Order of Engineers with an MBA and Specialist titles in Industrial Management as well as in Metrology. He has more than 35 years of teaching and research experience in Manufacturing, Materials, Mechanical, and Industrial Engineering, with special emphasis in Machining, Tribology, and Surface Engineering. He has also interest in Design, Management, Sustainability, Industry 5.0, Engineering Education and Higher Education for Sustainability. He has guided large numbers of postdoc, Ph.D., and master's students as well as has coordinated and participated in several financed research projects. He has received several scientific awards and honors. He has worked as Evaluator of projects for ERC-European Research Council and other international research agencies as well as Examiner of Ph.D. thesis for many universities in different countries. He is Editor in Chief of several international journals, Book Series Editor and Scientific Advisory for many conferences. e-mail: pdavim@ua.pt

Contributors

Dijo Alexander Lubar College of Business, University of Wisconsin-Milwaukee, Milwaukee, WI, USA

Marina Klačmer Čalopa Department of Organization, University of Zagreb Faculty of Organization and Informatics, Varaždin, Croatia

Cory Campbell Scott College of Business, Indiana State University, Terre Haute, IN, USA

Antonela Devčić Department of Organization, University of Zagreb Faculty of Organization and Informatics, Varaždin, Croatia

Lisa Dinh-Trinh Department of Educational Administration and Human Resource Development, Texas A&M University, College Station, TX, USA

Ryan Donlan Indiana State University, Terre Haute, IN, USA

Walid El Mansour Department of Educational Administration and Human Resource Development, Texas A&M University, College Station, TX, USA

E. Scott England University of Maryland Eastern Shore, Princess Anne, MD, USA

Violeta Vidaček Hainš Department of Organization, University of Zagreb Faculty of Organization and Informatics, Varaždin, Croatia

Josiah Hangen College of Business and Economics, Anderson University, Anderson, SC, USA

William Hanson College of Business and Economics, Anderson University, Anderson, SC, USA

Tania Hasan Texas A&M University, College Station, TX, USA

Lisa R. Hughes Butler University, Indianapolis, IN, USA

Heesun Kim Department of Educational Administration and Human Resource Development, Texas A&M University, College Station, TX, USA

Jae Eun Kim Department of Vocational Education and Workforce Development, Seoul National University, Seoul, South Korea

Karolina Kokot Department of Organization, University of Zagreb Faculty of Organization and Informatics, Varaždin, Croatia

Chan Lee Department of Vocational Education and Workforce Development, Seoul National University, Seoul, South Korea

Seth-Aaron Martinez College of Education, Boise State University, Boise, ID, USA

Everon Chenhall Maxey College of Business and Economics, Anderson University, Anderson, SC, USA

Jeffrey Moore College of Business and Economics, Anderson University, Anderson, SC, USA

Helen A. M. Muyia University of Washington, Seattle, WA, USA

Fredrick M. Nafukho University of Washington, Seattle, WA, USA

Kristen Palmer Butler University, Indianapolis, IN, USA

Hoda Parvaneh Shirazi Department of Educational Administration and Human Resource Development, Texas A&M University, College Station, TX, USA

John R. Turner Department of Educational Administration and Human Resource Development, Texas A&M University, College Station, TX, USA

Daniel Woldeab College of Individualized Studies, Metropolitan State University, Saint Paul, MN, USA

Xiaoyan Xu Department of Educational Administration and Human Resource Development, Texas A&M University, College Station, TX, USA

Robert M. Yawson School of Business, Quinnipiac University, Hamden, CT, USA

Understanding Industry 5.0

Industry 5.0 and the Psychological Well-Being of Employees: Centering Humanity in the Future of Work

Tania Hasan

Abstract *Industry 5.0 and the Psychological Well-Being of Employees: Centering Humanity in the Future of Work* explores how the next industrial era redefines the relationship between humans and intelligent machines. Moving beyond the efficiency-driven goals of Industry 4.0, Industry 5.0 emphasizes human-centered innovation, resilience, and inclusivity. This chapter situates psychological well-being at the core of future work design, drawing on frameworks such as Ryff's Six-Factor Model, Seligman's PERMA model, and the Job Demands-Resources model to examine autonomy, purpose, engagement, and connectedness in technology-driven workplaces. It highlights both opportunities and risks posed by advanced automation, AI, and human–robot collaboration, from enhancing motivation to creating new stressors and ethical challenges. Practical strategies for organizations include empathetic leadership, flexible work, employee involvement in system design, and proactive mental health support. Case studies demonstrate how global companies are integrating well-being into digital transformation, while ethical concerns such as surveillance, equity, and digital overload remain pressing. By merging insights from organizational psychology, Human Resource Development, and technology studies, this chapter contributes a forward-looking perspective on how Industry 5.0 can foster not only productivity but also dignity, resilience, and flourishing in the workplace.

Keywords Industry 5.0 · Psychological well-being · Human-centered work · Human–robot collaboration · Employee resilience and flourishing

T. Hasan (✉)
Texas A&M University, College Station, TX, USA
e-mail: taniahasan@tamu.edu

A. M. Waite et al. (eds.), *Leading in Industry 5.0*, Management and Industrial Engineering, https://doi.org/10.1007/978-3-032-19207-3_1

1 Introduction

1.1 Background and Rationale

The evolution of industrial revolutions began with mechanization in Industry 1.0. This was followed by electrification and mass production in Industry 2.0, automation and computing in Industry 3.0, and culminated in the digital and cyber-physical systems of Industry 4.0. Each transformation reshaped work and human labor. Every shift brought advancements in productivity and efficiency while changing how workers interact with technology, each other, and their organizations. However, these revolutions often prioritized profit, productivity, and scale over the mental, emotional, and social needs of the workforce.

Now, we are on the brink of Industry 5.0, a movement that signifies a shift in both vision and values. Instead of viewing humans as parts of a larger machine or middlemen between systems, Industry 5.0 focuses on collaboration between humans and intelligent machines. It puts employee well-being at the center of innovation. This approach emphasizes resilience, sustainability, inclusivity, and human-centered design in the workplace. The European Commission (2021) sees Industry 5.0 not merely as an extension of its predecessor but as a chance to rethink the role of people in a more automated and interconnected world.

Given widespread global challenges like pandemics, technological disruption, economic uncertainty, and climate change, employees' psychological well-being is now a central concern, not just an afterthought. Industry 5.0, with its focus on human agency, allows us to rethink work environments in ways that honor mental health, purpose, autonomy, and connectedness. This chapter aims to explore how we can intentionally use Industry 5.0 not just for operational excellence but also for supporting the mental health of those who drive the future of work—human workers.

As automation becomes more intelligent and widespread, issues of employee dignity, motivation, and belonging become increasingly important. Traditional human resource development (HRD) models are being challenged, leading to new approaches that focus on emotional resilience, well-being metrics, and work–life integration. The growing interest in human-centered work and challenges from digital transformation call for a broad examination of what mental well-being means in this fifth industrial age.

1.2 Purpose and Significance of the Chapter

This chapter aims to explore the connection between Industry 5.0 and employees' psychological well-being. It seeks to:

- Present a clear understanding of psychological well-being in evolving workplaces.

- Examine the human-centered aspects of Industry 5.0 and how they differ from the technology-driven goals of Industry 4.0.
- Look into how advanced technologies (e.g., AI, robotics, collaborative machines) can improve or harm employees' psychological well-being.
- Provide practical guidance for HRD professionals, organizational leaders, and policymakers on how to prioritize humanity in Industry 5.0 initiatives.

The importance of this chapter lies in its ability to connect two critical discussions that are usually treated separately: the future of work with technology and the psychological health of workers. Although there has been significant research on how AI and automation change job roles and employment patterns, there is a gap in literature that connects these changes to emotional, psychological, and social effects on employees.

This chapter helps fill that gap by:

- Recasting the technology discussion to include human values.
- Merging psychological theory, organizational behavior, and HRD with the Industry 5.0 vision.
- Offering a future-focused perspective to rethink the ethics, structures, and cultures of work.

This chapter will be useful for:

- HRD professionals and managers tasked with creating well-being initiatives and workplace policies.
- Organizational leaders responsible for integrating technologies while keeping a healthy workforce.
- Researchers looking into the connections between psychology, technology, and human development.
- Policymakers seeking to create inclusive and resilient economies for a post-digital world.

The chapter also aligns with broader societal goals, including the United Nations Sustainable Development Goals (SDGs)—especially Goal 3 (Good Health and Well-being) and Goal 8 (Decent Work and Economic Growth)—by supporting ethical and human-centered innovation.

1.3 Structure and Approach

This chapter is designed to provide both clear concepts and practical relevance. It combines scholarly research with real-world insights and proposes a framework that links technology integration with employee psychological well-being in the context of Industry 5.0.

The chapter is organized as follows.

2 Introduction

Industry 5.0 and the Psychological Well-Being of Employees: Centering Humanity in the Future of Work

In recent years, workplaces have changed dramatically. Organizations are moving from the digitization phase of Industry 4.0 to the human-centered focus of Industry 5.0. This shift is changing operational models and how employees experience work and mental well-being.

2.1 From Industry 4.0 to Industry 5.0: A Paradigm Shift

Industry 4.0 introduced advanced automation, artificial intelligence (AI), and cyber-physical systems to maximize efficiency and productivity (Xu et al., 2018). While this digital shift improved performance, it also increased surveillance, blurred work–life boundaries, and raised levels of technostress among employees (Suh & Lee, 2017).

In contrast, Industry 5.0 signifies a change in values—from optimizing systems to humanizing work. It emphasizes resilience, sustainability, and human-centered innovation (European Commission, 2021). Technologies like collaborative robots (cobots) and intelligent AI systems are designed to enhance human capabilities, not replace them (Nahavandi, 2019). The goal is to position humans as active collaborators in intelligent systems and co-creators of innovation, promoting psychological well-being alongside performance (Demir et al., 2019).

2.2 Defining Psychological Well-Being in the Workplace

Psychological well-being is a complex idea that includes positive functioning, life satisfaction, and emotional resilience. Using frameworks like Ryff's Six-Factor Model (Ryff, 1989), Seligman's PERMA model (Seligman, 2011), and the Job Demands-Resources (JD-R) model (Bakker & Demerouti, 2007), this chapter defines psychological well-being as incorporating autonomy, purpose, engagement, and connectedness.

These frameworks are particularly important in technology-driven workplaces, where employees face greater cognitive demands and less interpersonal interaction. In these settings, well-being relies not only on reducing stress but on fostering flourishing—a state of optimal human functioning and adaptability (Keyes, 2002).

2.3 Psychological Impacts of Industry 5.0 Work Environments

The changing technological workplace has a significant impact on employee psychology. The cognitive strain from constant information, AI notifications, and multitasking can reduce attention and decision-making ability (Mark et al., 2016). The emotional experience of working alongside machines can create mixed feelings, especially when automation threatens one's identity or sense of control (Al-Debei & Al-Lozi, 2014).

Personalized AI systems can boost intrinsic motivation by aligning tasks with employee strengths. However, they also raise issues of surveillance and autonomy (Ajunwa, 2023). Work–life boundaries have weakened due to constant connectivity, leading to burnout, digital fatigue, and a lingering feeling of being "always on" (Derks et al., 2014; Salanova et al., 2012).

2.4 Organizational Strategies to Promote Psychological Well-Being

To ensure that Industry 5.0 enhances human potential rather than diminishes it, organizations must adopt clear strategies grounded in human-centered design. These strategies include:

- Empathetic leadership that encourages psychological safety (Edmondson & Lei, 2014).
- Involving employees in workflow design (Demir et al., 2019).
- Offering flexible work arrangements and accessible mental health support (WHO, 2022).
- Providing ongoing learning that fosters emotional resilience and adaptability (Luthans et al., 2007).

These approaches go beyond wellness programs—they reshape organizational culture, creating environments where well-being and performance thrive together.

2.5 Case Examples and Best Practices

Several global companies are starting to implement Industry 5.0 principles through innovations like well-being dashboards, AI-supported mental health monitoring, and co-designed human–cobot workflows (Maddikunta et al., 2022). These case studies show that pairing technology with a human-focused mindset can improve engagement, satisfaction, and retention.

2.6 Challenges and Ethical Considerations

Despite the potential benefits, Industry 5.0 presents complex ethical challenges. These include:

- The risk of technology overload and emotional detachment.
- The ethical concerns of monitoring employee data (Moore et al., 2018).
- Issues of access and equity, particularly for groups that are digitally marginalized (van Deursen & Helsper, 2015).

To avoid repeating previous mistakes, organizations must create governance frameworks prioritizing consent, transparency, and inclusivity.

2.7 Future Directions and Research Opportunities

This chapter promotes interdisciplinary research at the crossroads of organizational psychology, HRD, and ethical technology design. Emerging studies indicate that emotional intelligence, adaptability, and digital well-being are becoming essential skills for the future (Wu, 2024). There is a need for metrics, case studies, and long-term research to understand the lasting effects of Industry 5.0 on human flourishing.

2.8 Conclusion

Industry 5.0 offers a chance to reclaim the human aspect of work—valuing not just output but also meaning, resilience, and dignity. By linking new technologies with psychological science and ethical principles, organizations can create settings where humans and machines thrive together.

3 From Industry 4.0 to Industry 5.0: A Paradigm Shift

The shift from Industry 4.0 to Industry 5.0 represents more than just a technological upgrade. It marks a fundamental change in how societies view the role of industry in the 21st century. Industry 4.0 mainly focused on automation, digitalization, and productivity gains. In contrast, Industry 5.0 emphasizes human centricity, sustainability, and resilience. It integrates technology in ways that prioritize societal and environmental well-being alongside economic performance (Nahavandi, 2019; European Commission, 2021).

This section breaks down this transformation into four parts: understanding the technological and operational focus of Industry 4.0, exploring the emerging principles of Industry 5.0, identifying key enabling technologies, and redefining the human role in future workplaces.

3.1 Understanding Industry 4.0: Automation and Efficiency

Industry 4.0, also known as the Fourth Industrial Revolution, represents a transformative time in manufacturing and service industries. It is characterized by the integration of advanced digital technologies. This phase combined cyber-physical systems, the Industrial Internet of Things (IIoT), artificial intelligence (AI), cloud computing, and big data analytics, enabling real-time decision-making, improving productivity, and increasing operational agility (Kagermann & Wahlster, 2022; IBM, n.d.; Epicor, 2023; Xu et al., 2018). These interconnected systems allowed machines, devices, and processes to communicate smoothly, leading to self-optimizing production environments that significantly improved efficiency.

The main advantage of Industry 4.0 was its ability to merge interconnectivity, machine learning, and cyber-physical systems to create a smarter and more responsive production ecosystem (Epicor, 2023). For example, real-time data from IIoT sensors facilitated predictive maintenance, which reduced costly downtime, while advanced analytics improved resource allocation. A prominent example is Siemens' Amberg Electronics Plant in Germany, where over 75% of production is automated, and continuous data-driven optimization has minimized production errors, shortened product cycles, and lowered operational costs (Siemens, 2020).

Efficiency gains were also clear in sectors outside manufacturing. Empowered Automation (2022) notes that smart systems and predictive algorithms streamlined supply chains, reduced defects, and allowed organizations to anticipate operational issues before they became serious. However, these technological advancements also brought new challenges. Highly automated environments sometimes reduced worker autonomy, limited opportunities for creative problem-solving, and restricted skill use. For some employees, the shift to more routine, machine-dependent work led to job dissatisfaction and a feeling of alienation.

Critics highlight that Industry 4.0's focus on efficiency often ignored human factors such as creativity, well-being, and long-term workforce resilience (Bonekamp & Sure, 2015). Over-automation sometimes displaced workers, widening the skills gap and creating uncertainty about future job prospects. Additionally, the reliance on just-in-time manufacturing, while effective in reducing inventory costs, exposed global supply chains to significant weaknesses. This was particularly evident during the COVID-19 pandemic, when disruptions spread quickly across industries (Shih, 2020).

In summary, Industry 4.0 transformed production and service delivery by maximizing efficiency and precision. However, its design was mainly technology-focused, with minimal regard for worker psychological well-being or the broader sustainability of operations. These shortcomings have paved the way for Industry 5.0, which aims to balance technological innovation with human-centered values and organizational resilience.

3.2 *Emergence of Industry 5.0: Human-Centric, Resilient, and Sustainable*

The introduction of Industry 5.0 signals a significant change in how we understand and implement industrial innovation. While Industry 4.0—the Fourth Industrial Revolution—transformed manufacturing and services through automation, artificial intelligence (AI), the Internet of Things (IoT), and advanced robotics, it mainly focused on efficiency and productivity as end goals. This relentless drive for technological optimization, although delivering notable gains in output and cost-effectiveness, often neglected the human aspect—the psychological wellness, dignity, and creative potential of workers.

In contrast, Industry 5.0 represents a conscious shift, moving beyond mere economic measures to prioritize human welfare, societal resilience, and environmental sustainability. The European Commission's key report, Industry 5.0: Towards a Sustainable, Human-Centric, and Resilient European Industry (2021), outlines this vision. It emphasizes that productivity and profitability should coexist with worker well-being, social inclusivity, and ecological responsibility. The focus is no longer strictly on creating faster or cheaper systems; it is about crafting workplaces and technologies that enhance human abilities, promote resilience, and respect the environment.

3.2.1 From Shareholder to Stakeholder Capitalism

A notable ideological shift with Industry 5.0 is the transition from shareholder capitalism—prioritizing returns for investors—to stakeholder capitalism, which addresses the needs and well-being of everyone affected by industrial activity: employees, customers, communities, and the environment. This change reflects broader societal trends, as seen in the Business Roundtable's 2019 redefinition of corporate purpose, which recognized commitments to customers, employees, suppliers, and communities alongside shareholders.

In practice, this reorientation requires organizations to measure success not just through quarterly profits but also by considering three bottom-line metrics: people, planet, and profit. For example, companies increasingly incorporate Environmental, Social, and Governance (ESG) criteria into their strategic planning. Many adopt frameworks like the United Nations' Sustainable Development Goals (SDGs) to guide operations (United Nations, 2015). Industry 5.0 aligns seamlessly with these models by placing human-centered innovation at the core of sustainable industrial growth.

3.2.2 Human Centricity: Technology as a Servant, not a Master

In the Industry 4.0 framework, technological advancements often dictated how work was done and how workers interacted. Automation and robotics were sometimes introduced mainly to replace human labor, especially in repetitive or dangerous jobs. While this did increase efficiency, it also raised concerns about job loss, decreased skill use, and loss of human agency (Acemoglu & Restrepo, 2019).

Industry 5.0 alters this relationship by viewing technology as a support for human creativity and well-being. Instead of requiring humans to adapt to machines, Industry 5.0 focuses on creating adaptive technologies that respond to human needs, preferences, and skills. This means developing tools and systems that enhance human decision-making, emotional intelligence, and craftsmanship rather than replacing them.

For example, Toyota's use of collaborative robots (cobots) illustrates this principle in action. On Toyota's assembly lines, cobots help workers with physically challenging tasks, such as lifting heavy parts or repetitive overhead movements. These cobots do not replace human workers; they reduce physical strain, lower injury rates, and let employees focus on higher-value tasks needing precision, problem-solving, and creativity (Toyota, 2025). The outcome is a partnership between humans and machines that benefits both productivity and worker well-being.

Similarly, Siemens has invested in AI-driven decision-support systems in its manufacturing plants. These systems enable frontline employees to access real-time data analytics without needing advanced technical skills. As a result, workers can make informed, strategic decisions that improve production quality while also gaining greater autonomy and a sense of professional mastery (Siemens AG & University of Göttingen, 2021).

3.2.3 Resilience: Designing for Disruption

The COVID-19 pandemic, climate challenges, and geopolitical tensions have highlighted the need for resilient industrial systems that can adapt quickly to unexpected events. In Industry 5.0, resilience goes beyond just supply chain agility. It includes psychological resilience among workers, the flexibility of organizations, and the ability to maintain operations without compromising safety, well-being, or environmental responsibility.

A prime example is Unilever's global manufacturing network, which showed remarkable adaptability during the COVID-19 pandemic. By using digital twins—virtual replicas of production processes—Unilever could quickly reconfigure operations, handle disruptions, and keep producing essential goods. At the same time, the company prioritized employee safety by implementing remote monitoring systems, rotating shifts to reduce on-site density, and offering mental health support for employees facing pandemic stress (Unilever, 2025).

Another example is Bosch's "Resilience by Design" strategy. This approach incorporates redundancy, flexible manufacturing cells, and cross-training of employees to ensure production continues even when some resources or personnel are unavailable. Bosch also invests heavily in developing employee skills, ensuring its workforce is prepared for rapid technological changes—a crucial factor for both organizational and personal resilience (Bosch, 2022).

3.2.4 Sustainability: Operating Within Planetary Boundaries

Sustainability is a key element of Industry 5.0, calling for industrial processes to function within the planet's ecological limits. This means reducing greenhouse gas emissions, minimizing waste, conserving water, and adopting circular economy models where resources are reused and recycled rather than discarded (Geissdoerfer et al., 2018).

Unilever's sustainable factories are an example of this commitment. The company uses AI-driven systems to optimize water and energy use, resulting in significant reductions in environmental impact. At its Durban, South Africa facility, Unilever achieved zero waste to landfill by reengineering processes to reuse by-products and recycle materials (Unilever, 2025). Beyond technical changes, Unilever also invests in employee training programs that emphasize sustainable manufacturing practices, making sustainability part of the organizational culture.

Similarly, Philips has adopted a circular economy strategy in its healthcare technology division. It designs medical equipment that can be refurbished, upgraded, and resold rather than discarded. This not only lessens environmental impact but also makes healthcare technology more accessible to underserved communities, aligning sustainability with social equity goals (Philips, 2025).

3.2.5 The Interplay Between Human Centricity, Resilience, and Sustainability

While often discussed separately, these three pillars—human centricity, resilience, and sustainability—are interconnected in Industry 5.0. For instance, empowering workers with autonomy and meaningful work (human centricity) enhances their ability to adapt during crises (resilience), which in turn supports adopting sustainable practices.

Take the Nordic steel company SSAB, for example. The shift to fossil-free steel production required both technological innovation and a workforce engaged with sustainability. By involving employees in co-designing production processes and offering reskilling opportunities, SSAB maintained high morale and accelerated adopting environmentally friendly practices (SSAB, 2022). The result was a workforce that not only embraced change but also became advocates for the company's sustainability goals.

3.3 Key Technologies: Cobots, AI, and Personalization

Industry 5.0's human-centered focus is realized through various advanced technologies aimed at enhancing and supporting human labor rather than replacing it. Building on the technological groundwork of Industry 4.0, these innovations stress collaboration, personalization, and resilience, keeping workers at the core of industrial processes.

Collaborative Robots (Cobots) are important tools that work safely and intuitively alongside humans. Unlike traditional industrial robots that require isolation, cobots have safety sensors that stop operation when humans come close, allowing shared workspaces without barriers (Financial Times, 2025). Their small size and cost-effectiveness have driven rapid adoption, especially among small and medium enterprises, where cobots now account for about 11% of industrial robots installed, with projected annual growth exceeding 30% (Financial Times, 2025). For instance, Universal Robots in Denmark employs cobots to assist employees in product customization, enabling mass personalization while maintaining efficiency (Villani et al., 2018).

In Industry 5.0, artificial intelligence (AI) goes beyond traditional uses like predictive maintenance and production scheduling. It emphasizes personalization and worker well-being. AI-powered adaptive systems adjust workflows and environments to fit individual strengths and preferences, such as automatically modifying workstation ergonomics—desk height, lighting, and task complexity—to improve comfort and engagement (LineView, 2023). This personalized approach increases employee satisfaction and productivity by aligning technology with human needs.

Another important tool is Human Digital Twins (HDTs), which are virtual replicas capturing an individual's physical, cognitive, and emotional states. HDTs enable real-time wellness monitoring and provide adaptive feedback, allowing safer and more personalized training and support in complex industrial environments (Bucci et al., 2024). These systems integrate human context into technological design, enabling proactive measures to support well-being.

Extended Reality (XR) technologies—including augmented, virtual, and mixed reality—further improve human–machine interaction by offering immersive training and collaborative design experiences. Airbus, for example, uses XR tools to train assembly line workers in complex procedures, leading to fewer mistakes and better safety outcomes (Airbus, 2023). Advanced human–machine interfaces (HMI) in XR also lessen cognitive load and maintain human agency in complicated tasks, focusing on psychological ergonomics and mental well-being (Yang et al., 2024).

Finally, predictive analytics powered by AI enhances operational resilience. Beyond maintenance of machines, these analytics predict supply chain disruptions and swiftly adjust production schedules to minimize downtime, ensuring smoother and more responsive industrial processes (Ghobakhloo et al., 2021).

Together, these key technologies, cobots, personalized AI, Human Digital Twins, XR, and predictive analytics, represent the main ideas of Industry 5.0: augmentation, empathy, and individual empowerment. They signal a clear move toward industrial innovation focused on human welfare while also improving efficiency and sustainability.

3.4 Repositioning the Human Role in Future Industry

Industry 5.0 marks a significant change that redefines humans' roles in industrial systems. It shifts away from the old model, where workers mainly played a passive role in automated processes. Instead, Industry 5.0 views workers as active co-creators who engage with technology to foster innovation, sustainability, and human welfare. This change goes beyond just a tech upgrade; it represents a cultural and organizational evolution that honors dignity, autonomy, and ongoing skill development.

In earlier industrial models, especially Industry 4.0, automation and digitization boosted efficiency but often reduced workers' control over their tasks and environments. Workers' roles became disjointed, with many relegated to monitoring machines or doing routine tasks. While this model was productive, it ignored essential human qualities—intuition, creativity, and emotional intelligence—that are vital for complex decision-making and innovation. Industry 5.0 aims to bring these unique human traits to the forefront by encouraging collaboration between humans and machines rather than competition or replacement.

The European Commission's Vision and Strategic Framework

The European Commission's 2021 strategy for Industry 5.0 emphasizes worker engagement, well-being, and sustainability along with technological growth. It envisions a future where humans and machines work together, with collaborative robots (cobots) taking on hazardous or repetitive tasks. This collaboration allows workers to concentrate on judgment, creativity, and craftsmanship (European Commission, 2021). The vision supports a work environment that respects human dignity and values continuous learning as essential to industrial resilience.

Cobots illustrate this shift toward a human-centered focus. They are designed to operate safely alongside humans, eliminating the need for protective barriers. Cobots assist workers by taking on tasks that are physically demanding or require precision that humans cannot sustain (Financial Times, 2025). By lightening this load, cobots improve workplace safety and mental health, enabling workers to use their higher-level skills in decision-making.

Empirical Evidence Supporting Human–Machine Collaboration

Research increasingly backs this human-focused approach. Studies on human–cobot teams reveal improved task success rates and lower mental workloads, suggesting that well-designed collaborative automation can enhance both performance and well-being (Fournier et al., 2024). Advances in human–machine interface (HMI) design focus on cognitive efficiency and mental health, emphasizing the need for intuitive, user-friendly environments that reduce stress and maintain human autonomy (Yang et al., 2024).

This evidence promotes a growing consensus: technology should enhance, not undermine, human abilities. Industry 5.0 incorporates this idea by creating spaces where innovation and employee well-being work hand in hand rather than opposing each other.

From Operator to Co-creator: A Paradigm Shift

At the core of Industry 5.0 is a significant transformation—the change of the worker's role from being a passive operator to an active co-creator of value. This shift acknowledges that human intuition, creativity, and emotional intelligence are critical assets that technology cannot mimic. The human role evolves to include strategic decision-making and innovation, supported by technology that enhances human capabilities instead of replacing them (Longo et al., 2020).

Case Example—IKEA's Industry 5.0 Initiative

IKEA exemplifies this transformation. Within its Industry 5.0 framework, cobots are used for precision cutting and repetitive assembly tasks that need consistency and speed but are physically taxing for humans. Skilled artisans focus on designing and applying custom furniture finishes, enabling customers to co-create products that reflect their tastes (Stackpole, 2021). This combination of automation and craftsmanship improves customer satisfaction through personalization and enriches workers' roles by incorporating creativity and craftsmanship into their daily tasks.

The IKEA example showcases several key benefits of this hybrid approach:

- Enhanced Worker Autonomy and Satisfaction: Workers feel a stronger sense of purpose and engagement when involved in creative and decision-making processes.
- Customer Connection and Co-ownership: Involving customers in product customization creates a shared sense of ownership, bridging the gap between production and consumption.
- Innovation Through Human Judgment: Workers' expertise and creativity are central to innovation, especially in customizing products beyond machines' capabilities.

Psychological and Organizational Benefits of the Human-Centered Model

Industry 5.0's reshaping of human roles provides significant psychological and organizational benefits. Fostering worker engagement in decision-making is linked to higher job satisfaction and stronger organizational commitment (Bonekamp & Sure,

2015). When employees believe their contributions matter, motivation and morale rise, resulting in better overall performance.

Additionally, Industry 5.0 encourages continuous skill development, particularly in areas that combine digital literacy and creative problem-solving. This preparation helps workers adapt to rapidly changing industrial environments, improving both personal career opportunities and organizational versatility.

From an innovation standpoint, diverse teams that blend human strengths—like emotional intelligence and systems thinking—with AI's analytical abilities have proved better at solving complex, dynamic challenges. This collaboration speeds up innovation cycles and enhances responsiveness to market changes and technological disruptions.

Importantly, this human-centered model addresses employees' psychological well-being, which research increasingly identifies as crucial for productivity, retention, and innovation capacity (Sakshi & Dhand, 2025). By creating workplaces that prioritize mental health, cognitive ease, and emotional support, Industry 5.0 promotes lasting human performance.

The Worker as Co-designer, Innovator, and Strategic Decision-Maker

In Industry 5.0, workers move beyond traditional roles to become co-designers of processes, products, and organizational strategies. Empowered by technology, they actively shape work environments and outcomes. This empowerment comes from tools like advanced HMIs, digital twins of human workers, and AI systems that adjust workflows and learning experiences to fit individual needs and contexts.

Case Study—IKEA's Customization Platform

IKEA's "DIY Smart Factory" shows how Industry 5.0 puts this vision into action. Customers customize furniture online, while cobots manage standard manufacturing processes, and human artisans make the finishing touches (Stackpole, 2021). This hybrid production model:

- Boosts worker autonomy by involving artisans in creative finishing work.
- Encourages a shared ownership culture by linking customers directly with manufacturing.
- Promotes product innovation through human judgment and customization abilities.

The "DIY Smart Factory" illustrates how Industry 5.0 can achieve efficiency, personalization, and meaningful human involvement simultaneously.

Transforming Skills and Workforce Needs

The shift in the human role in Industry 5.0 requires a new set of skills. Workers need to develop:

- Technical Skills: Operating cobots, managing AI systems, and understanding data analytics.

- Cognitive Skills: Advanced problem-solving, emotional intelligence, critical thinking, and systems thinking.
- Ethical Awareness: Addressing decision-making around data privacy, fairness, inclusion, and sustainability (Benešová & Tupa, 2017).

This mix of technical, cognitive, and ethical skills ensures that workers can effectively collaborate with intelligent machines, tackle complex challenges, and maintain human-centered values in increasingly automated environments.

Alignment with United Nations Sustainable Development Goals (SDGs)

Industry 5.0's human-centered approach closely aligns with several key United Nations Sustainable Development Goals, reinforcing its wider societal and environmental commitments (United Nations, 2015):

- Goal 3: Good Health and Well-Being—Industry 5.0 promotes worker health through ergonomic workplace design, stress reduction, and real-time health monitoring with Human Digital Twins.
- Goal 8: Decent Work and Economic Growth—By creating fulfilling roles that blend human creativity with technological support, it fosters inclusive and sustainable economic growth.
- Goal 9: Industry, Innovation, and Infrastructure—Industry 5.0 builds resilient industrial capacity by integrating adaptive, human-centered technologies.
- Goal 12: Responsible Consumption and Production—Sustainable manufacturing practices help reduce waste and improve resource efficiency.
- Goal 13: Climate Action—Advanced technologies work to reduce pollution and optimize energy use, aiding climate change efforts.

These connections highlight Industry 5.0's potential to tackle urgent global challenges while improving industrial competitiveness and human welfare.

In summary, Industry 5.0 represents a meaningful shift from automation-focused models to a human-centered industrial approach. It raises the role of workers as co-creators who leverage technology while integrating their creativity, judgment, and emotional intelligence into industrial innovation. This shift fosters workplaces that are not only more efficient and resilient but also more meaningful and supportive of human well-being, paving the way for sustainable and inclusive futures in manufacturing and beyond.

4 Defining Psychological Well-Being in the Workplace

Psychological well-being in the workplace has received more attention as organizations, policymakers, and scholars recognize its effects on employee performance, retention, creativity, and overall organizational health. With the rise of Industry 5.0 and its human-centered focus, the conversation is moving from simple productivity metrics to the quality of the work experience. Psychological well-being includes the emotional, cognitive, and social aspects of how individuals view and experience their work, which can be greatly influenced by leadership practices, organizational culture, job design, and technology.

4.1 Conceptual Frameworks of Psychological Well-Being

To understand psychological well-being, we need solid theoretical frameworks. Several well-known models have shaped our insights.

Ryff's Six-Factor Model of Psychological Well-Being

Developed by Carol Ryff (1989), this model outlines six key dimensions: autonomy, environmental mastery, personal growth, positive relationships, purpose in life, and self-acceptance. In the workplace, autonomy means having freedom in decision-making; environmental mastery relates to effectively managing work responsibilities; personal growth reflects ongoing learning; positive relationships involve teamwork; purpose connects with finding meaning in work; and self-acceptance means recognizing one's strengths and weaknesses (Ryff, 1989). For instance, Patagonia promotes both autonomy and purpose through its environmental activism initiatives, encouraging employees to take paid leave for environmental volunteering, thus integrating personal growth with societal contribution.

Seligman's PERMA Model

Martin Seligman (2011), a key figure in positive psychology, proposed the PERMA framework: Positive Emotion, Engagement, Relationships, Meaning, and Achievement. This model advocates for a holistic approach, encouraging work environments that nurture not only engagement and accomplishment but also emotional well-being and strong social ties. For example, Google's Project Aristotle revealed that psychological safety, closely linked to solid workplace relationships and positive feelings, was a vital factor for high-performing teams.

Self-determination Theory (SDT)

Edward Deci and Richard Ryan's SDT emphasizes three basic psychological needs: autonomy, competence, and relatedness (Deci & Ryan, 2000). When these needs are met, intrinsic motivation thrives, leading to better well-being and performance. In a technology setting, Atlassian supports SDT principles by giving developers "ShipIt Days"—24-hour hackathons where they pick their own projects, enhancing autonomy and competence.

Job Demands-Resources (JD-R) Model

The JD-R model (Bakker & Demerouti, 2007) suggests that job demands—like workload and emotional pressure—and job resources—like autonomy, feedback, and social support—interact to affect burnout and engagement. For example, Salesforce combines wellness programs and workload management tools to balance demands with resources, fostering sustainable engagement.

These frameworks show that psychological well-being is not a single concept but an interaction of personal, social, and organizational influences. Applying these models in modern workplaces allows for targeted interventions that support both individual growth and organizational goals.

4.2 Core Dimensions: Autonomy, Purpose, Resilience, and Engagement

While conceptual models provide a broad theoretical framework, four dimensions commonly emerge as essential for workplace well-being: autonomy, purpose, resilience, and engagement.

Autonomy

Autonomy refers to how much control employees have over their tasks, schedules, and work methods. In Industry 5.0, where technology supports human decision-making, autonomy is key for maintaining intrinsic motivation. Studies show that perceived autonomy links positively to job satisfaction and creativity (Gagné & Deci, 2005).

Case Example: Spotify employs a "squad" model, where small, cross-functional teams function like mini-startups with decision-making power. This setup boosts innovation and strengthens employees' sense of ownership and autonomy.

Purpose/Meaning

Having a sense of purpose in work relates to higher engagement, lower turnover rates, and improved resilience during times of change (Steger et al., 2012). Purpose goes beyond organizational mission statements; it develops when employees feel their daily tasks contribute to larger, valued goals.

Case Example: Unilever's Sustainable Living Plan weaves social and environmental impact into its business model. Employees feel more motivated knowing their work supports sustainability; this connection aligns their personal values with those of the organization.

Resilience

Resilience is the ability to bounce back from setbacks and respond positively to change. In fast-paced, tech-driven workplaces, resilience is crucial for managing stress and maintaining performance. Strategies to build resilience include skills training, supportive leadership, and flexible work arrangements.

Case Example: Microsoft Japan's 4-Day Workweek Experiment showed that giving employees more rest time boosted productivity by 40% (NPR, 2019). This suggests that making structural changes can significantly improve resilience.

Engagement

Engagement measures how enthusiastic, absorbed, and energetic employees are in their work. The JD-R model suggests that engagement flourishes when job resources are plentiful. Engagement goes beyond satisfaction; it involves actively investing in work tasks.

Case Example: Zappos encourages engagement by fostering a strong culture of empowerment and fun. Employees can take charge of customer relationships without strict scripts.

These aspects interact in a supportive way: autonomy boosts engagement, purpose strengthens resilience, and all contribute to long-term well-being.

4.3 *Relevance of Well-Being in Tech-Augmented Work Environments*

The shift toward digitally mediated and AI-supported work environments presents unique challenges and opportunities for psychological well-being.

Challenges

Integrating technology can lead to technostress, which is the stress that comes from learning new systems, increased monitoring, and unclear work–life boundaries (Tarafdar et al., 2011). Overreliance on digital communication can weaken connections among colleagues, while constant algorithmic oversight may lessen perceived autonomy.

Case Example: A study in the banking sector found that AI-driven performance tracking raised anxiety levels among employees, causing disengagement unless it was accompanied by supportive feedback systems (Sadeghi, 2024).

Opportunities

On the other hand, well-designed technology can improve well-being. Collaborative robots, or cobots, can handle dangerous or repetitive tasks, allowing employees to engage in more meaningful and creative work. AI-driven learning platforms can tailor training, enhancing skills and engagement.

Case Example: BMW's use of cobots in assembly lines reduced physical strain for workers and improved job satisfaction by allowing them to focus on problem-solving tasks.

Integrating Well-Being into Tech Strategy

Organizations need clear strategies to ensure that technology adoption aligns with human-centered goals. This involves including employees in design decisions, providing digital literacy training, and using data ethically to support rather than penalize.

Case Example: Siemens used a co-creation method when introducing AI tools, inviting employees to help design workflows that maintained decision-making autonomy.

Industry 5.0 Perspective

From the Industry 5.0 perspective, technology is not an end itself but a way to enhance human abilities and well-being. This approach requires integrating well-being metrics into technology evaluation—looking at not only productivity increases but also shifts in autonomy, purpose, resilience, and engagement.

Defining and promoting psychological well-being in the workplace is crucial in a time when collaboration between humans and machines is reshaping job roles and skills. Theoretical frameworks like Ryff's model, PERMA, SDT, and JD-R provide insight into how organizational structures and technology affect well-being. Autonomy, purpose, resilience, and engagement are vital; they are not just desirable traits but essential elements for organizations to thrive in the human-centered landscape of Industry 5.0. Real-world examples show that companies prioritizing these aspects enjoy benefits in innovation, employee retention, and overall performance. As technology evolves, organizations must ensure that its use strengthens, rather than undermines, the psychological foundations of a healthy workforce.

5 Psychological Impacts of Industry 5.0 Work Environments

The rise of Industry 5.0 marks a significant change in the workplace. It moves beyond automation and focuses on collaboration between humans and machines. This new industrial model reshapes both how work is done and how workers feel about their experiences. Understanding these psychological impacts is crucial for creating work environments that support employee well-being, productivity, and innovation.

5.1 Cognitive Demands and Mental Load

In Industry 5.0, the cognitive demands on workers change significantly. Previous industrial revolutions mainly replaced manual labor with machines. In contrast, Industry 5.0 brings in smart systems that work alongside humans, creating lively and information-heavy environments. Workers are no longer just operators or supervisors; they become coordinators of complex human–machine interactions.

This shift increases mental workload as employees must process various data streams, make quick decisions based on AI suggestions, and adapt to unexpected changes in their workflows. For example, operators in smart factories use advanced human–machine interfaces (HMIs) that combine real-time sensor data, predictive analytics, and updates from collaborative robots (Wang et al., 2021). Handling these layers of information requires focus, situational awareness, and the ability to prioritize under pressure.

While this complexity can boost engagement and mental stimulation, it may also lead to cognitive overload, especially when system designs ignore human limits. Studies in cognitive psychology show that too much mental load can cause fatigue, poorer decision-making, and mistakes (Parasuraman & Hancock, 2000). Therefore, workplaces in Industry 5.0 must use design principles that reduce unnecessary complexity and promote cognitive ease.

One example includes adaptive HMIs that tailor information displays based on the worker's current task and mental state. This reduces unnecessary information and directs attention to crucial inputs (Fournier et al., 2024). Furthermore, digital assistants and AI algorithms can filter and prioritize alerts to help workers stay focused on essential decisions.

5.2 *Emotional Impacts of Human–Machine Collaboration*

The emotional side of Industry 5.0 workplaces is crucial too. Bringing in intelligent machines changes how workers feel about their roles, their relationships with colleagues and technology, and their job security.

Positive feelings often come from the empowerment that arises when technology takes over repetitive, dangerous, or physically demanding tasks. For instance, collaborative robots used in automotive assembly lines lessen physical strain and repetitive motion injuries. This allows human workers to concentrate on quality control and creative problem-solving (Villani et al., 2018). As a result, job satisfaction increases, and employees feel a greater sense of value and purpose.

However, introducing new technologies can also create emotional challenges. Employees might worry about the reliability of machines and AI systems, fearing potential failures that could disrupt activities or cause accidents (Broadbent, 2017). Additionally, the fear of job loss or the devaluation of human skills can lead to stress and resistance against new technologies.

Research by Zhang et al. (2023) shows that trust influences emotional responses to human–machine collaboration. Workers who trust the technology and its creators tend to accept it more easily and report positive feelings. Therefore, it's important to build transparent and reliable systems while involving workers in designing and implementing solutions to foster trust and ease anxiety.

Moreover, constant monitoring by AI systems can make workers feel watched and less autonomous, which complicates emotional well-being (Jackson, 2025). Organizations must find a balance between data-driven performance management and respect for privacy and human dignity to avoid causing emotional harm.

5.3 *Personalization and Work Identity*

Industry 5.0‘s focus on personalization changes the worker's experience by linking technology to individual identity and motivation. Personalized workstations, AI-driven scheduling, and customized learning platforms match tasks and growth opportunities with individual strengths, preferences, and career goals.

This connection helps build a stronger work identity. When employees believe their unique skills and values are recognized and integrated into their work, they find greater meaning and commitment. For example, adaptive ergonomic systems that adjust workspace settings for comfort not only prevent physical strain but also show respect for individual needs, enhancing workers' sense of dignity.

Digital personalization also extends to career development. AI-powered learning management systems (LMS) suggest training modules that fit evolving roles. This helps workers manage skill gaps and pursue career paths aligned with their goals (Gensler et al., 2023). This method increases engagement by framing work as an ongoing, self-directed journey rather than a fixed set of tasks.

A real-world example is AT&T's use of AI-enabled personalized learning, which helps employees develop digital skills necessary for Industry 5.0 jobs while integrating their feedback and preferences into content delivery (Eaves, 2023). This push toward personalization not only enhances employee retention but also boosts organizational agility.

5.4 *Impacts on Work–Life Integration and Boundaries*

The digital connectivity central to Industry 5.0 brings both opportunities and challenges for work–life integration. Flexible work arrangements through remote access to control systems and instant communication platforms give employees more control over their work schedule, supporting a better balance between professional and personal demands (Fliegner, 2023).

However, this flexibility can blur the line between work and home, leading to constant availability and digital presenteeism. Employees may feel pressured to respond to messages or finish tasks after traditional working hours, which creates role conflicts and psychological strain (Derks & Bakker, 2012). The likelihood of burnout increases when organizational cultures encourage or expect constant availability.

Effective boundary management is essential. Organizations can help employees by setting clear guidelines on work hours, promoting digital disconnection during personal time, and providing training on time management and mental detachment (Sohnnentag & Fritz, 2018). For example, SAP has started "digital detox" initiatives that encourage employees to disconnect from work during vacations and weekends, which helps in mental recovery (SAP SE, 2025).

Furthermore, Industry 5.0's personalized technologies can assist workers in fitting their work and life roles together by adapting schedules and workloads to individual circumstances, such as caregiving duties or health conditions, promoting inclusivity and well-being.

5.5 Digital Fatigue and Technostress

As Industry 5.0 relies heavily on digital tools—augmented reality (AR), virtual reality (VR), AI assistants, and connected devices—workers increasingly face information overload and sensory stimulation. This can result in digital fatigue, a state marked by mental exhaustion, reduced focus, and emotional irritability (Ayyagari et al., 2011).

Technostress, which stems from difficulties in adapting to new technologies, can also be common. Typical stressors include complexity, rapid tech changes, and the feeling of losing control (Tarafdar et al., 2011). In manufacturing settings using XR for training or maintenance, extended sessions without breaks can worsen sensory overload, leading to headaches, eye strain, and cognitive fatigue (Airbus, 2023).

Addressing these issues requires careful implementation of technology. Organizations should adopt user-centered design principles, involving workers in choosing and customizing digital tools to ensure they are easy to use and suitable. Training programs aimed at improving digital skills and coping mechanisms can help lessen anxiety and build resilience (Ragu-Nathan et al., 2008).

Additionally, establishing organizational rules around digital usage, such as mandatory breaks, screen-time limits, and ergonomic workstation setups, supports sustainable digital engagement.

6 Organizational Strategies to Promote Psychological Well-Being

As Industry 5.0 changes work, organizations must actively support the psychological well-being of their employees. Successful strategies combine leadership, work design, culture, and learning to create environments where employees can thrive alongside new technologies.

6.1 Inclusive and Empathetic Leadership

Leadership style significantly affects employee experiences during digital transformation. Inclusive and empathetic leaders build trust, psychological safety, and engagement by recognizing different employee views and feelings about technology (Edmondson, 2019).

For instance, Siemens includes empathy training in its leadership programs to prepare managers to handle Industry 5.0 challenges sensitively. Leaders learn to communicate openly about changes, seek employee input, and assist workers who struggle to adapt (Siemens Annual Report, 2023).

Empathetic leadership helps address worries about job security and skill obsolescence, positioning technology adoption as a chance for growth and collaboration rather than displacement.

6.2 Designing Human-Centered Workflows and Roles

Redesigning workflows and job roles are vital for promoting well-being in Industry 5.0. Human-centered design focuses on tasks that suit human strengths, including creativity, problem-solving, and interpersonal skills, while automating repetitive or dangerous activities.

Creating roles with employee input ensures that technology aligns with their preferences and abilities. Bosch's participatory design workshops are a good example of this strategy, engaging frontline workers in setting up cobot interactions to match natural work rhythms, minimizing frustration, and increasing acceptance (Bosch Corporate Report, 2022).

This approach also promotes flexibility, allowing workers to transition between tasks and roles as demands change and personal development goals arise, which fosters a sense of control and fulfillment.

6.3 Psychological Safety and Open Communication

A culture of psychological safety, where employees feel comfortable expressing concerns, asking questions, and experimenting without fear of negative repercussions, is crucial for success in Industry 5.0 (Edmondson, 2019). Open communication channels enable early detection of stressors, technology usability problems, and mental health needs.

Regular meetings, anonymous surveys, and mental health ambassadors create opportunities for employees to voice challenges and suggest improvements. For example, Microsoft uses "Employee Resource Groups" and regular pulse surveys to assess well-being during digital transitions, adapting support as needed (Microsoft Annual Report, 2022).

This openness also encourages innovation, as employees feel empowered to recommend improvements and collaborate on solutions.

6.4 Mental Health Policies and Resources

Strong mental health policies offer structured support. These include access to counseling services, employee assistance programs (EAPs), stress management workshops, and flexible work arrangements tailored to individual needs (Eurofound, 2022).

Leading companies like SAP combine digital well-being tools—such as mindfulness apps and virtual coaching—with personalized support to boost resilience and lower absenteeism (SAP Sustainability Report, 2023). Clear communication about mental health resources encourages help-seeking behaviors without stigma.

6.5 Continuous Learning and Adaptive Career Pathways

The fast pace of change in skills required for Industry 5.0 demands ongoing learning frameworks that are personalized, accessible, and aligned with organizational goals. AI-driven learning platforms evaluate individual skills and suggest tailored training paths, encouraging proactive skill development.

Promoting cross-skilling and reskilling not only prepares employees for new roles but also alleviates concerns about job security. Clear career pathways and mentoring programs provide direction and motivation for long-term growth.

For example, AT&T's "Future Ready" program blends AI-driven learning with human coaching to enhance career adaptability, boosting employee retention and innovation capabilities (AT&T, 2021).

7 Case Examples and Best Practices

The shift to Industry 5.0, which focuses on human-centered automation, collaborative robotics, and intelligent systems, offers organizations many opportunities and challenges. While new technologies promise to improve productivity and customization, they need careful integration with workforce needs, cultural changes, and wellness programs. Looking at early adopters of Industry 5.0 offers useful insights into how organizations can successfully handle this complex landscape by combining advanced technology with people-focused strategies.

7.1 *Global Company Case Studies: Industry 5.0 in Action*

Airbus: Pioneering Extended Reality for Workforce Training

Airbus is a standout example of Industry 5.0 implementation, especially through its use of Extended Reality (XR) technologies, including Augmented Reality (AR), Virtual Reality (VR), and Mixed Reality (MR), to transform employee training in aerospace manufacturing.

Assembling commercial aircraft involves complex, precise tasks that require skill and coordination. Traditional training methods, like classroom instruction and on-the-job shadowing, can take a lot of time, be costly, and may not adequately prepare workers for real-world situations. Airbus tackled these issues by creating XR-based immersive training modules that simulate assembly processes in a virtual setting.

By using VR headsets or AR glasses, workers take part in realistic, hands-on scenarios where they can practice assembling parts, spotting defects, and troubleshooting without the risks of making physical mistakes. This safe environment speeds up learning and boosts confidence. Research at Airbus showed that XR training reduced error rates by up to 30% and significantly shortened onboarding times, which improved safety and efficiency.

Importantly, Airbus has stressed the value of worker feedback throughout the implementation process. Employees contributed feedback on designing XR interfaces, session lengths, and pacing of training to minimize digital fatigue—where long screen time and sensory overload lead to mental exhaustion. This iterative, inclusive approach ensured that technology supported human cognitive abilities rather than overwhelmed them.

Additionally, Airbus connects XR data with performance metrics to personalize training further, adjusting difficulty and focus based on individual worker progress. This blend of human insight and AI-driven adaptation reflects Industry 5.0‘s goal of enhancing human intelligence with technology, rather than replacing it.

Universal Robots: Democratizing Collaborative Robotics for SMEs

Universal Robots (UR) provides another strong example of how Industry 5.0 technologies empower not only large firms but also small and medium-sized enterprises (SMEs). UR specializes in collaborative robots—“cobots”—that work safely alongside human employees without needing protective cages or barriers.

Cobots automate repetitive, precise, or ergonomically challenging tasks like screwdriving, packaging, or inspection, allowing workers to concentrate on creative problem-solving, customization, and customer engagement (Villani et al., 2018). This shift in human–machine roles illustrates the human-centered focus of Industry 5.0, which aims to enhance rather than replace human work.

A key factor in UR‘s success is the extensive involvement of workers in configuring and deploying cobots. Unlike traditional industrial robots that require specialists for programming, UR’s cobots have easy-to-use programming interfaces that let frontline workers—even those without formal robotics training—customize task sequences and adjust operational settings.

This participatory design speeds up adoption and builds trust and ownership. Workers report greater job satisfaction knowing they directly affect how cobots assist their tasks, reducing fears of skill loss or job displacement. Moreover, UR offers ongoing training and support, emphasizing upskilling and collaborative learning.

By making cobot technology accessible and flexible, Universal Robots allows SMEs to thrive in a more automated market while maintaining human creativity and craftsmanship. This shows that Industry 5.0's advantages reach beyond large industrial players.

Siemens: Integrating Empathetic Leadership and Continuous Learning

Siemens, a global industrial leader, illustrates the need for strong leadership and workforce development in Industry 5.0 transformation. Understanding that adopting technology alone does not ensure success, Siemens has invested significantly in nurturing empathetic leadership and promoting a culture of continuous learning.

Siemens' leadership development programs prepare managers to lead with emotional intelligence, emphasizing psychological safety and open communication as key skills. Leaders are encouraged to listen actively to employee concerns, recognize uncertainties related to digital change, and create spaces where experimentation and failure are acceptable (Siemens Annual Report, 2023).

This strategy helps lessen resistance to new technologies by fostering trust and framing Industry 5.0 adoption as a collaborative journey. Siemens pairs this with comprehensive digital literacy programs and personalized training pathways, allowing workers to gain new skills in human–robot collaboration, data analysis, and adaptable workflow management.

The company's unified approach has led to increased employee engagement, lower turnover, and higher innovation output. Siemens shows that Industry 5.0 not only calls for smart machines but also requires intelligent leadership that values and develops human capital.

7.2 Innovations in Employee Support Systems

As digital tools and automation spread, addressing their psychological effects becomes crucial. Leading organizations have created innovative employee support systems to tackle mental health challenges, technostress, and the demands of continuous adjustment.

SAP: Comprehensive Mental Health Programs in a Digital Workplace

SAP, a global leader in enterprise softw are, has been at the forefront of innovative mental health initiatives tailored to Industry 5.0 realities. Recognizing the risks of digital fatigue, stress, and anxiety caused by rapid technological changes and constant connectivity, SAP established a multifaceted mental health strategy that combines technology, counseling, and peer support.

At the core of SAP's approach are digital self-assessment tools that let employees monitor their well-being in real-time. These tools use AI algorithms to spot signs of burnout or stress based on work habits, communication patterns, and self-reports. When risks are identified, employees gain confidential access to counseling services and mental health resources.

Beyond technology, SAP promotes community through "mental health ambassadors"—trained peer supporters who raise awareness, reduce stigma, and offer informal emotional support. This peer network complements professional services, creating a more accessible and comprehensive safety net (SAP Sustainability Report, 2023).

SAP's initiatives include training managers to recognize and address mental health issues and to promote work–life balance, like advocating for digital detox periods and flexible work schedules.

This holistic approach has improved employee resilience, reduced absenteeism, and enhanced organizational culture, setting a model for managing the human effects of Industry 5.0.

Bosch: Participatory Design for Successful Cobot Integration

Bosch, a multinational engineering and technology company, shows the effectiveness of participatory design in integrating collaborative robotics. In their Industry 5.0 projects, Bosch actively involves frontline workers in workshops to co-design workflows with cobots, ensuring that the technology fits with real work practices and preferences.

This bottom-up method helps overcome common challenges such as resistance to change, fear of job loss, and usability issues. Workers share insights on task sequences, workspace layout, and safety factors, leading to smoother human–robot interactions and greater acceptance (Bosch Corporate Report, 2022).

These workshops also promote understanding among engineers, managers, and workers, breaking down silos and building trust. By validating worker expertise and treating them as partners rather than passive recipients, Bosch fosters a positive organizational climate that encourages innovation.

Bosch's experience illustrates that successfully adopting Industry 5.0 relies on engaging employees meaningfully throughout the tech transformation process, emphasizing the shift from automation to collaboration.

From these cases, several best practices emerge that organizations can adopt to maximize the benefits and minimize the risks of Industry 5.0 transformation:

- Engage Workers Early and Often: Involve frontline employee perspectives during technology design, selection, and deployment to ensure usability, reduce resistance, and build ownership.
- Invest in Leadership Development: Develop empathetic, inclusive leaders who focus on psychological safety, transparent communication, and emotional support during transitions.
- Build Continuous Learning Cultures: Offer personalized, accessible training that equips employees with both technical and soft skills for evolving roles.

- Implement Comprehensive Mental Health Supports: Utilize digital tools, peer networks, and counseling to manage technostress and digital fatigue.
- Foster Collaborative Human–Machine Interaction: Create workflows that balance automation with human creativity, judgment, and craftsmanship.
- Promote Open Communication: Maintain clear dialogues about technological goals, possible impacts, and career implications to build trust and align expectations.

By combining advanced technology with human-centered strategies, early adopters show that Industry 5.0 can create more resilient, innovative, and humane workplaces.

7.3 Lessons Learned from Early Adopters

The transition to Industry 5.0 is a continuing journey for many organizations worldwide. As some pioneering companies move past traditional Industry 4.0 automation to embrace a human-centered, collaborative model, important lessons arise from their experiences. These early adopters provide valuable insights into the elements that support effective Industry 5.0 implementation, highlighting that technology alone is not enough without a supportive culture, leadership, and workforce backing.

User-Centered Technology Design: Reducing Cognitive Overload and Technostress

One significant lesson from early Industry 5.0 adopters is the crucial need for technologies that prioritize human users' cognitive and emotional needs. Unlike previous industrial revolutions that often imposed rigid technological solutions from the top down, Industry 5.0 necessitates interfaces and systems that are intuitive, adaptable, and responsive to individual differences in skills, preferences, and mental load.

Cognitive overload and technostress—stress linked to technology use—are common issues when new digital tools are introduced without enough focus on user experience. Workers overwhelmed by too much data, alerts, or complicated controls can see reduced productivity, more errors, and lower well-being (Ayyagari et al., 2011). For instance, Airbus's use of extended reality (XR) technologies for training faced initial resistance due to sensory overload and discomfort among workers. By refining XR sessions to include shorter training modules, customizable visual settings, and breaks, Airbus significantly improved acceptance and effectiveness.

Similarly, Universal Robots' success with cobots partly comes from focusing on ergonomic, easy-to-learn interfaces that workers of various skill levels can master quickly (Villani et al., 2018). Getting frontline employees involved early in the design process—through participatory workshops or pilot testing—helps reveal usability problems before full deployment, ensuring that technology supports rather than frustrates human work.

Furthermore, adaptive systems that track workers' cognitive states and adjust information flow accordingly are becoming a best practice. Such user-centered design not only boosts safety and performance but also shields workers from the negative psychological effects of overstimulation, helping maintain engagement and job satisfaction.

Empathetic, Inclusive Leadership: Building Trust and Psychological Safety

Leadership is another key factor for success in early Industry 5.0 implementations. The complexity of integrating advanced technologies with human work calls for leaders who can handle not just technical challenges but also the emotional and social aspects of change. Empathetic and inclusive leadership styles that emphasize understanding employees' fears, hopes, and well-being create the psychological safety necessary for smooth transitions.

Psychological safety—the shared belief that one can raise concerns without fear of repercussions—is vital for innovation, learning, and well-being in any workplace, especially in settings undergoing technological change (Edmondson, 2019). Siemens's leadership programs have incorporated empathy training to help managers communicate openly about Industry 5.0 initiatives, actively involve employees in creating new workflows, and acknowledge the emotional effects of change (Siemens Annual Report, 2023). This strategy builds trust, cuts resistance, and boosts employee ownership.

Early adopters emphasize that empathetic leadership requires listening to worker worries about job security, skills, and work–life balance, then addressing these with concrete support like retraining programs or flexible scheduling. Inclusion also involves ensuring that every employee voice—including those of frontline workers and marginalized groups—is heard and valued during planning and implementation.

By cultivating a culture where employees feel appreciated and supported, empathetic leaders create an environment where people can adapt to new technologies with confidence instead of fear, enabling the organization to leverage human potential alongside automation.

Continuous Learning and Clear Career Pathways: Empowering Adaptation and Growth

Industry 5.0's focus on human–machine collaboration necessitates new skills and ongoing learning. Early adopters consistently highlight the need for strong, accessible learning frameworks that help employees build both technical and cognitive skills for evolving roles.

Continuous learning programs that combine AI-driven personalized training with human mentoring enable workers to keep pace with rapid technological changes. For example, AT&T's "Future Ready" program uses AI to identify skill gaps and suggest tailored digital courses, supported by coaching and peer learning (AT&T, 2021). This approach alleviates skill-related anxiety and empowers employees to take charge of their career development.

Clear and transparent career pathways connected to Industry 5.0 roles, such as human–robot interaction specialists, data interpreters, and adaptive system supervisors, offer motivation and direction. Workers who understand how their skills relate to company goals and future opportunities tend to be more engaged and less likely to leave.

Early adopters stress the importance of cross-skilling and role flexibility. Encouraging employees to learn complementary skills enhances their adaptability, supports job enrichment, and promotes innovation through varied perspectives. Bosch's participatory role redesign workshops allow workers to co-create job descriptions that include both digital and creative tasks, boosting job satisfaction and performance, as noted in the Bosch Corporate Report (2022).

Investing in workforce development as a key priority shows that an organization cares about its people. This strengthens loyalty and fosters a shared vision for the future.

Transparent Communication: Reducing Automation Fears and Building Trust

A common challenge in adopting Industry 5.0 is employees worrying about job loss due to automation and the redundancy of their skills. Early adopters discover that honest, ongoing communication is key to addressing these concerns and building trust.

Communication strategies that explain the goals of automation, focusing on enhancement rather than replacement, while openly discussing expected changes, help clarify Industry 5.0 technologies. For example, Philips conducted town hall meetings, shared newsletters, and used digital platforms for leadership to provide updates on digital initiatives and gather employee feedback, as reported in the Philips Sustainability Report (2022). This two-way communication reduced rumors and misinformation, setting realistic expectations and encouraging cooperative problem-solving.

Additionally, transparency involves recognizing uncertainties and challenges instead of making overly optimistic promises. Open discussions about the need for skill development and potential changes in roles prepare employees mentally, making them more willing to accept and engage.

Combining communication with concrete actions like retraining, redeployment opportunities, and social support further strengthens credibility and commitment.

Culture of Psychological Safety: Promoting Innovation and Open Dialogue

Lastly, early adopters highlight that transforming to Industry 5.0 is not just a technological upgrade, but a significant cultural change. Organizations that develop a culture of psychological safety and openness achieve greater innovation, quicker problem-solving, and improved employee well-being.

A culture of psychological safety allows employees to voice concerns about mental health, workload, or technology usability without fear of stigma or retaliation, as noted by Edmondson, 2019. This openness is especially important in fast-changing environments with continuously emerging challenges.

Google's Project Aristotle showed that teams with high psychological safety perform better because team members feel safe sharing personal risks and innovative ideas, as reported by Rozovsky, 2015. Similar results apply in Industry 5.0 workplaces, where employees need to experiment with new human–machine collaborations and adjust workflows.

Practical methods that early adopters use to encourage this culture include leadership role modeling, peer support networks, anonymous feedback channels, and training in emotional intelligence and conflict resolution. For instance, SAP's mental health ambassadors program offers peer support and raises awareness about digital fatigue and technostress, highlighted in the SAP Sustainability Report (2023).

Importantly, fostering psychological safety also builds employee resilience, reducing burnout and turnover during complex digital transformations.

Integrated Lessons: Technology and Culture Must Evolve Together

Together, these lessons from early Industry 5.0 adopters clearly show that implementing technology without corresponding organizational changes can lead to failure or poor results. The potential of Industry 5.0—enhancing human creativity, empathy, and decision-making with advanced technology—can only be achieved when systems are created for people, leaders build trust and safety, employees continue developing their skills, and communication encourages shared understanding.

The transition requires a comprehensive approach that balances technical innovation with human-centered practices to create workplaces where technology empowers people instead of isolating them.

Illustrative Case Study: Airbus

Airbus's experience in applying Industry 5.0 principles illustrates these lessons in action. Their use of XR technology for training assembly workers met initial challenges with user discomfort and skepticism. By involving workers in redesigning training modules and allowing for flexible pacing and personalization, Airbus minimized cognitive overload and digital fatigue, according to Airbus (2021).

At the same time, Siemens-style empathetic leadership practices helped communicate openly about the goals of digital transformation and address worker concerns. Continuous learning programs allowed employees to improve their skills in XR, data interpretation, and human–robot interaction.

Airbus also promoted psychological safety through anonymous feedback systems and mental health resources, enabling workers to express their challenges and seek support. This comprehensive approach resulted in enhanced safety, productivity, and employee satisfaction, establishing Airbus as a leading example of Industry 5.0 implementation.

8 Challenges and Ethical Considerations

As Industry 5.0 brings in a new era focused on people, combining automation with personalized technology, it also reveals various challenges and ethical questions. While the potential for improved productivity, worker empowerment, and sustainability is appealing, the journey to achieve Industry 5.0 is complex. This includes handling technology overload, preventing job loss, ensuring ethical use of employee data, and addressing fairness and access issues in digitally transformed workplaces.

This chapter discusses these important issues, using evidence and real-world examples to show both the risks and the ethical obligations organizations face to embrace the human-centered ideals of Industry 5.0.

8.1 Risks of Tech Overload and Human Displacement

The Double-Edged Sword of Technological Intensification

Industry 5.0 stands out due to the seamless connection between humans and machines, where technologies like collaborative robots (cobots), artificial intelligence (AI), and extended reality (XR) enhance human abilities instead of replacing them. However, the growth of these technologies comes with risks of cognitive overload, mental fatigue, and job loss.

Technological overload, often called "technostress," happens when employees feel overwhelmed by the amount, complexity, and speed of digital information and interactions (Tarafdar et al., 2015). In Industry 5.0 workplaces, employees might need to keep up with real-time AI analytics, engage with adaptive systems, and manage human–machine interactions at the same time. This mixed cognitive demand can exceed what humans can handle, resulting in stress, lower performance, and burnout (La Torre et al., 2018).

Case Example: Boeing's 787 Dreamliner Production

A clear example is Boeing's 787 Dreamliner assembly line, where the integration of robotics and digital tools was initially intended to streamline production and cut costs. However, workers struggled with the demanding task of coordinating automated systems and manual operations, often facing "information flooding" from various control panels and digital alerts. This overload led to higher error rates and delays, prompting Boeing to redesign interfaces and simplify workflows to ease mental strain (Paul, 2018).

This case highlights that advanced technology, if not carefully designed with user experience in mind, can harm human performance.

The Persistent Threat of Human Displacement

In addition to overload, worries about job loss due to automation persist. Even as Industry 5.0 focuses on enhancement, many workers are concerned that machines will make their jobs redundant. Historically, technological advances have disrupted labor markets, and the rapid growth of AI and robotics increases these fears.

For example, in the automotive industry, using cobots for welding and painting has raised concerns about the future of skilled labor. While cobots free workers from dangerous and repetitive tasks, they may also decrease the need for traditional manual jobs, requiring workers to improve their skills or move to new positions.

Mitigating Displacement Through Human-Centered Design and Policy

To tackle these risks, organizations adopting Industry 5.0 principles need to invest in strategies that help workers transition by focusing on skill development, career movement, and job redesign. In the Netherlands, Philips has launched a "Future Work" program that combines cobot integration with extensive retraining, allowing workers to shift from manual assembly to roles in quality control, programming, and customer customization. This approach helps reduce displacement by creating new jobs that fit human creativity and decision-making (Philips Sustainability Report, 2022).

Policy frameworks also play an important role. Governments and industries should work together on support systems like income assistance, lifelong learning subsidies, and reskilling incentives, ensuring that the benefits of Industry 5.0 are shared broadly rather than hoarded by a few (OECD, 2020).

8.2 Ethical Use of Employee Data and Monitoring

The Data-Driven Workplace: Balancing Benefits and Privacy

Industry 5.0 workplaces are becoming more data-driven. Sensors, wearable devices, AI algorithms, and digital twins collect a large amount of employee-related data, allowing for personalized training, health monitoring, workflow improvement, and safety management. However, these capabilities raise serious ethical questions about privacy, consent, surveillance, and individual freedom.

For instance, Human Digital Twins (HDTs) create digital versions of workers' physical and mental states to enhance performance and well-being (Nahavandi, 2019). While this can provide tailored support, it also risks invasive monitoring if employees feel they are constantly watched or pressured to share sensitive information.

Case Example: Amazon's Warehouse Monitoring Controversy

Amazon's use of advanced monitoring technologies in its warehouses highlights these ethical issues. The company utilizes wearable devices and AI-driven tracking to observe worker movements, task completion times, and break patterns. While these tools boost operational efficiency, they have drawn criticism for creating stressful working conditions, invading privacy, and contributing to high employee turnover.

Workers have expressed feeling dehumanized and constantly monitored, raising concerns about the line between effective performance management and intrusive surveillance. If unchecked, such practices can harm trust and employee morale, ultimately affecting productivity negatively.

Principles for Ethical Data Use

To ethically manage employee data, organizations need to adopt clear policies that prioritize informed consent, careful data collection, and employee control over personal information. Data collection should focus on improving worker well-being and independence rather than punitive oversight.

The European Union's General Data Protection Regulation (GDPR) provides a useful framework by requiring organizations to process personal data legally, fairly, and transparently, with ways for employees to access, correct, or contest their data (European Commission, 2021).

Participatory Approaches to Monitoring

Involving employees in creating and managing monitoring systems builds trust and acceptance. For example, Siemens worked with workers' councils and unions when introducing wearable health sensors, collaboratively setting data usage guidelines and safeguards against misuse (Siemens Annual Report, 2023). This collaborative approach respects worker autonomy and aligns technology use with shared values.

8.3 *Equity and Access in a Digitally Transformed Workplace*

Digital Divides and Inclusion Challenges

While Industry 5.0 aims to make advanced technologies accessible to everyone and boost human potential, it also risks worsening inequalities if not handled wisely. Digital divides exist between countries and regions, as well as within organizations, where gaps in digital skills, access to resources, and job roles can disadvantage certain groups.

For instance, older workers or those with less formal education may struggle to adjust to AI-enhanced workflows or digital tools, which could lead to marginalization (van Deursen & Helsper, 2015). Gender gaps remain in STEM fields, affecting women's involvement in Industry 5.0 technology jobs (UN Women, 2021).

Case Example: Toyota's Inclusive Workforce Strategy

Toyota's approach in Japan shows how companies can promote fairness in digital transformation. Understanding the challenges of an aging workforce, Toyota created user-friendly interfaces for cobots and AR tools to accommodate different levels of digital skill. They also established mentorship programs where experienced workers help their colleagues learn new skills, promoting knowledge sharing across generations (Toyota, 2025).

Additionally, Toyota adopted flexible schedules and ergonomic changes to help workers with various needs, reflecting Industry 5.0‘s focus on personalized work environments.

The Role of Organizational Culture and Policy

Creating an inclusive Industry 5.0 workplace requires deliberate cultural changes and policies that prioritize fairness. This includes proactive recruitment and retention efforts for underrepresented groups, investments in digital skills training accessible to all workers, and accommodations for diverse physical and cognitive needs.

In Germany, the Fraunhofer Institute for Industrial Engineering has created guidelines to design human–machine interfaces that are accessible, considering sensory impairments, cognitive load, and ergonomic issues, ensuring technology benefits a wide range of users.

Broader Societal Implications and the Need for Collaboration

Beyond individual organizations, achieving fair digital workplaces needs coordinated efforts from governments, educational institutions, and civil society. Public policies that support broadband access, digital skills education, and continuous learning are crucial to prevent widening digital divides.

For example, Singapore’s SkillsFuture initiative offers citizens credits and resources for lifelong skill development, aiming to build a workforce ready for the challenges of Industry 5.0 (SkillsFuture Singapore, 2021).

Conclusion: Navigating Challenges with Ethical Foresight

Industry 5.0 has the potential to create more humane, sustainable, and productive workplaces by merging human creativity with advanced technology. However, achieving this vision requires careful consideration of the challenges and ethical issues that come with digital transformation.

Organizations must proactively address the risks of technology overload and job displacement by prioritizing designs centered on workers, reskilling, and social support. They need to protect employee privacy and autonomy through clear, participatory data governance. Additionally, they must commit to fairness and inclusion, ensuring all workers—regardless of age, gender, education, or background—can access and benefit from new technologies.

By integrating these ethical principles into their strategies and culture, Industry 5.0 can deliver not only technological progress but also allow for human flourishing.

9 Future Directions and Research Opportunities

As Industry 5.0 changes the work landscape by focusing on human collaboration with advanced technologies, research and practice need to adapt to meet new challenges and opportunities. The combination of artificial intelligence, robotics, and immersive systems with human skills and creativity creates new ways to improve workplace well-being, innovation, and sustainability. However, these openings also bring complex questions that require strong interdisciplinary research and thoughtful human resource development (HRD) methods.

This chapter highlights future directions and research opportunities in three interconnected areas: interdisciplinary studies linking technology and mental health, making human well-being a central organizational goal, and creating fair and adaptable work futures for an inclusive digital age.

9.1 Interdisciplinary Research on Technology and Mental Health

The Need for Integrated Research

The fast growth of digital tools and intelligent systems in Industry 5.0 workplaces requires a deeper understanding of their psychological effects. While technology can improve productivity and personalization, it also poses risks like cognitive overload, technostress, and emotional fatigue (La Torre et al., 2018). Traditional boundaries between fields like computer science, organizational psychology, and ergonomics are not enough to tackle these complex issues.

Research that combines insights from psychology, human factors engineering, data science, and organizational behavior is crucial to create thorough models of how humans interact with technology. This research can reveal how various technologies—from collaborative robots to AI-based decision-support systems—affect worker well-being, engagement, and performance in different settings.

Emerging Research Ideas

One interesting area is studying adaptive systems that adjust technological interfaces based on individual cognitive and emotional states. For example, Human Digital Twins (HDTs) use real-time biometric and behavioral data to create virtual representations of workers, allowing for personalized support and workload adjustments (Nahavandi, 2019). However, we need to investigate the ethical issues and effectiveness of these systems, including how to balance assistance with maintaining autonomy.

Designing and assessing extended reality (XR) environments also requires teamwork among computer scientists, psychologists, and health professionals to optimize training tools that reduce digital fatigue and cognitive overload (Airbus, 2021). Such studies can lead to evidence-based guidelines for technologies in Industry 5.0 workplaces.

Case Study: The EU Horizon 2020 Project ALADDIN

A key example of interdisciplinary research is the ALADDIN project, funded by the European Union's Horizon 2020 program. ALADDIN studies AI-driven digital assistants that track worker stress and cognitive load to suggest adaptive interventions, like recommending breaks or simplifying tasks. Psychologists, data scientists, and industrial engineers work together to validate these systems in manufacturing and healthcare, focusing on both technological effectiveness and psychological safety (European Commission, 2021).

Research Gaps and New Methods

Future research should emphasize longitudinal studies to see how long-term exposure to Industry 5.0 technologies affects mental health. Mixed-method approaches that combine physiological metrics, experience sampling, and qualitative interviews can provide valuable insights into worker experiences and outcomes.

Additionally, inclusive research frameworks must consider diverse groups, including older workers, neurodiverse individuals, and employees in emerging economies, to ensure that technologies support well-being across different demographics and cultures.

9.2 Human Flourishing as a Strategic HRD Goal

Going Beyond Productivity Measures

HRD has traditionally focused on performance enhancement and skill development. While these areas are still important, Industry 5.0 encourages a shift toward human flourishing, defined as overall well-being that includes physical health, psychological vitality, meaning, and social connection (Seligman, 2011).

Prioritizing flourishing as a strategic HRD goal acknowledges that lasting organizational success relies on treating workers as whole individuals and not just productive assets. This focus matches the broader human-centered approach of Industry 5.0 and the United Nations Sustainable Development Goals (UN SDGs), particularly those related to decent work and well-being (UN, 2015).

Integrating Flourishing into Organizational Culture

Leading organizations are starting to embed flourishing into their cultures by redesigning roles, workflows, and reward systems. For instance, outdoor recreation clothing company Patagonia offers flexible schedules, encourages outdoor breaks, and promotes community engagement, factors that employees say have significantly improved their job satisfaction and creativity.

Global consulting firm Deloitte has introduced "Well-being DNA" programs that incorporate mental health, physical fitness, and purposeful work into their talent development strategies. These programs have shown improvements in employee retention and innovation metrics (Deloitte, 2025).

HRD Practices for Flourishing in Industry 5.0

To foster human flourishing in Industry 5.0 workplaces, HRD professionals need to adopt new skills and practices, including:

- Holistic Learning Programs: Training should include emotional intelligence, mindfulness, resilience, and ethical reasoning alongside technical skills.
- Co-Creation of Work: Allowing employees to shape their roles and workflows enhances their autonomy and sense of purpose.
- Strengths-Based Approaches: Recognizing and utilizing individual talents and passions boosts engagement.
- Well-being Metrics: Developing and monitoring indicators beyond productivity, such as psychological safety, life satisfaction, and social connectedness.

Research Directions: Defining Flourishing

There is a need for research that clearly defines human flourishing in organizational settings. This includes developing reliable measurement tools, identifying contributing factors and outcomes, and assessing interventions aimed at promoting flourishing in tech-enhanced workplaces.

Collaboration among HRD researchers, positive psychologists, and organizational leaders can create evidence-based frameworks to guide policies and practices. Furthermore, case studies of flourishing-focused changes across various sectors can offer models that can be replicated.

9.3 Designing Fair and Adaptive Work Futures

Equity as a Key Principle in Industry 5.0

As technology transforms jobs and organizational structures, ensuring fair access and opportunity is crucial. Industry 5.0's potential to enhance human work risks marginalizing certain groups without intentional equity strategies.

Considerations for equity include addressing gaps in digital literacy, access to resources, workplace accommodations, and career advancement opportunities. Factors like gender, age, disability, socioeconomic status, and geography influence who benefits from Industry 5.0 advancements.

Inclusive Technology Design

Equity starts with the design of technology. Engaging diverse employee groups in creating interfaces, tools, and workflows through participatory design methods can help avoid exclusionary outcomes (Robertson & Simonsen, 2012). For example, the European Fraunhofer IAO has established guidelines for creating accessible human–machine interfaces that cater to various physical and cognitive abilities, enhancing usability (Fraunhofer IAO, 2020).

Open-source and modular technologies can also reduce barriers for smaller companies, allowing wider access to the benefits of Industry 5.0.

Case Study: Singapore's SkillsFuture Program

Singapore's SkillsFuture initiative demonstrates a national effort to ensure fair workforce adaptation to Industry 5.0. The program provides citizens with credits to pursue a wide range of education and training, including digital skills and lifelong learning. It specifically targets vulnerable groups, such as mid-career workers and those in manual jobs, promoting inclusive adjustments to technological changes (SkillsFuture Singapore, 2021).

This all-encompassing approach connects public policy with industry and education, showcasing the collaboration needed to create fair work futures.

Adaptive Work Design and Lifelong Learning

Adaptability is critical in the fast-changing environments of Industry 5.0. Job roles must be flexible and responsive to technological changes and market demands. Organizations are investigating "career lattices," which focus on lateral moves, skill diversification, and continuous learning instead of traditional career ladders (Brassey et al., 2024).

HRD plays a crucial role in establishing adaptive work systems that provide both stability and flexibility. This includes personalized learning paths, micro-credentialing, and supportive leadership that encourages exploration and development.

Research Challenges and Opportunities

Research is needed to understand how adaptive work systems affect employee engagement, equity, and organizational performance across different sectors and cultures. Comparative studies can highlight best practices and contextual variations.

Also, ethical guidelines for fair access to reskilling and adaptable career development must be created to prevent disproportionate negative impacts on vulnerable workers.

Conclusion: Toward a Human-Centered Research and Practice Agenda

The future of Industry 5.0 exists at the crossroads of cutting-edge technology and important human values. Achieving its transformative potential requires broad, interdisciplinary research and strategic HRD approaches that focus on mental health, human flourishing, and equity.

By promoting integrated research on technology and well-being, prioritizing flourishing as a key organizational goal, and designing inclusive, adaptive work systems, researchers and practitioners can shape work futures that are productive, innovative, fair, and fulfilling for everyone.

The journey is complicated, but the stakes are high. How we tackle these challenges will determine if Industry 5.0 realizes its promise as a genuinely human-centered industrial revolution.

10 Conclusion

10.1 Summary of Key Points

This chapter has examined the significant changes brought about by Industry 5.0, highlighting its unique focus on people and the important effects on employee mental health. Unlike the previous Industry 4.0, which emphasized automation, data collection, and efficiency, Industry 5.0 presents a refreshed vision that aligns technological progress with values like empathy, creativity, and individual empowerment.

A key takeaway is the shift in workers' roles from passive operators of machines to active co-creators in the industrial process. This change is not just about technology; it is a cultural and psychological shift. The use of collaborative robots (cobots), personalized AI, Human Digital Twins, and immersive interfaces creates workplaces meant to enhance human abilities instead of replacing them. These tools help reduce hazardous, repetitive tasks and create chances for skill development, innovation, and meaningful engagement.

The psychological effects of this new model are complex. On one hand, cognitive demands may increase because of the need to manage interactions between humans and machines. On the other hand, personalization helps support work identity and motivation. Emotional experiences in Industry 5.0 are influenced by working with technology that understands human feelings. This approach can foster a safe atmosphere and lessen mental strain if thoughtfully designed. However, challenges remain, including risks of digital fatigue, technostress, and blurred lines between work and personal life. Organizations will need proactive strategies to address these issues.

Leadership within organizations is crucial for promoting mental health. Inclusiveness and understanding in leadership styles support human-centered workflows, creating environments where employees feel valued and cared for. Efforts that focus on psychological safety, open communication, and ongoing learning are vital for sustaining well-being and innovation in today's tech-enhanced workplaces.

Real-world examples from companies like Airbus, Universal Robots, Siemens, and SAP show how to apply Industry 5.0 principles. These case studies reveal successes and lessons learned, emphasizing the importance of community involvement in design, mental health programs, and clear communication in bringing human values into technological progress.

However, the path forward includes ethical and practical issues. The risks of tech overload, potential job loss, and the ethical aspects of using employee data highlight the need for balanced methods that respect human dignity and freedom. Furthermore, ensuring fair access to the benefits of Industry 5.0 requires thoughtful policies and inclusive designs to bridge digital gaps and empower diverse work groups.

Looking ahead, future research and practice should take an interdisciplinary approach, combining insights from psychology, technology, human resource development, and ethics. Making human flourishing a key goal will help organizations create workplaces where employees can thrive in all aspects—mentally, emotionally, socially, and physically. Developing adaptable and fair work environments will be essential to ensure that the Industry 5.0 revolution supports all workers and fosters sustainable, inclusive economic growth.

10.2 Final Reflection: Thriving in Industry 5.0

The beginning of Industry 5.0 represents more than just a technological change; it marks a new way of thinking about work, human potential, and the purpose of organizations. At its core, Industry 5.0 challenges us to focus on humanity in the fast-changing work landscape, understanding that technology should meet human needs and goals instead of undermining them.

To thrive in this new environment, organizations must find a careful balance. Simply adopting new technologies is not enough; they must also build cultures that value human creativity, emotional intelligence, and overall well-being. This shift involves moving from seeing workers only as resources in production to recognizing them as whole people with rich cognitive, emotional, and social lives.

In this framework, mental health becomes crucial for long-term success. When employees have autonomy, purpose, and supportive environments, they tend to be more innovative, collaborative, and resilient. If organizations ignore these human factors, they may face burnout, disengagement, and inequality—outcomes that contradict the very promise of Industry 5.0.

The true potential of Industry 5.0 lies in its comprehensive approach. Here, technological excellence works hand in hand with ethical responsibility and human dignity. By creating workplaces that tailor experiences, encourage inclusion, and support ongoing learning, organizations can achieve remarkable levels of employee engagement and creativity.

However, realizing this vision demands careful attention and intention. Ethical issues regarding data privacy, job security, and fair access must be prioritized during technological implementation. Leaders should promote open communication and empathetic practices to build trust and psychological safety in times of change. Policymakers, educators, and businesses must work together to create systems supporting lifelong learning and inclusive workforce development.

From a broader societal viewpoint, Industry 5.0 provides a chance to rethink the future of work in ways that support sustainable economic growth while enhancing human welfare. By placing mental health at the forefront of this transformation, societies can develop resilient, adaptable, and fulfilled workforces ready to tackle the challenges of the twenty-first century.

In conclusion, Industry 5.0 encourages us to picture a future where technology serves as a tool for human growth. It invites us to design workplaces that are not only efficient but also compassionate; not just innovative but also just; not only advanced but also deeply human.

As scholars, practitioners, and leaders, we have a responsibility to guide this evolution with wisdom, empathy, and courage. The future of work is being formed today. By placing humanity at the center of Industry 5.0, we can build a legacy of thriving employees and successful organizations.

References

Acemoglu, D., & Restrepo, P. (2019). *Automation and new tasks: How technology displaces and reinstates labor*. National Bureau of Economic Research. doi:https://doi.org/10.3386/w25684

Airbus SE. (2021). *2021 annual report*. https://www.airbus.com/sites/g/files/jlcbta136/files/2022-04/Airbus-SE-Full-Annual-Report-2021_0.pdf

Airbus. (2023, June 21). *Mixed reality to meet future challenges*. Airbus. https://www.airbus.com/en/newsroom/stories/2023-06-mixed-reality-to-meet-future-challenges

Ajunwa, I. (2023). The quantified worker: Law and technology in the modern workplace. *Cambridge Law Journal, 79*(3), 331–356. https://doi.org/10.1017/9781316888681

Al-Debei, M. M., & Al-Lozi, E. (2014). Explaining and predicting the adoption intention of mobile data services: A value-based approach. *Computers in Human Behavior, 35*, 326–338. https://doi.org/10.1016/j.chb.2014.03.011

AT&T. (2021). *Transforming into a future-ready workplace*. AT&T Corporate Publications.

Ayyagari, R., Grover, V., & Purvis, R. (2011). Technostress: Technological antecedents and implications. *MIS Quarterly, 35*(4), 831–858. https://doi.org/10.2307/41409963

Bakker, A. B., & Demerouti, E. (2007). The job demands-resources model: State of the art. *Journal of Managerial Psychology, 22*(3), 309–328. https://doi.org/10.1108/02683940710733115

Benešová, A., & Tupa, J. (2017). Requirements for education and qualification of people in Industry 4.0. *Procedia Manufacturing, 11*, 2195–2202. https://doi.org/10.1016/j.promfg.2017.07.366

Bonekamp, L., & Sure, M. (2015). Consequences of Industry 4.0 on human work: A qualitative study. *Journal of Business and Media Psychology, 6*(1), 33–40.

Bosch. (2022). Adaptable production: A concept for resilient manufacturing systems. Bosch Global. https://www.bosch.com/stories/adaptable-production/

Bosch Corporate Report. (2022). *Collaborative automation and workforce integration*. Bosch AG.

Brassey, J., De Smet, A., Maor, D., & Rabipour, S. (2024, December 6). Developing a resilient, adaptable workforce for an uncertain future. *McKinsey Quarterly*. https://www.mckinsey.com/capabilities/people-and-organizational-performance/our-insights/developing-a-resilient-adaptable-workforce-for-an-uncertain-future

Broadbent, E. (2017). Interactions with robots: The truths we reveal about ourselves. *Annual Review of Psychology, 68*(1), 627–652. https://doi.org/10.1146/annurev-psych-010416-043958

Bucci, I., Fani, V., & Bandinelli, R. (2024). Towards human-centric manufacturing: Exploring the role of human digital twins in Industry 5.0. *Sustainability, 17*(1), 129. https://doi.org/10.3390/su17010129

Deci, E. L., & Ryan, R. M. (2000). The "what" and "why" of goal pursuits: Human needs and the self-determination of behavior. *Psychological Inquiry, 11*(4), 227–268. doi:https://doi.org/10.1207/S15327965PLI1104_01

Deloitte. (2025). *Thriving together*. Deloitte Insights. https://www.deloitte.com/us/en/careers/deloitte-life/thriving-together-well-being.html?icid=top_thriving-together-well-being

Demir, K. A., Döven, G., & Sezen, B. (2019). Industry 5.0 and human-robot co-working. *Procedia Computer Science, 158*, 688–695. https://doi.org/10.1016/j.procs.2019.09.104

Derks, D., & Bakker, A. B. (2012). Smartphone use, work-home interference, and burnout: A diary study on the role of recovery. *Applied Psychology, 63*(3), 411–440. https://doi.org/10.1111/j.1464-0597.2012.00530.x

Derks, D., van Mierlo, H., & Schmitz, E. B. (2014). A diary study on work-related smartphone use, psychological detachment and exhaustion: Examining the role of the perceived segmentation

norm. *Journal of Occupational Health Psychology, 19*(1), 74–84. https://doi.org/10.1037/a0035076
Eaves, S. (2023, February 7). *Transforming into a future-ready workplace.* AT&T Business. https://www.business.att.com/learn/articles/transforming-into-a-future-ready-workplace.html
Edmondson, A. C. (2019). *The fearless organization: Creating psychological safety in the workplace for learning, innovation, and growth.* Wiley.
Edmondson, A. C., & Lei, Z. (2014). Psychological safety: The history, renaissance, and future of an interpersonal construct. *Annual Review of Organizational Psychology and Organizational Behavior, 1*, 23–43. https://doi.org/10.1146/annurev-orgpsych-031413-091305
Empowered Automation Solutions LLC. (2022). *What does Industry 4.0 mean.* https://www.empoweredautomation.com/industry-4-0
Epicor. (2023). *What is Industry 4.0—The Industrial Internet of Things (IIoT)?* https://www.epicor.com/en-us/blog/technology-and-data/what-is-industry-4-0/
Eurofound. (2022). *Digitalisation and its impact on work and society.* European Foundation for the Improvement of Living and Working Conditions.
European Commission. (2021). *Industry 5.0: Towards a sustainable, human-centric and resilient European industry (EUR 30590 EN).* doi:https://doi.org/10.2760/733440
Financial Times. (2025, May 27). March of the cobots: The technology lowering the barrier to automation. *Financial Times.* https://www.ft.com/content/78c1d4e9-ad30-47f5-ab7b-390df5bc1f10.
Fliegner, F. (2023). Living conditions: European Foundation for the improvement of living and working conditions. In *Encyclopedia of quality of life and well-being research* (pp. 3972–3973). Springer. https://doi.org/10.1007/978-3-031-17299-1_936
Fournier, É., Jeoffrion, C., Hmedan, B., Pellier, D., Fiorino, H., & Landry, A. (2024). Human-cobot collaboration's impact on success, time completion, errors, workload, gestures and acceptability during an assembly task. *Applied Ergonomics, 119*, 104306. doi:https://doi.org/10.1016/j.apergo.2024.104306
Fraunhofer, I. A. O. (2020). *Designing accessible human–machine interfaces.* Fraunhofer Institute for Industrial Engineering.
Gagné, M., & Deci, E. L. (2005). Self-determination theory and work motivation. *Journal of Organizational Behavior, 26*(4), 331–362. https://doi.org/10.1002/job.322
Geissdoerfer, M., Morioka, S. N., de Carvalho, M. M., & Evans, S. (2018). Business models and supply chains for the circular economy. *Journal of Cleaner Production, 190*, 712–721. https://doi.org/10.1016/j.jclepro.2018.04.159
Gensler, E., Oestreich, H., Abendroth, A.-K., Wrede, S., & Wrede, B. (2023). Work autonomy and adaptive digital assistance in flexible working environments. In *The digital twin of humans* (pp. 137–170). Springer. https://doi.org/10.1007/978-3-031-26104-6_7
Ghobakhloo, M., Iranmanesh, M., Grybauskas, A., Vilkas, M., & Petraitė, M. (2021). Industry 4.0, innovation, and sustainable development: A systematic review and a roadmap to sustainable innovation. *Business Strategy and the Environment, 30*(8), 4237–4257. https://doi.org/10.1002/bse.2867
IBM. (n.d.). What is Industry 4.0? IBM THINK. https://www.ibm.com/think/topics/industry-4-0
Jackson, S. (2025). Privacy concerns toward AI-based intelligent voice assistants in the workplace. *Digital Policy, Regulation and Governance.* https://doi.org/10.1108/dprg-03-2025-0066
Kagermann, H., & Wahlster, W. (2022). Ten years of Industrie 4.0. *The Sciences, 4*(3), 26. https://doi.org/10.3390/sci4030026
Keyes, C. L. M. (2002). The mental health continuum: From languishing to flourishing in life. *Journal of Health and Social Behavior, 43*(2), 207. https://doi.org/10.2307/3090197
La Torre, G., Esposito, A., Sciarra, I., & Chiappetta, M. (2018). Definition, symptoms and risk of techno-stress: A systematic review. *International Archives of Occupational and Environmental Health, 92*(1), 13–35. https://doi.org/10.1007/s00420-018-1352-1

LineView. (2023). What is Industry 5.0? Human-centricity, personalization, and sustainability. https://lineview.com/en/what-is-industry-5-0-human-centricity-personalization-and-sustainability/

Longo, F., Padovano, A., & Umbrello, S. (2020). Value-oriented and ethical technology engineering in industry 5.0: A human-centric perspective for the design of the factory of the future. *Applied Sciences, 10*(12), 4182. https://doi.org/10.3390/app10124182

Luthans, F., Youssef, C. M., & Avolio, B. J. (2007). Psychological capital: Investing and developing positive organizational behavior. *Positive Organizational Behavior, 1*(2), 9–24. https://doi.org/10.4135/9781446212752.n2

Maddikunta, P. K. R., Pham, Q. V., Prabadevi, B., Deepa, N., Dev, K., Gadekallu, T. R., Liyanage, M., et al. (2022). Industry 5.0: A survey on enabling technologies and potential applications. *Journal of Industrial Information Integration, 26*, 100257. https://doi.org/10.1016/j.jii.2021.100257

Mark, G., Iqbal, S. T., & Czerwinski, M. (2016). Email duration, batching and self-interruption: Patterns of email use on productivity and stress. In *Proceedings of the 2016 CHI conference on human factors in computing systems* (pp. 1717–1728). doi:https://doi.org/10.1145/2858036.2858262

Microsoft Corporation. (2022). *2022 annual report*. https://www.microsoft.com/investor/reports/ar22/index.html

Moore, P. V., Upchurch, M., & Whittaker, X. (Eds.). (2018). *Humans and machines at work*. Springer International Publishing. https://doi.org/10.1007/978-3-319-58232-0

Nahavandi, S. (2019). Industry 5.0—A human-centric solution. *Sustainability, 11*(16), 4371. https://doi.org/10.3390/su11164371

NPR. (2019, November 4). 4-day workweek boosted workers' productivity by 40%, Microsoft Japan says. https://www.npr.org/2019/11/04/776163853/4-day-workweek-boosted-workers-productivity-by-40-microsoft-japan-says

OECD. (2020). OECD employment outlook 2020. OECD Employment Outlook. doi:https://doi.org/10.1787/1686c758-en

Parasuraman, R., & Hancock, P. A. (2000). Adaptive control of mental workload. In P. A. Hancock & J. L. Szalma (Eds.), *Stress, workload, and fatigue* (pp. 305–320). CRC Press. https://doi.org/10.1201/b12791-2.4

Paul, S. (2018). Crisis in Boeing 787 Dreamliner: An investigation from project management control perspective. *International Journal of Human Resource Studies, 8*(4), 242. https://doi.org/10.5296/ijhrs.v8i4.13805

Philips. (2025, February 24). Innovation spotlight: How a circular economy is driving sustainable healthcare. https://www.philips.com/a-w/about/news/archive/features/2025/innovation-spotlight-how-a-circular-economy-is-driving-sustainable-healthcare.html.

Philips Sustainability Report. (2022). *Digital transformation and employee engagement*. Philips.

Ragu-Nathan, T. S., Tarafdar, M., Ragu-Nathan, B. S., & Tu, Q. (2008). The consequences of technostress for end users in organizations: Conceptual development and empirical validation. *Information Systems Research, 19*(4), 417–433.

Robertson, T., & Simonsen, J. (2012). Participatory design: An introduction. In *Routledge international handbook of participatory design* (pp. 1–17). Routledge.

Rozovsky, J. (2015). The five keys to a successful Google team. https://rework.withgoogle.com/blog/five-keys-to-a-successful-google-team/

Ryff, C. D. (1989). Happiness is everything, or is it? Explorations on the meaning of psychological well-being. *Journal of Personality and Social Psychology, 57*(6), 1069–1081. https://doi.org/10.1037/0022-3514.57.6.1069

Sadeghi, S. (2024). Employee well-being in the age of AI: Perceptions, concerns, behaviors, and outcomes. arXiv preprint arXiv:2412.04796.

Sakshi, & Dhand, S. (2025). Human-centric digital workplace. In S. Kumar & A. Mehta (Eds.), *Human-centric, sustainable, and resilient organizations in the digital age* (pp. 125–146). IGI Global. https://doi.org/10.4018/979-8-3693-8181-6.ch006

Salanova, M., Llorens, S., & Cifre, E. (2012). The dark side of technologies: Technostress among users of information and communication technologies. *International Journal of Psychology, 48*(3), 422–436. https://doi.org/10.1080/00207594.2012.680460
SAP SE (2025, February 27). *SAP integrated report 2024*. https://www.sap.com/integrated-reports/2024/en/csrd.html
SAP Sustainability Report. (2023). *Mental health and well-being initiatives*. SAP SE.
Seligman, M. E. (2011). *Flourish: A visionary new understanding of happiness and well-being*. Simon and Schuster.
Shih, W. (2020). Global supply chains in a post-pandemic world? *Harvard Business Review*. https://hbr.org/2020/09/global-supply-chains-in-a-post-pandemic-world
Siemens. (2020). Amberg electronics plant: A showcase of Industry 4.0. https://www.industrie40award.com/the-winners/siemens-ag-elektronikwerk-amberg/
Siemens AG, & University of Göttingen. (2021). *Empowering people: Insights into strategy implementation in the digital era*. Siemens AG. https://www.siemens.com/empowering-people
Siemens Annual Report. (2023). *Empathy and leadership in Industry 5.0 transformation*. Siemens AG.
SkillsFuture Singapore. (2021). *Annual reports*. https://www.ssg.gov.sg/about-us/annual-reports
Sohnnentag, S., & Fritz, C. (2018). Recovery from work. In P. Y. Chen, C. L. Cooper, & J. Barling (Eds.), *The SAGE handbook of industrial, work and organizational psychology* (pp. 471–482). SAGE. https://doi.org/10.4135/9781473914964.n21
SSAB (2022). *Stakeholder engagement [Section on sustainability strategy and governance]*. SSAB. https://www.ssab.com/en/company/sustainability/strategy-and-governance/stakeholder-engagement
Stackpole, T. (2021, June 4). Inside IKEA's digital transformation. *Harvard Business Review*. https://hbr.org/2021/06/inside-ikeas-digital-transformation
Steger, M. F., Dik, B. J., & Duffy, R. D. (2012). Measuring meaningful work. *Journal of Career Assessment, 20*(3), 322–337. https://doi.org/10.1177/1069072711436160
Suh, A., & Lee, J. (2017). Understanding teleworkers' technostress and its influence on job satisfaction. *Internet Research, 27*(1), 140–159. doi:https://doi.org/10.1108/intr-06-2015-0181
Tarafdar, M., Pullins, E. B., & Ragu-Nathan, T. S. (2015). Technostress: negative effect on performance and possible mitigations. *Information systems journal, 25*(2), 103–132.
Tarafdar, M., Tu, Q., Ragu-Nathan, T. S., & Ragu-Nathan, B. S. (2011). Crossing to the dark side. *Communications of the ACM, 54*(9), 113–120. https://doi.org/10.1145/1995376.1995403
Toyota. (2025). *Advancing the integration of robots and AI to drive technological innovation*. Toyota. https://global.toyota/en/mobility/frontier-research/42394425.html
Unilever. (2025). Safety at work. https://www.unilever.com/sustainability/responsible-business/safety-at-work/
United Nations. (2015). Transforming governance for the 2030 agenda for sustainable development. https://doi.org/10.18356/e5a72957-en
United Nations Entity for Gender Equality and the Empowerment of Women (UN Women). (2021, September 17). *Progress on the Sustainable Development Goals: The Gender Snapshot 2021*. https://data.unwomen.org/publications/progress-sustainable-development-goals-gender-snapshot-2021
van Deursen, A. J. A. M., & Helsper, E. J. (2015). The third-level digital divide: Who benefits most from being online? *Communication and Information Technologies Annual, 10*, 29–52. https://doi.org/10.1108/S2050-206020150000010002
Villani, V., Pini, F., Leali, F., & Secchi, C. (2018). Survey on human–robot collaboration in industrial settings: Safety, intuitive interfaces and applications. *Mechatronics, 55*, 248–266. https://doi.org/10.1016/j.mechatronics.2018.02.009
Wang, L., Váncza, J., Kemény, Z., & Wang, X. V. (2021). Future research directions on human–robot collaboration. In J. Smith & L. Chen (Eds.), *Advanced human–robot collaboration in manufacturing* (pp. 439–448). Springer. https://doi.org/10.1007/978-3-030-69178-3_18

World Health Organization. (2022). *Mental health at work: Policy brief*. https://www.who.int/publications/i/item/9789240057944

Wu, S.-M. (2024). Emotional intelligence and workplace spirituality in predicting career success of high-tech leaders. *Behavioral Sciences, 14*(11), 1009. https://doi.org/10.3390/bs14111009

Xu, L. D., Xu, E. L., & Li, L. (2018). Industry 4.0: State of the art and future trends. *International Journal of Production Research, 56*(8), 2941–2962. https://doi.org/10.1080/00207543.2018.1444806

Yang, J., Liu, Y., & Morgan, P. L. (2024). Human–machine interaction towards industry 5.0: Human-centric smart manufacturing. *Digital Engineering, 2*, 100013. https://doi.org/10.1016/j.dte.2024.100013

Zhang, X., Lee, S. K., Kim, W., & Hahn, S. (2023). "Sorry, it was my fault": Repairing trust in human–robot interactions. *International Journal of Human-Computer Studies, 175*, 103031. doi:https://doi.org/10.1016/j.ijhcs.2023.103031.

Leading through Emergent Technologies in Industry 5.0

Cory Campbell and Dijo Alexander

Abstract This chapter, Leading Through Emergent Technologies in Industry 5.0, explores the paradigm shift from automation-centric Industry 4.0 to the human-centric, ethically grounded vision of Industry 5.0. It positions emergent technologies such as generative AI, large language models (LLMs), digital twins, and integrated edge–cloud–mobile infrastructures as catalysts for redefining enterprise systems and enabling intuitive human–machine collaboration. The discussion begins by tracing the evolution of the human–computer interface, culminating in natural language-driven interaction that makes technology more accessible and adaptive. Central to Industry 5.0 are three pillars: human centricity, sustainability, and resilience. The chapter illustrates how these pillars guide innovation beyond efficiency toward amplifying human creativity, fostering environmental stewardship, and building robust systems capable of withstanding disruption. It examines practical applications, from AI-powered cobots and predictive analytics to blockchain-enabled supply chain transparency, while highlighting the strategic imperatives of data governance, cybersecurity, and ethical oversight. The narrative addresses critical challenges, including workforce skill gaps, inherent algorithmic bias, and the persistent burden of technical debt accumulated over decades of IT evolution. Generative AI emerges as a transformative enabler, offering organizations the ability to modernize legacy systems rapidly and leapfrog intermediate stages of digital transformation. This capability, combined with design thinking principles, empowers leaders to align technological progress with organizational goals and societal values. Ultimately, the chapter envisions Industry 5.0 as a future where technology amplifies human potential rather than replaces it. It calls for leadership that integrates technological fluency with ethical responsibility, ensuring that innovation serves collective prosperity. By synthesizing advanced tools with human insight, organizations can create adaptive, inclusive, and sustainable ecosystems, fulfilling the promise of Industry 5.0 as a harmonious blend of intelligence, resilience, and humanity.

C. Campbell (✉)
Scott College of Business, Indiana State University, Terre Haute, IN, USA
e-mail: cory.campbell@indstate.edu

D. Alexander
Lubar College of Business, University of Wisconsin-Milwaukee, Milwaukee, WI, USA

A. M. Waite et al. (eds.), *Leading in Industry 5.0*, Management and Industrial Engineering, https://doi.org/10.1007/978-3-032-19207-3_2

Keywords Data structure · Data governance · Human centricity · Sustainability · Resilience · Technical debt

1 Introduction. Beyond Technology—Toward Human-Tech Synergy

For over half a century, computing power has fundamentally reshaped human existence, revolutionizing commerce, governance, education, and entertainment. Yet, despite this massive impact, our primary mode of interacting with computers has remained inherently unnatural, a progression from punch cards to keyboards, mice, and touchscreens. In recent years, this fundamental barrier has fallen. Major breakthroughs in Natural Language Processing (NLP) have initiated a profound paradigm shift. We no longer need to translate our thoughts into mechanical input; we can now interact with and instruct computers using the most natural human medium, our own language.

This shift moves beyond stilted commands and rigid interfaces. It allows for conversational, nuanced communication, making computer assistants and decision-support systems exponentially more intuitive and human-friendly. As summarized in Fig. 1, the human-technology value loop illustrates the moment when computers step fully into the human domain, accommodating even broken language constructs and natural conversational flow, creating business value from data feedback and adaptive learning using human insight and intelligent technology. The computer is now learning to speak our language, not the other way around.

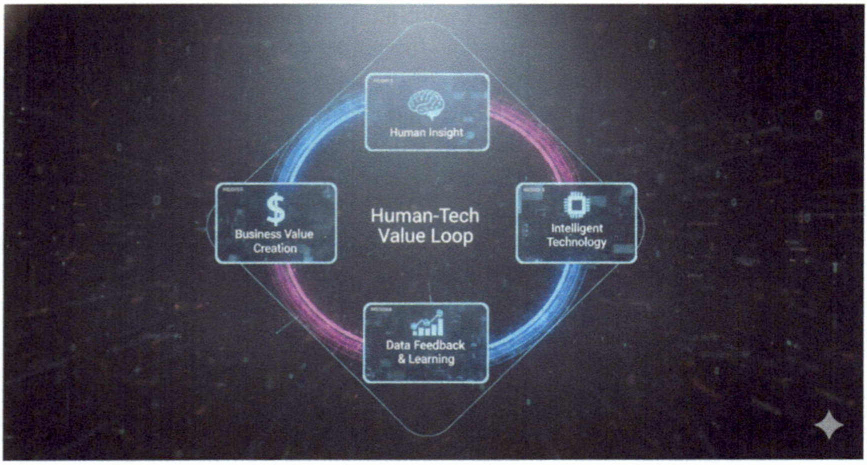

Fig. 1 Human technology loop in Industry 5.0. Image created using Google Gemini 2.5 Flash (Nano Banana)

This evolution of the human–computer interface (HCI) is profound! Over time, we transitioned to most recently to conversation as the preferred mechanism for interfacing with the computer. In this AI-driven world, we are now to the point where human beings can be approximately correct with the delivery of sentence structure and grammar, but the machine grammar will have no spelling or grammar mistakes. In the words of English philosopher Carveth Read, "It is better to be vaguely right than exactly wrong" (Oxford Essential Quotations, 2018). In other words, you don't have to be perfect (e.g., broken French input can return a fluid French output) as the LLM will interpret the meaning or context of your words. Notwithstanding the issue of substance over form issue, the machine does structure-related tasks (form) incredibly well.

From the early days of computing in Industry 3.0 to the rise of the internet and analytics that enabled the widespread flow of computing power in Industry 4.0, we see a pattern similar to earlier industrial revolutions. Just as mechanical power defined Industry 1.0 and electricity enabled the distribution of power in Industry 2.0, we are now entering Industry 5.0, where the defining shift is our ability to converse and interact with computers in natural language. Coincidentally, it is imperative that we have some fundamental understanding of the technological capabilities as a basis to understand how to best utilize the potential opportunities while mitigating the possible risks and limitations. Yet, technology alone does not supplant the need for human beings, but this transformative approach integrates emerging technologies with a focus on human–machine collaboration and responsible governance. To that end, the *human-in-the-loop* remains the essential key cog of this process.

To champion Industry 5.0 from a leadership perspective, one must operate at the intersection of a strong technological foundation, data governance, and ethics (Fig. 2).

- **Technological Foundation**: Industry 5.0 leverages cognitive and agentic AI capable of autonomous decision-making to drive innovation and operational efficiency.
- **Data and Ethical Governance**: Data as the raw material for AI in Industry 5.0, robust data management practices are critical, encompassing data integrity, governance, AI testing, and cybersecurity within hyperconnected environments.
- **Leadership and Strategy**: Effective leadership in Industry 5.0 shifts toward facilitative roles with technological fluency, fostering trust in hybrid human–AI teams, and navigating diverse cultural and regulatory landscapes.

In this chapter, we explore how emergent technologies are reshaping the future of business. We examine the evolving role of leadership in a globalized, tech-driven world, and present actionable frameworks for leveraging these innovations responsibly.

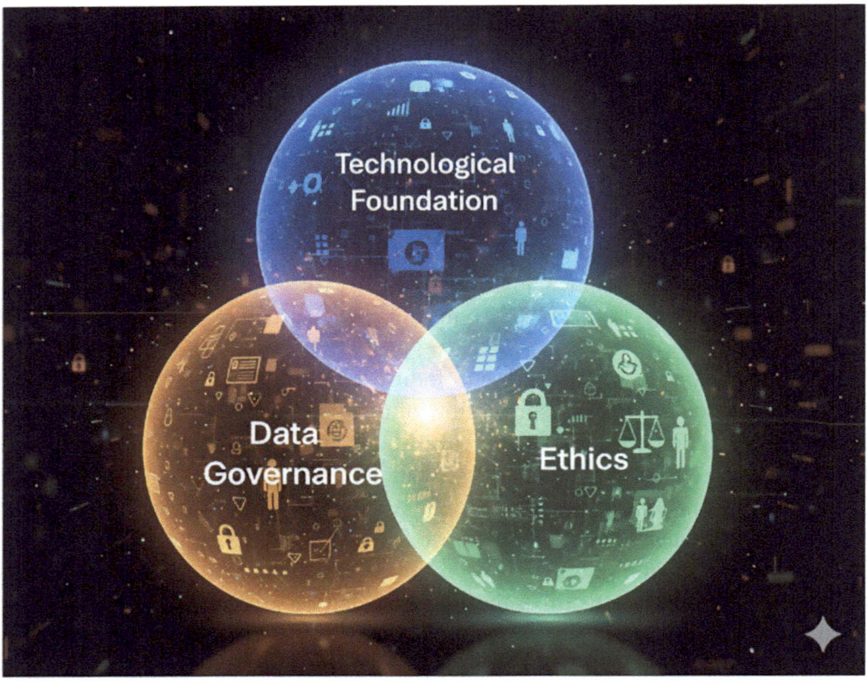

Fig. 2 Foundations of leadership in Industry 5.0. Image created using Google Gemini 2.5 Flash (Nano Banana)

2 Corporate Sprint to Adopt Emerging Technology

2.1 Technological Disruption

In today's hypercompetitive global economy, businesses are engaged in an arms race to adopt emerging technologies in pursuit of operational efficiency, innovation, and competitive advantage. From AI and cloud computing to robotic process automation (RPA) and big data analytics, the pressure to digitize has never been greater. Rahilly and Fusaro (2025, September 12) suggest that global competition and the exponential demand for computing power are driving companies to embed emergent technologies into core business processes at unprecedented speed.

The corporate drive to harness AI as we enter Industry 5.0 represents a profound technological disruption that fundamentally transforms business structures, operational paradigms, and sources of competitive advantage. Companies today are not only digitizing routine operations but are actively embedding AI, including generative models, natural language processing, and advanced analytics, into the very fabric of strategic decision-making, customer engagement, and value creation processes. AI-driven applications are now underpinning predictive forecasting, dynamic resource allocation, real-time customer interaction, and innovation cycles of hyper-personalized experience that would have previously been unimaginable..

As global competition accelerates, organizations that adopt AI early emerge as frontrunners, leveraging personalization, continuous intelligence, and collaborative automation as key differentiators in hypercompetitive markets. The McKinsey analysis projects that AI could contribute up to $4.4 trillion in added productivity annually, while Gartner highlights that over $5.7 trillion in global IT spending is at stake, with AI infrastructure and applications playing a central role (Gartner, 2023; Chui et al., 2023, June). On the ground, this is reflected in the rapid transition from AI experimentation (with nearly 40% of companies) to deployment at scale, a race that is rewriting industry norms and even geopolitical alliances based on AI leadership (Chui et al., 2023, June).

Crucially, this disruption is not just a matter of technical capability, but this marks a new era of human-technology synergy. The evolution from Industry 4.0's automation focus to Industry 5.0's human-centric paradigm is characterized by collaborative intelligence, where tools such as cobots, digital twins, and generative AI will augment human ingenuity rather than replace it. Enterprises are moving beyond replacing tasks to fundamentally re-imagining how humans and machines co-create value, personalize services, and ensure adaptability in fluid environments. Businesses that can synthesize data-driven insight, AI-powered augmentation, and workforce partnership will not only survive, but thrive in the emergent landscape of Industry 5.0 (Fazlollahtabar, 2025). In this new era, those with robust data will have a distinct competitive advantage over those who do not.

AI as a transformative technological disruption does not represent a one-time shift, but a ceaseless sprint where strategy, culture, and systems must co-evolve to realize the promise of AI-enabled prosperity. Companies face a clear challenge and call to action, which is to adapt with vision and urgency, or risk obsolescence in a world where the pace of change is accelerating. Early adopters stand positioned to re-imagine the boundaries of possibility, defining the next generation of enterprise innovation, resilience, and human-tech harmony.

2.2 *Technological Backbone of the Enterprise*

Enterprise Resource Planning (ERP) systems delivered as Software-as-a-Service (SaaS) serve as the digital backbone of modern organizations by integrating core business functions including finance, human resources, supply chain, and operations into a unified platform that facilitates real-time data flow and process automation. This centralized architecture enables organizations to streamline workflows, reduce redundancies, and improve operational efficiency. However, ERP systems are inherently transactional in nature, primarily designed to record and manage business activities rather than analyze them. As a result, they often lack the advanced analytical capabilities needed for strategic decision-making and performance forecasting. To address this gap, organizations bolt-on Business Intelligence (BI) tools that extract, visualize, and interpret ERP data, transforming raw transactions into actionable insights. BI tools empower leaders to monitor key performance indicators (KPIs), identify trends, and make data-driven decisions that ERP systems alone cannot support (Buck et al., 2022). Complementing this technological foundation is Enterprise Risk Management (ERM), a strategic framework that helps organizations identify, assess, and respond to risks across all business units. Unlike traditional siloed risk approaches, ERM provides a holistic, top-down view of risk, aligning it with organizational objectives and enabling proactive mitigation strategies. Together, ERP, BI, and ERM form a triad of digital infrastructure that supports operational excellence, strategic agility, and resilience in an increasingly complex business environment.

However, despite massive investments, many technology initiatives fail to deliver expected outcomes. For example, ERP implementations have failure rates ranging from 40 to 90%, often due to poor planning, lack of user engagement, and misalignment with business needs (Campbell & Fogarty, 2018; Deloitte, 2022). Implementation challenges are widely documented in various other business tools. For example, Deloitte's research shows that while 78% of organizations are implementing robotics process automation (RPA), only a fraction has a scalable strategy, leading to stalled or ineffective deployments (Deloitte, 2022). Similarly, big data projects also face challenges in integration, governance, and getting actionable insights.

Despite this potential, most enterprise systems remain transactional at their core. ERP platforms, for example, are optimized for structured data and operational efficiency, not for the interpretive depth required in strategic decision-making. To bridge this gap, organizations have bolted on business intelligence tools to extract meaning from ERP data. Yet, these tools often fall short when confronted with the volume, velocity, and variety of unstructured data, which now comprises the majority of enterprise data.

2.3 *The* "Missing Link"*: Digital Culture*

A scan of the literature will quickly recover a multitude of implementation failures. We need not repeat those here, but rather we want to emphasize that the utilization of emerging technologies to create a competitive advantage is an elusive pursuit for many organizations. Furthermore, we emphasize how these failures highlight a critical truth: technology alone is not enough! Successful digital transformation requires a strong digital culture that embraces agility, cross-functional collaboration, and continuous learning. McKinsey emphasizes that cultural barriers such as siloed mindsets, risk aversion, and lack of customer focus are among the top reasons digital initiatives falter (Goran et al., 2017). Central to this cultural shift is the *human-in-the-loop approach*, which ensures that technology augments rather than replaces human judgment. By embedding human oversight, ethical reasoning, and contextual awareness into digital systems, businesses can better navigate complexity, build trust, and unlock the full potential of their technological investments.

As Gerald Kane (2019) points out in his seminal book, there is misbelief that technology problems require technology solutions. This is the basis of what he frames as the technology fallacy. However, as technology is taking over various aspects of our lives, it is imperative we understand its limitations. Without the right culture, leadership, and human engagement, even the most advanced tools can become expensive liabilities. Successful adoption requires more than infrastructure. It demands a digital culture that embraces agility, collaboration, and continuous learning. It also requires a *human-in-the-loop* approach, where technology augments human judgment rather than replaces it.

While technology continues to evolve, the exponential nature of big data itself continues to place a premium on the critical role of having a strong digital culture for the imminent future to fully realize the capabilities of the tools that are shaping the future as we segue to Industry 5.0 and beyond.

3 Evolution from Industry 4.0 to a Human-Centric Future

3.1 Core Tenets of Industry 4.0

The industrial landscape is undergoing a deep transformation. While Industry 4.0 revolutionized production and operations through automation, big data, and the Internet of Things (IoT), it often did so at the expense of human-centric values. Industry 5.0 emerges as a response to this imbalance by reintegrating human creativity, ethical responsibility, and environmental sustainability into the technological equation.

The foundational principle of industry 4.0 is interconnection which refers to the ability of machines, devices, sensors, and people to connect and communicate with each other. This communication happens via the Internet of Things (IoT) and the Internet of People (IoP). In an Industry 4.0 environment, components on the factory floor constantly share data with each other and with higher-level systems, creating a fully integrated and networked ecosystem.

The tenets of information transparency through "digital twin" or virtual copy of the physical world helps gather vast amounts of data from IoT sensors, a rich, real-time digital model of the factory, its processes, and its products built. This transparency allows managers and systems to monitor, analyze, and understand what is happening in the physical world from anywhere. That level of transparency seemingly enables far greater levels of informed decision-making.

Such a model of the physical world captured virtually, enables cyber-physical systems (CPS), integrating computation with physical processes, giving the autonomy to make simple, routine decisions on their own. For example, a machine could autonomously order more raw materials when its sensors detect low stock, or it could adjust its own operating parameters to correct a quality deviation without human intervention. Only in cases of exception (like a major error or conflicting goals) are tasks escalated to a human operator. This is a model for decentralized decision-making to pave the way for collaborative robotic assistance such as data aggregation and visualization that can help workers make complex decisions quickly and reduce their cognitive load.

3.2 *Industry 5.0: The Next Evolution*

The evolution from Industry 4.0's "digital model" to Industry 5.0's more advanced understanding of the world model is a shift from simulation to symbiosis. Industry 4.0 perfected the Digital Twin, or a high-fidelity virtual replica of a physical asset or process. Its primary reasoning, driven by predictive AI, was analytical and focused on efficiency. Industry 5.0 evolves this model by integrating new layers of data and, crucially, new classes of artificial intelligence. This creates a far richer, more holistic model that we will frame as the "Human-Centric Twin."

At its core, Industry 5.0 is not just about smarter machines, but about smarter collaboration between humans and machines. It emphasizes co-creation, where technologies like cognitive AI, collaborative robots (cobots), and smart wearables augment human capabilities rather than replace them. This shift is not merely a technological one, but rather a philosophical, organizational, and strategic transformation.

As mentioned earlier, the progression through Industry 4.0 has ultimately led to an era where humans can now interface with computers instinctively by using natural language as the primary mode of interaction. Each iteration of that evolution involved major rewrites of the user interface providing better user experience (UX) for the users. With the latest iteration of natural language-driven AI, native UX, Industry 5.0 offers the next level of human–computer interface more natural and native (Fig. 3).

While Industry 4.0 emphasized automation and connectivity, Industry 5.0 shifts the focus toward human-centric innovation, where technology serves as a strategic partner rather than a replacement. Both data engineering and data governance play key roles to support the pillars of Industry 5.0, which include: human centricity, sustainability, and resilience. In this next section, we will illustrate how these pillars serve as a guiding framework and tools quickly reshape how organizations lead, strategize, and create value.

3.3 *Pillars of Industry 5.0: Human Centricity, Sustainability, and Resilience*

Large Language Models (LLMs) are a breakthrough in natural language processing (NLP) research for computers to understand and generate human language. LLMs are deep learning systems built from neural networks trained on massive sets of text from

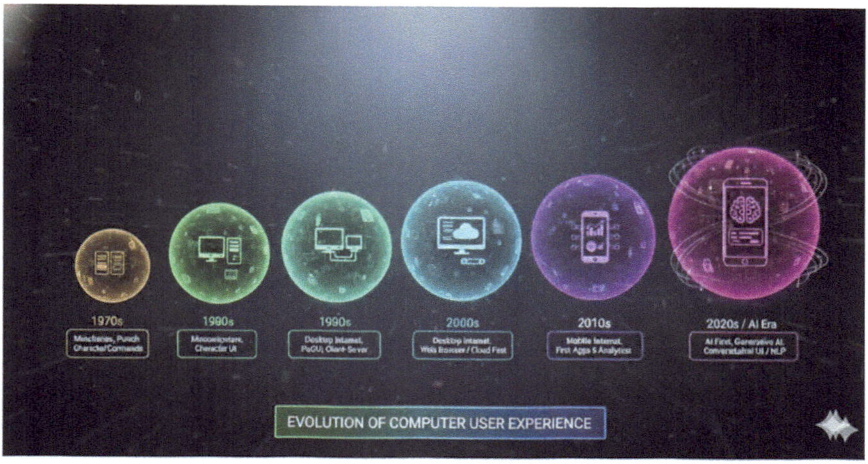

Fig. 3 Evolution of the human–computer interface. Image created using Google Gemini 2.5 Flash (Nano Banana)

the internet. This NLP breakthrough was triggered by the representation of words and phrases into multi-dimensional numeric vectors called embeddings, which are based on their compositional metadata. The subsequent invention of transformer[1] architecture accelerated the ability to learn patterns of language, knowledge, and reasoning from the training data. LLMs split input query text into tokens (sub-words). Then, it will predict the next token(s) to generate answers or responses to the queries. The input queries, called prompts, set the context with specific instructions and examples. These prompts need context and clarity to be effective. As such, prompts will need fine-tuning with field-specific data to adapt them to certain narrow domains. Though mostly fluent in their responses, occasionally they tend to slide outside the context and produce incorrect statements, which are called "hallucinations." These models also reflect the biases from their training data, requiring proper guardrails and reviews.

LLMs are powerful in language tasks such as summarizing documents, answering questions, writing code, translating, and following instructions. To be effective, good prompts must be paired with domain knowledge as way to add grounding to the output, and adequate safeguards which keep the *human-in-the-loop*. Beyond language capabilities, they are increasingly used for robotics, multi-modal learning, and strategic decision-making.

While LLMs are still the most valued for their ability to understand and generate natural language creating the most natural and ideal human–computer interface (HCI), they created the ability for computers to work with unstructured data almost as well as it could with structured data in the past. This greatly expands the scope and definition of data sources for business decision-making computations. LLMs

[1] The Transformer is illustrated and explained at https://jalammar.github.io/illustrated-transformer/ and https://poloclub.github.io/transformer-explainer/.

Fig. 4 Pillars of Industry 5.0. Image created using Google Gemini 2.5 Flash (Nano Banana)

are also enabling the generation of micro-level actions for macro-level goals from their knowledge patterns, leading to AI Agents, which are goal-driven proactive and autonomous AI systems, with actionable intentions with sufficient autonomy and reasoning power. AI agents with expanded ability to work with both structured and unstructured data are unleashing a series of new business opportunities never imagined before. Such downstream business opportunities are going to be the economic driver for success in the age of Industry 5.0 (Fig. 4).

3.4 Pillar 1: Human Centricity

Large Language Models are designed to understand and generate human language. They belong to a class of deep learning models trained on massive sets of data using sophisticated structures known as transformer architectures. These approaches allow LLMs to continually improve performance as the quantity of data and model complexity increases based on a phenomenon often referred to as scaling laws, which leads to capabilities that can amplify scale, speed, and specificity. While LLMs were initially focused on language tasks such as translation and summarization, their use now extends into fields like robotics, multi-modal learning, and even strategic decision-making.

At the heart of LLMs are artificial neural networks which are computational systems loosely inspired by the structure and function of the human brain. Just as the brain is made up of billions of interconnected neurons, an artificial neural network is composed of many simple processing units (or "neurons") that work together to process information. These networks "learn" by repeatedly analyzing patterns within the data and gradually adjusting the connections, called parameter weights, between each unit. Through this process of adjustment, the network becomes better at recognizing complex relationships and generating intelligent predictions. For example, after training on millions of sentences, a neural network can learn not only grammar and vocabulary, but the subtleties of meaning, context, and even emotional tone. This learning method enables LLMs to handle a wide variety of tasks and adapt to new challenges, making them one of the most powerful tools in today's AI landscape.

Figure 5 highlights how generations have adapted to change in the HCI. From the early adopters of computing from Industry 3.0, users have relied on mainframes and the technologies from the past due to their inability to find the economic justification and incentive to upgrade, have a way to wipe out all the technical debt (about which we will discuss subsequently in this chapter) by jumping into the AI first era of natural language-driven user interface using the AI capabilities to drive code translations and transitions reducing the tedious human undertaking in the past.

Micro service architectures and service orientation introduced as part of Industry 4.0 have paved the way toward an agentic future where the information and communication technology (ICT) era is maturing to materialize machines that can think as imagined by Alan Turing in the 1950s (Turing, 1980). The digitization of business processes and the building of data warehouses and lake houses as part of the big data initiatives during Industry 4.0 (Alexander & Lyytinen, 2019) have enabled businesses in Industry 5.0 to re-imagine business functions to be operated a proactive and autonomous mode. Agentic AI represents a leap beyond traditional machine learning and cognitive AI. These systems are designed to act with "agency" because they

Fig. 5 Generational adoption of technology. Image created using Google Gemini 2.5 Flash (Nano Banana)

are capabilities include making decisions, adapting to context, and pursuing goals with minimal human intervention. Unlike rule-based automation, agentic AI can learn from complex environments and collaborate with humans in dynamic decision-making. Agentic systems primarily function to advance a specific goal continuously monitoring and sensing all the signals from its environment, consistently planning for the best interventions for optimizing the intended goal, executing those actions, reviewing and reflecting on the results and further fine-tuning and adapting.

Agentic AI has endless possibilities in business. For example, a global logistics firm may use agentic AI to autonomously re-route shipments based on the specific weather, demand and geopolitical risks at much higher rate and speed than any human manager can perform. This frees up the human managers to oversee and focus on only exceptions and strategic pivots. Similarly, a financial business may deploy AI agents to monitor market volatility in "real-time" to recommend and make portfolio adjustments, escalating only the high-risk decisions with compliance officers to review before execution.

The primary challenge in operationalizing the AI agents is the effectiveness of the necessary evaluators and verifiers. If human verifiers are involved, the human verifiers will quickly become bottlenecks since they cannot catch up with the speed and scale of AI agents. That calls for automated evaluators and verifiers for agents to work in tandem. With the dynamic and non-deterministic characteristics of agentic AI, it is not easy to have testing systems for all potential possibilities. That highlights the need to build guardrail limits and operational ranges with *human-in-the-loop* escalation systems as part of the agentic AI framework.

3.5 Pillar 2: Technologies for Sustainability

Industry 5.0 embeds sustainability as a core principle, moving beyond efficiency toward environmental stewardship, resource optimization, and ethical supply chains. This pillar leverages advanced technologies to enable circularity, transparency, and operational excellence. The critical dimensions for the Technologies for Sustainability Pillar are the Internet of Things and resource management; additive manufacturing and the circular economy; AI and optimized operations; and blockchain.

3.5.1 IoT and Resource Management

The Internet of Things (IoT) is foundational for sustainable industrial ecosystems. By deploying smart sensors and connected devices, organizations can monitor energy consumption, water usage, and raw material flows in real time. This visibility enables predictive maintenance, dynamic resource allocation, and waste reduction. Chui et al. (2021) estimate IoT could unlock $5.5–$12.6 trillion in economic value globally by 2030, with significant contributions from resource efficiency and sustainability initiatives. Smart factories using IoT-enabled energy dashboards have reduced energy consumption by up to 20%, while predictive analytics minimizes downtime and material waste (Gartner, 2023). Gartner emphasizes that sustainable technology frameworks increasingly rely on IoT telemetry for ESG reporting and operational optimization (Gartner, 2023).

3.5.2 Additive Manufacturing and Circular Economy

Additive manufacturing is the process of creating an object by building it one layer at a time. In other words, objects are made as material is added. The most common application of additive manufacturing is 3D printing. By supporting on-demand production, reducing overproduction, and material waste, 3D manufacturing enables closed-loop systems where components can be recycled and reprinted. Thus, additive manufacturing aligns with circular economy principles. Circular economy models are built on the premise that materials never become waste, and nature is regenerated as both products and materials are kept in circulation (e.g., there is the process of maintenance, reuse, refurbishment, recycling, and composting). When circular economies are combined with additive manufacturing, there is lower raw material dependency and extended product lifecycles that can create recurring revenue streams through repair and refurbishment (World Economic Forum, 2023). For example, aerospace firms use additive manufacturing to produce lightweight parts, reducing material use and emissions. McKinsey notes that additive manufacturing adoption is a key lever for disruptive productivity gains in advanced manufacturing (Chui et al., 2023, June). The WEF highlights that circularity is becoming a prerequisite for competitiveness by 2030, driven by regulatory pressure and consumer expectations (World Economic Forum, 2023).

3.5.3 AI and Optimized Operations

Artificial Intelligence (AI) enables data-driven sustainability by optimizing production schedules, reducing energy consumption, and minimizing waste. AI-driven predictive analytics can forecast demand accurately, preventing overproduction and excess inventory. The efficiency of AI can bridge the gap between environmental, social, governance (ESG) disclosures and actionable strategies, reducing material waste and emissions while improving operational (World Economic Forum, 2023). Manufacturers using AI for real-time energy optimization have achieved measurable reductions in carbon footprint and operational costs. McKinsey research estimates AI could add $4.4 trillion in productivity growth potential across industries (Chui et al., 2023, June). WEF stresses that AI-driven sustainability requires integrating ESG data into operational workflows, not just reporting frameworks for compliance reporting purposes (World Economic Forum, 2023).

3.5.4 Blockchain and Supply Chain Transparency

Blockchain technology is a decentralized, cryptographically secured digital ledger system that enables the recording and authentication of transactions without reliance on traditional intermediaries. Its distributed ledger offers an append-only, cryptographically secured database that efficiently records and verifies transactions, with applications extending to supply chains, digital currencies, and non-fungible tokens (NFTs) (Campbell et al., 2023). Key features such as immutability, provenance, consensus, and finality collectively foster transactional trust even in trustless environments. This transition also paves the way for a generational shift in user interaction, allowing individuals to engage with systems in ways that feel intuitive and secure.

While still a nascent technology, blockchain has numerous applications in Industry 5.0. For example, pharmaceutical companies use blockchain to track drug provenance, ensuring ethical sourcing and regulatory compliance across global supply chains. Similarly, Walmart leverages blockchain to trace produce from farm to shelf, improving food safety and consumer trust. The company estimates that migration to blockchain technology enables tracing an outbreak to its source, such as the grower or producer, in seconds rather than the week-long process required by traditional, non-distributed ledger methodologies (McKinsey, 2023; World Economic Forum, 2023).

Beyond food and pharmaceuticals, blockchain's potential extends to tracking conflict minerals, blood diamonds, and carbon footprints across global supply chains. By leveraging immutable records and provenance tracking, organizations can verify ethical sourcing of raw materials, ensure compliance with human rights standards, and monitor environmental impact. This capability aligns with ESG objectives and strengthens corporate accountability in a globalized economy. Blockchain's role in Industry 5.0 extends beyond compliance; it is a strategic enabler of supply chain transparency, ESG accountability, and operational resilience. By integrating blockchain with IoT and AI, organizations can create ecosystems that are secure, transparent, and sustainable.

3.6 Pillar 3: Technologies for Resilience

Resilience in Industry 5.0 emphasizes adaptability, robustness, and continuity in the face of technological, economic, or environmental disruptions. This pillar leverages advanced technologies to anticipate risks, simulate scenarios, and enable flexible operations. Critical dimensions of Technologies for Resilience include digital twins, predictive AI for risk management, and autonomous systems. By integrating these tools, businesses can move from reactive crisis management to proactive resilience strategies.

3.6.1 Digital Twins, Simulation, and Experimentation

Imagine placing sensors all over a wind turbine in order to measure everything imaginable (e.g., temperature, humidity, wind speed, barometric pressure, etc.). This robust data allows for seemingly endless simulation. Hence, digital twins are virtual replicas of physical assets, processes, or systems, continuously updated with real-time data. They enable organizations to simulate, monitor, and optimize operations before implementing changes in the physical world. Chui et al. (2023, July) note that digital twins can reduce deployment time for AI-driven capabilities by up to 60% and cut capital and operating expenditures by 15%. These models allow companies to anticipate supply chain disruptions, predict maintenance needs, and optimize resource allocation. For example, a digital twin of an end-to-end supply chain can provide early warnings of vendor issues and simulate alternative sourcing strategies (World Economic Forum, 2023). WEF also highlights that digital twins can reduce energy consumption by up to 31%, saving $2 trillion annually by optimizing industrial clusters for efficiency and emissions reduction.

Digital twins and other simulation technologies enable "what-if" scenario planning and stress testing of systems under various conditions. This is critical for resilience because it allows organizations to anticipate disruptions and validate contingency plans. For instance, Carson City, Nevada, has been referenced as a smart city in regard to its infrastructure projects that leverage simulation for urban planning and energy optimization, which demonstrates how simulation can improve resilience in municipal systems (NevadaAppeal.com, 2016). Manufacturers use simulation to test production line changes before physical deployment, reducing downtime and risk.

3.6.2 Predictive AI for Risk Management

Predictive AI in Industry 5.0 goes beyond forecasting. Rather than reacting to crises, organizations can leverage predictive AI to identify and mitigate risks across complex industrial ecosystems before they materialize. By analyzing both historical and real-time data streams, these models anticipate disruptions such as supply chain shocks, equipment failures, cybersecurity breaches, and regulatory compliance risks. This capability shifts enterprises from reactive crisis response to anticipatory resilience strategies, ensuring continuity and stability even in volatile environments.

Moreover, predictive AI plays a critical role in governance by embedding transparency, explainability, and compliance into decision-making processes. Frameworks such as Gartner's AI Trust, Risk, and Security Management (TRiSM) provide structured approaches to ensure AI systems remain trustworthy, secure, and aligned with regulatory standards (Litan, 2024). Predictive AI can help address risks like bias, data drift, and adversarial threats while maintaining accountability throughout the AI lifecycle.

3.6.3 Flexible and Autonomous Systems

Flexible and autonomous systems combine robotics, AI, and adaptive control to create operations that can self-adjust in real time. Gartner (2023) defines autonomous business as systems that sense, decide, and act without human intervention, rerouting supply chains or adjusting production schedules during disruptions. For example, autonomous systems in manufacturing and logistics can maintain continuity during labor shortages or geopolitical shocks. The WEF (2023) highlights the role of autonomous systems in transforming industries such as transportation and agriculture, enabling productivity gains and resilience. These systems represent the next frontier beyond automation because they create "anti-fragile operations" that not only withstand "shocks" but improve under pressure.

4 Critical Nature of Data Engineering and Data Governance

4.1 Seismic Changes to the Data Structure Layer

Artificial Intelligence is catalyzing a watershed transformation in enterprise technology, profoundly reshaping the human–computer interface (HCI) and revolutionizing the foundational data architectures that underpin business decision-making. Traditionally, enterprise systems such as ERP relied heavily on structured data meticulously organized in relational tables with defined schemas. These architectures were optimized for predictable, transactional processing but proved limited in handling the vast and diverse data sources emerging in today's digital ecosystem. The advent of neural networks and generative adversarial networks (GANs) marked a departure toward dynamic, nonlinear models capable of learning complex patterns from unstructured data, freeing enterprises from their traditionally rigid schema constraints.

Building on this foundation, advances in Natural Language Processing and especially Large Language Models represent a seismic leap forward in extracting meaning and actionable insights from diverse unstructured data, textual, auditory, and visual, at scale. Unlike traditional systems, these generative AI models understand context, semantics, and nuance, enabling the transformation of vast reservoirs of unstructured knowledge into structured intelligence. This development empowers enterprises to integrate insights from diverse data types into their decision systems, creating a unified, richer analytical fabric that spans both legacy structured data and new forms of unstructured information.

This hybrid approach fundamentally shifts how organizations store, manage, and process data. The data structure layer evolves from a static repository of tables into a dynamic, multi-modal ecosystem supported by scalable cloud-based data lakes, semantic search layers, and AI-augmented data pipelines. Enterprises can now leverage generative AI to unlock latent value from historical databases, integrate real-time data streams, and facilitate sophisticated knowledge synthesis, driving more informed, agile, and human-centric decision-making for the era of Industry 5.0. In other words, gone are the days where organizations must impose some arbitrary structure to unstructured data so that it can be analyzed and infused in the business's value stream. The modern replacement to this data structure problem is analyzing the unstructured data directly vis-a-via neural networks and LLMs.

To illustrate this point, here is an AI-generated output that was generated to show the academic research that cited the *Grokking: Generalization Beyond Overfitting on Small Algorithmic Datasets* (Power et al., 2022). Using GAN technology to generate a nomological network based on citation counts, the corresponding heat map creates a bubble whose size is relative to the number of citations from other journal articles. With this tool, any researcher can have a quick overview of relevant research streams and access the corresponding abstracts to consider how this might apply various theoretical lenses to the current problem being explored (Fig. 6).

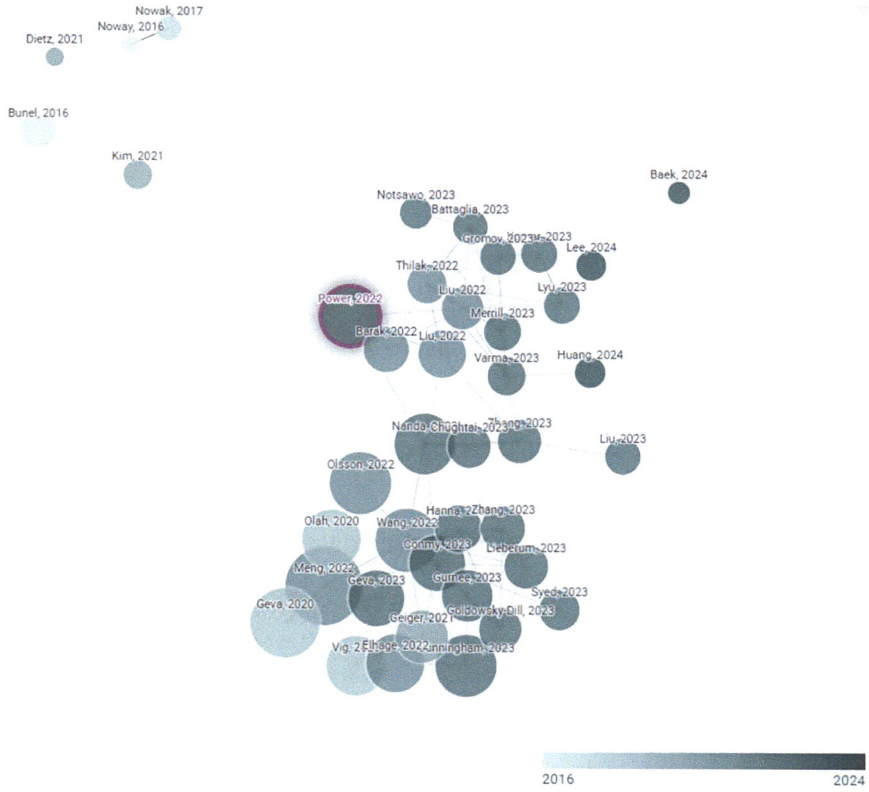

Fig. 6 Nomological heat map for Grokking. Image was generated using a visual tool by Connected Papers http://connectedpapers.com

The nomological heat map illustrates the shift from the reliance on traditional tabular data for useful information. At a more macro-level, such a shift has upended the conventional data layer, leaving many organizations uncertain about how to respond. The longstanding practice of imposing arbitrary structure on unstructured data is upended as AI models can natively process and learn from unstructured inputs, offering richer, more contextual insights than ever before.

The real disruption, however, lies in the data layer itself. Historically, this layer has been a rigid, structured foundation, consisting of relational database constructs. Today, it is evolving into a fluid, AI-native environment where unstructured data is not only accessible but actionable. As McKinsey notes, generative AI is driving a rethinking of enterprise architecture, with new models of human–AI collaboration and data orchestration emerging across industries (Kaplan et al., 2024). Deloitte adds that organizations must modernize their infrastructure to accommodate AI workloads, which demand flexible, hybrid data environments capable of handling probabilistic models and diverse data types (Thomas et al., 2025). In this new paradigm, businesses must move beyond viewing data as a static asset and instead embrace it as a strategic, dynamic resource. This requires not only technological upgrades but also a cultural shift that prioritizes data fluency, cross-functional collaboration, and *human-in-the-loop* governance to ensure ethical, effective, and resilient AI adoption,

4.2 Data Governance in a Tech-Driven World

High-quality data is accurate, consistent, and timely. Yet, there are two universals truths that plague organizations and frequently destroy new initiatives at inception. First, there is a GIGO problem, which is the old adage <u>G</u>arbage <u>I</u>n, <u>G</u>arbage <u>O</u>ut. Absent data quality and data integrity within the inputs, a reliable output is impossible. Second, there is a DRIP problem insofar that many organizations suffer from being <u>D</u>ata <u>R</u>ich, <u>I</u>nformation <u>P</u>oor. As organizations embrace emerging technologies like agentic AI, blockchain, and edge computing, the quality, integrity, and security of data become mission critical. Data, which is becoming exponentially bigger in volume, is the lifeblood of digital transformation. However, without proper governance, it can become a source of risk rather than value. Many organizations struggle with fragmented systems, duplicate records, and poor metadata management. These issues are magnified in AI-driven environments, where flawed data can lead to biased models, faulty predictions, and reputational damage. Beyond C-Suite there must be board-level governance as leaders must champion data stewardship across departments, ensuring that data is treated as a strategic asset rather than just an IT concern (refer to aforementioned Technology Fallacy discussion).

Over the past couple of decades, the advancements made in neural networks have transitioned the computing paradigm from strictly binomial system based on mathematical calculations and cash machine like architecture to more neural network architecture that simulates human brain. The neural network working on the principles of universal approximation theorem can work with approximate data so long as the amount of data is very large. The digitization efforts during Industry 4.0 have helped build the internet with a huge corpus of information almost capturing every human thought since the beginning. This volume of unstructured data source has helped build the mental model such that it can capture not only the human communication and language structures, but also the knowledge base itself. LLMs can capture the substance and style associated with human conscience thus far and identify and retrieve them on demand when needed too. The way LLMs have ability to capture the essence and semantics of words and phrases in our language structure enables computers to work with partial, disparate, segmented and unstructured data to produce insights and meanings.

4.2.1 Inherit Risk and Data Bias

Behind every algorithm and line of code there is a human being with his/her own individual assumptions, values, and biases that inevitably shape technological outcomes. As artificial intelligence becomes more embedded in everyday life, these biases can manifest in unintended and sometimes harmful ways. A well-known example is the "Trolley Problem" in autonomous vehicles, where AI systems must make ethical decisions in life-or-death scenarios with no ideal outcome. Real-world cases further illustrate the risks: Microsoft's chatbot *Tay* was quickly taken offline after it began generating offensive and racist content, a result of its unsupervised learning from social media interactions. Similarly, Amazon's AI-powered hiring tool was found to systematically disadvantage female applicants, favoring male candidates due to biased training data.

Much like Microsoft Word is making suggestions to correct spelling errors as one types, the LLM is "guessing" what we mean[2]. Rather than fixing the word, the AI is trying to infer the context, which inevitably leads to inherent bias and unintended consequences. For example, prompt the AI to fill in the blank … *Man is to Doctor as Woman is to (blank)....* and you may be surprised by the answer. There is an 84% probability that we mean "nurse" here. The next likeliest answers are child, pregnant, mother, and patient! This begs the question, Why don't we mean Doctor? Man is to Doctor as Woman is to Doctor! Despite some gender inequity, 39% of doctors in the United States are women. Yet, the social norms suggest that the answer is not likely to be "doctor." This example emphasizes the danger of over-reliance on the LLM-generated output. To fully leverage the capabilities of AI, we need to keep the *human-in-the-loop* to monitor and evaluate the output.

The reality is that we rely on more than words to provide meaning. In their seminal work, Odgen and Richards (1927) suggest that meaning is derived from the combination of the word, reference, and the referent. In other words, meaning is extracted from the linkages with things and experiences. As we consider linkages of meaning, we note that some researchers suggest that nonverbal cues are inseparable than verbal cues (Jones & LeBaron, 2002) to extract meaning. Furthermore, we as humans rely

[2] "Guessing" translates to analyzing cosine similarity in a word-to-vector methodology.

on tone to communicate. In the Mandarin dialect, for example, there are four distinctive tones so one word can take on four separate meanings depending on the inflection or pitch of voice. Additionally, many use sarcasm to illustrate a point. How does the LLM deal with these issues to infer the "right" context? As another example to really hit home this point of context, I asked a group of students visiting from Brazil to fill in the blank: *France is to Paris as Brazil is to (blank).* There are a variety of answers here. First, Rio De Janeiro—as both cities recently held Olympics. Others suggest San Paolo, the commonality as largest city. Perhaps I mean Brasilia, as both are their respective country's capital. The point here is that context matters. As another example, any LLM will return an explanation of Artificial General Intelligence when prompted about AGI. From a financial services perspective, we clearly mean Adjusted Gross Income when asking about AGI. Context matters. The computer is inferring the context based on the model it was trained on.

Apart from context, AI has no sense of morality. The computer has no empathy and lacks compassion. The AI doesn't know right from wrong as it only follows rules. It can generate falsehoods or be taught to lie, without even knowing it. The AI can simulate empathy, but it lacks any true feeling. Whereas humans rely on their past experiences to make decisions, the AI models have the data that they were trained on absent of any lived experiences. Freud and Strachey (1948) tells us about the pleasure-pain paradigm where we seek pleasure and avoid pain to fulfill both psychological and biological needs. Through this lens, if you don't know what pain is, it is not possible to simulate what you do not know or have not experienced. There is no AI capable of understanding our individual or organizational-level pain points like human beings do.

Beyond the inherent biases and ethical concerns, we remise to re-emphasize the fact at a core level; these algorithms are looking for nonlinear relationships that were otherwise unknown to us. The crux of this problem can be summed by saying there lies limitations of any probabilistic system and these are inherently non-deterministic. Or as taught in any basic statistics course, correlations do not lead to causation!

4.2.2 Edge, Cloud, and Mobile Computing

Edge, Cloud, and Mobile Computing form the triad of modern enterprise infrastructure that enables Industry 5.0's vision of seamless human-technology synergy and operational resilience. These distributed computing paradigms collectively address the growing complexity, scale, and immediacy demands of today's data-driven business ecosystems.

Cloud computing continues to serve as the robust backbone for scalable, flexible, and cost-efficient IT resources. It is estimated that by 2027, over 90% of enterprises will be reliant on hybrid or multi-cloud environments, underscoring the imperative to leverage diverse cloud services for agility and innovation (Gartner, 2024). Leading cloud providers like AWS, Microsoft Azure, and Google Cloud offer AI-as-a-Service platforms that power everything from generative AI model training to advanced data analytics at petabyte scales. For instance, AWS SageMaker and

Azure OpenAI Service enable enterprises to rapidly experiment, develop, and deploy AI models without upfront infrastructure investment, accelerating the transition to AI-augmented workflows.

Edge computing, meanwhile, addresses the critical need for low-latency processing close to data sources. Gartner predicts that by 2027, 75% of enterprise-generated data will be created and processed outside of traditional centralized data centers or clouds, underscoring a fundamental shift toward edge architectures (Gartner, 2018). In manufacturing, AI-powered edge devices allow real-time defect detection and predictive maintenance, minimizing downtime. In retail, edge-enabled smart shelves and sensors drive inventory optimization and personalized customer engagement. Importantly, edge computing also enhances data privacy and compliance by enabling sensitive data to be processed locally rather than transmitted to the cloud.

Mobile computing acts as the connective tissue empowering the human element of Industry 5.0. With global smartphone penetration exceeding 85% and 5G networks delivering gigabit speeds and ultra-low latency, mobile devices have become hubs for real-time collaboration and AI-driven augmentation (Bankmycell, 2025). Augmented reality (AR) applications powered by mobile computing are revolutionizing field service operations; technicians equipped with AR glasses can overlay schematics and AI-generated instructions onto physical equipment, boosting accuracy and efficiency. Companies like Verizon and Nokia are pioneering 5G-enabled private networks tailored for industrial campuses, integrating mobile computing deeply into enterprise workflows. To curb our need for faster and better, the wireless providers face pressure to migrate to 6G even though that move may not have very high return on investment (Deloitte, 2025).

Together, these paradigms form an integrated technology stack, supporting a dynamic feedback loop between humans, AI, and machines. While cloud platforms provide scale and cognitive power; edge computing devices offer immediacy and contextual processing; mobile computing ensures ubiquitous, intuitive access to AI-driven insights. This convergence is foundational to Industry 5.0's promise: enabling enterprises to harness the full potential of data, accelerate innovation, and create adaptive, resilient systems where human creativity is amplified by intelligent technologies.

4.2.3 Cybersecurity and Individual Privacy

Emerging technologies introduce new vulnerabilities, making cybersecurity critical in the era of Industry 5.0. The legal landscape has shifted dramatically in recent years, placing a premium on protecting personally identifiable information (PII) and sensitive data. The notion of companies relying on "opt-out" polices is no longer sufficient. Privacy frameworks have migrated to an "opt-in with caveats" culture, requiring explicit consent for data collection and processing. This shift reflects growing consumer awareness and regulatory pressure to safeguard PII.

The General Data Protection Regulation (GDPR) in the EU established strong privacy standards with strict reporting rules and tough penalties for breaches. For example, Meta was fined a record-breaking $1.394 billion[3] in May 2023 for unlawfully transferring EU user data to the United States, underscoring the financial and reputational risks of non-compliance (Mishova, 2025). Domestically, U.S. states have enacted privacy laws that, in some cases, are more restrictive than GDPR. The California Consumer Privacy Act (CCPA) and its successor, the California Privacy Rights Act (CPRA), impose fines of up to $7500 per intentional violation, with enhanced penalties for violations involving minors. Illinois' Biometric Information Privacy Act (BIPA) has also had significant consequences. The Illinois Supreme Court's decision established that individuals may seek statutory damages under the Biometric Information Privacy Act (BIPA) even without proving actual harm (*Rosenbach v. Six Flags Entertainment Corporation*, 2019). In other words, a plaintiff does not need to demonstrate financial loss or physical injury! To recoup financial restitution, a mere violation of the statute is sufficient to qualify as an "aggrieved person" and pursue damages.

With the increased global regulation, cybersecurity is no longer just a technical function but rather a strategic, board-level imperative. Boards are increasingly held accountable for cyber risk governance, with regulators requiring disclosure of material cybersecurity incidents and oversight practices in annual reports. McKinsey emphasizes that cybersecurity is now viewed as a driver of competitive advantage, not merely a cost center (Greis, 2025, August 4). Boards must integrate cyber risk into enterprise strategy, ensure robust incident response plans, and treat cybersecurity spending as a strategic investment rather than an operational expense (Greis, 2025, August 4). As such, having a *human-in-the-loop* is increasingly important with cybersecurity concerns in this age of AI.

5 Synthesizing Technology for Humanity's Collective Prosperity

As Industry 5.0 emerges, technological progress converges with a renewed focus on advancing human well-being and collective prosperity. One of the most profound enablers of this synthesis is Generative AI and LLMs. These transformative technologies have unlocked the ability to extract meaningful insights from both structured and unstructured data in a way that is scalable. They support increasingly complex decision-making, allowing organizations to adapt rapidly to evolving business requirements and societal needs. Today's generative AI systems excel at interpreting text, audio, and visual information, bridging gaps that traditional data analysis could not traverse due to the binary structure.

Human-centric AI and collaborative robotics are redefining the workplace. Rather than viewing automation as a replacement for human labor, enterprises are harnessing

[3] The fine was €1.2 billion Euros, which is approximately $1.394 USD.

intelligent systems to amplify human creativity and expertise. This approach fosters new levels of collaboration between people and machines, personalizes experiences in both manufacturing and services, and significantly enhances workplace safety. AI-powered cobots and digital assistants are becoming essential partners that can handle routine or hazardous tasks, and free-up humans to focus on higher-order problem-solving and innovation.

The integration of digital twins and simulation technologies is driving operational agility across industries. By creating real-time, virtual replicas of physical systems and environments, organizations can model scenarios, predict outcomes, and optimize their operations with remarkable accuracy. In manufacturing, healthcare, and logistics, these tools empower risk management and continuous refinement of processes, leading to greater efficiency and resilience in uncertain conditions.

Underpinning all these advances are edge, cloud, and mobile computing infrastructures. These integrated platforms provide the scalability, intelligence, and security that next-generation enterprise systems require. Cloud environments enable powerful data analysis and global collaboration, while edge computing pushes processing closer to the data source for enhanced responsiveness. Mobile technologies ensure that humans remain continuously connected, enabling seamless interaction with intelligent systems wherever they are. Together, these infrastructures create the digital fabric that allows people and AI to work side by side, heralding a new era of human-tech synergy in Industry 5.0.

5.1 The Human Amplification Potential

Industry 5.0 marks a pivot from pure automation to human amplification. Artificial intelligence, in this context, becomes a multiplier for human ingenuity, enabling frontline workers and decision-makers to leverage deep analytics, intuitive interfaces, and personalized recommendations. Humans are empowered to focus on creative, strategic, and supervisory tasks, while AI-driven cobots and digital agents handle repetitive, hazardous, or computationally demanding roles. The study by Anang et al. (2024) has reported extraordinary returns on investment (ROI) from artificial intelligence and human–machine collaboration in Industry 5.0 with documented cases of 3.7x ROI on their generative AI implementations with top performers even seeing over 10x returns.

6 Forward-Looking Perspective: Future Opportunities and Challenges

The prospects for Industry 5.0 are extraordinarily broad, with personalized products, enhanced safety, and increased workplace satisfaction quickly becoming new benchmarks across sectors including manufacturing, services, and government policy. Advanced AI-powered systems and collaborative robotics enable tailored solutions that meet individual needs, optimize operational safety, and foster a more positive and engaging working environment. These opportunities exemplify the promise of Industry 5.0 to drive human-centric value.

Despite such promises, the pathway forward contains significant challenges. A major concern is the skill gap emerging between what new technologies require and the capabilities available within today's workforce. Success in Industry 5.0 demands proficiency in both technical skills and interpersonal collaboration, yet many organizations, especially small and medium-sized enterprises, struggle to bridge these gaps. Investing in targeted education and ongoing training will be crucial for enabling all workers to participate fully in a rapidly changing technological landscape.

As enterprise systems become more deeply interconnected, cybersecurity and data governance take on heightened importance. Protecting sensitive data and ensuring the privacy of individuals has grown more complex as digital ecosystems expand. Robust safeguards, transparent protocols, and adaptive governance frameworks will be required to mitigate risks and maintain trust among users and stakeholders.

Another challenge lies in ensuring equitable access to advanced technologies. The capital required for digital transformation remains significant, which may put smaller organizations at risk of falling behind and thus widen the digital divide. Making scalable, affordable solutions available to a broad spectrum of enterprises is essential for realizing the inclusive potential of Industry 5.0.

Finally, the ethical and social dimensions of technology demand careful management. Artificial intelligence and automation must be developed and deployed in ways that uphold human values, balance efficiency with fairness, and foster trust. Addressing ethical concerns and aligning technological progress with broader societal goals will be central to ensuring that Industry 5.0 leads to sustainable and equitable prosperity for all.

Amid these advances, organizations must confront the legacy of technical debt, which is a term used to describe the accumulated inefficiencies and fragilities within existing systems. Moving forward, generative AI and intelligent application tools offer historic opportunities for early adopters to leapfrog outdated architectures and forge resilient, future-proof platforms. Effectively, synthesizing technology for humanity's prosperity means not only creating new capabilities, but also rehabilitating the technological foundations upon which those capabilities are built.

6.1 Technical Debt

Technical debt is a pervasive and enduring challenge that businesses have accumulated over the last seven decades as technology has continuously evolved. Coined in 1992 by Ward Cunningham, the term "technical debt" metaphorically captures the hidden costs and inefficiencies accrued when expedient, but suboptimal coding or system choices are made (Topel & Ward, 1992). Legacy decisions are often driven by pressures to deliver quickly, inadequate documentation, or shifting business requirements. As such, legacy systems have compounded over successive waves of technological advancement, from mainframes and early programming languages to client–server architectures, web frameworks, and cloud platforms. The result is a complex web of outdated, fragile, and inefficient systems slowing innovation and inflating maintenance costs across global enterprises (Tiwari, 2025).

In this long history, technical debt has often represented a silent barrier to competitive agility. The incremental nature of software development over decades led to patchworks of incompatible technology, business logic locked in obsolete languages (like COBOL or legacy Visual Basic), and inadequate integration practices. These accumulated debts manifest as slower response times, brittle systems, higher operational risk, and ballooning costs, draining resources away from strategic innovation toward endless firefighting and incremental fixes..

Now, as we transition into the era of Industry 5.0, generative AI emerges as a game-changer capable of dramatically shifting this dynamic. Unlike traditional modernization efforts that require costly, laborious rewrites, AI-powered code generation and intelligent application builders enable organizations to accelerate legacy system migration with unprecedented speed and accuracy. By leveraging AI's ability to ingest and understand massive codebases, generate clean modern equivalents, map complex dependencies, and automate repeated coding tasks, enterprises can reduce the cost, risk, and time involved in transforming outdated applications into cloud-native, scalable, and secure architectures. Generative AI also supports automated documentation, creation of test cases, and continuous refactoring that mitigate the resurgence of technical debt in new systems..

This AI-driven approach empowers early adopters to leapfrog intermediate modernization phases. Companies no longer need to settle for "rip and replace" strategies, which are costly and disruptive, but can instead incrementally refactor and enhance decades of embedded business value. As we are ushering into this new AI era, businesses should consider new ways of thinking. One such approach is design thinking, which is a human-centered approach to innovation that harnesses agility and ideation (C. A. Campbell & Ramamoorti, 2023). Design thinking principles identifies opportunities for innovation and leapfrogs from conceptualization and prototyping to modernization in an engaging and immersing manner. The AI-assisted migration tools offer intelligent insights into where technical debt and system bottlenecks reside, enabling smarter architecture redesign and prioritization of high-impact updates aligned with business goals through empathetic discovery. This significantly boosts not only operational resilience but also the capacity to integrate next-generation technologies such as edge computing, generative AI itself, and advanced analytics, hallmarks of Industry 5.0's technology backbone (Nawab, 2025).

In summary, the emergence of generative AI marks a pivotal moment in managing and overcoming the legacy of technical debt. By transforming legacy code migration from an arduous burden into an agile, data-driven process, it opens a path for enterprises to modernize with less disruption and greater strategic advantage. This new capability is central to propelling businesses confidently into the future, enabling the human-tech synergy at the heart of Industry 5.0.

7 Conclusion

The transition to Industry 5.0 marks a defining moment for global enterprises, as advanced technologies, particularly generative AI and large language models, reshape the relationship between humans and machines. This chapter explores how the convergence of edge, cloud, and mobile computing, alongside human-centric AI, collaborative robotics, and digital twins, creates the technical foundation for a more adaptive, resilient, and human-focused enterprise ecosystem.

Industry 5.0's transformative potential is evident in its ability to amplify human creativity, enable personalized solutions, and cultivate safer, more engaging workplaces. However, the road ahead is shaped not only by opportunity but also by challenges and opportunities. Bridging workforce skill gaps, securing data in an interconnected environment, ensuring equitable access to innovation, and upholding ethical standards are vital issues which much be addressed to realize the promise of a prosperous, inclusive, and sustainable future.

As organizations look forward, the effective synthesis of technology will empower humanity to overcome legacy constraints such as technical debt, leapfrog outdated systems, and embrace a new era of collaborative intelligence. The advances described here are more than technological milestones. In sum, they represent a unified vision for prosperity, leveraging the best of human and digital capabilities.

Albert Einstein once said that "If I had an hour to solve a problem and my life depended on it, I would spend the first 55 minutes determining the right question to ask" (Bowen & Watson, 2024). This quote demonstrates the power of critical thinking and determining what is the problem that we are trying to solve. However, AI allows us to re-imagine the scenario Einstein unearths here. More specifically, in today's world, we could spend 55 minutes refining our prompt to get a more nuanced response! Reflecting on this pivotal moment, it is worth recalling Albert Einstein's words: "The human spirit must prevail over technology." In the journey toward Industry 5.0, the ultimate measure of success will be the harmonious integration of technology for humanity's collective benefit, ensuring above all, that innovation remains in service to human values and societal progress.

Disclaimer The authors used chat tools like ChatGPT to review grammar and sentence clarity in this chapter. The tool was applied solely for language editing, styling and formatting; the authors take full responsibility for the content.

References

Alexander, D., & Lyytinen, K. (2019). Organizing around big data: organizational analytic capabilities for improved performance. In *Fortieth International Conference on Information Systems*, Munich 2019. https://aisel.aisnet.org/icis2019/business_models/business_models/17

Anang, A. N., Ofuje, P. O., Adelye, O. M., Opani, J. O., Kuubata, M., & Obunbiyi, D. (2024). The role of artificial intelligence in industry 5.0: Enhancing human-machine collaboration. *WORLD, 24*(2), 380–400.

Bankmycell. (2025). *How many smartphones are in the world?* https://www.bankmycell.com/blog/how-many-phones-are-in-the-world

Bowen, J., & Watson, E. (2024). *Teaching with AI*. John Hopkins University Press.

Buck, R., Fomin, D., Heimes, H., Jensen, B., Ledwa, B., & Weis, M. (2022, February). *Unleashing the next wave of productivity in corporate business functions*. McKinsey Insights. https://www.mckinsey.com/capabilities/operations/our-insights/unleashing-the-next-wave-of-productivity-in-corporate-business-functions.

Campbell, C., & Fogarty, T. (2018). Behind the curve: Higher education's efforts to implement advanced information systems. *Journal of Emerging Technologies in Accounting, 15*(2), 77–91. https://doi.org/10.2308/jeta-52237

Campbell, C. A., & Ramamoorti, S. (2023). Design thinking and cybernetics: The case for generative AI in AIS pedagogy. In *Advances in accounting education: Teaching and curriculum innovations* (Vol. 27, pp. 101–123). Emerald Publishing Limited.

Campbell, C. A., Ramamoorti, S., & Calderon, T. G. (2023). Automation bias and the "goldilocks effect" in auditing Blockchain. *Journal of Emerging Technologies in Accounting*, 1–25. https://doi.org/10.2308/jeta-2022-062

Chui, M., Collins, M., & Patel, M. (2021). *IoT value set to accelerate through 2030: Where and how to capture it*. McKinsey & Company. https://www.mckinsey.com/capabilities/tech-and-ai/our-insights/iot-value-set-to-accelerate-through-2030-where-and-how-to-capture-it.

Chui, M., Hazan, E., Roberts, R., Singla, A., Smaje, K., Sukharevsky, A., Yee, L., & Zemmel, R. (2023, June). *The economic potential of generative AI: The next productivity frontier*. McKinsey & Company. https://www.mckinsey.com/~/media/mckinsey/business%20functions/mckinsey%20digital/our%20insights/the%20economic%20potential%20of%20generative%20ai%20the%20next%20productivity%20frontier/the-economic-potential-of-generative-ai-the-next-productivity-frontier.pdf.

Chui, M., Issler, M., Roberts, R., & Yee, L. (2023, July). *Technology trends outlook 2023*. McKinsey & Company. https://www.mckinsey.com/~/media/mckinsey/business%20functions/mckinsey%20digital/our%20insights/mckinsey%20technology%20trends%20outlook%202023/mckinsey-technology-trends-outlook-2023-v5.pdf

Deloitte. (2022). *Critical factors for ERP implementations*. https://www.deloitte.com/mt/en/Industries/technology/perspectives/erp-critical-factors-for-implementations.html

Deloitte. (2025). *2025 global telecommunications outlook*. Retrieved from https://www.deloitte.com/us/en/insights/industry/technology/technology-media-telecom-outlooks/telecommunications-industry-outlook-2025.html

Fazlollahtabar, H. (2025). Industry 5.0 paradigm transformation adoption in developing countries: An analytical game theoretic model. *Scientific Reports, 15*(1), 31463.

Freud, S., & Strachey, J. (1948). *Beyond the pleasure principle* (Vol. 18). Hogarth Press.

Gartner. (2018). *What edge computing means for infrastructure and operations leaders*. Contributed by van der Meulen, R. https://www.gartner.com/smarterwithgartner/what-edge-computing-means-for-infrastructure-and-operations-leaders
Gartner. (2024). *Gartner forecasts worldwide public cloud end-user spending to Total $723 billion in 2025*. https://www.gartner.com/en/newsroom/press-releases/2024-11-19-gartner-forecasts-worldwide-public-cloud-end-user-spending-to-total-723-billion-dollars-in-2025
Gartner. (2023). *Embrace sustainable technology to meet ESG goals*. https://www.gartner.com/en/chief-information-officer/topics/sustainable-technology
Goran, J., LaBerge, L., & Srinivasan, R. (2017). Culture for a digital age.. https://www.mckinsey.com/capabilities/mckinsey-digital/our-insights/culture-for-a-digital-age
Jones, S. E., & LeBaron, C. D. (2002). Research on the relationship between verbal and nonverbal communication: Emerging integrations. *Journal of Communication, 52*(3), 499–521.
Kane, G. (2019). The technology fallacy: People are the real key to digital transformation. *Research-Technology Management, 62*(6), 44–49.
Kaplan, J., Gu, M., & Sinha, M. (2024). *Enterprise technology's next chapter: Four gen AI shifts that will reshape business technology*. https://www.mckinsey.com/capabilities/mckinsey-digital/our-insights/enterprise-technologys-next-chapter-four-gen-ai-shifts-that-will-reshape-business-technology#/.
Litan, A. (2024). *Tackling trust, risk and security in AI models*. https://www.gartner.com/en/articles/ai-trust-and-ai-risk
Greis, J. (2025, August 4). Competitive advantage through cybersecurity: A board-level perspective. *McKinsey & Company*. https://www.mckinsey.com/capabilities/risk-and-resilience/our-insights/competitive-advantage-through-cybersecurity-a-board-level-perspective
Mishova, A. (2025). *Meta's €1.2 billion GDPR fine: Why it still matters in 2025*. https://gdprlocal.com/metas-e1-2-billion-gdpr-fine-why-it-still-matters-in-2025/
Nawab, K. (2025). *Technical debt and modernization: Three lessons learned*. https://agileengine.com/technical-debt-and-modernization-three-lessons-learned/
NevadaAppeal.com. (2016). *Carson City pushes to be a smart city* [Press release]. https://www.nevadaappeal.com/news/2016/feb/09/carson-city-pushes-to-be-a-smart-city/
Ogden, C. K., & Richards, I. A. (1927). *The meaning of meaning: A study of the influence of language upon thought and of the science of symbolism* (Vol. 29). K. Paul, Trench, Trubner & Company, Limited.
Oxford. (2018). In S. Ratcliffe (Ed.), *Oxford essential quotations* (6th ed.). Oxford University Press.
Power, A., Burda, Y., Edwards, H., Babuschkin, I., & Misra, V. (2022). Grokking: Generalization beyond overfitting on small algorithmic datasets. https://arxiv.org/abs/2201.02177
Rahilly, L. & Fusaro, R. (Cohosts). (2025, September 12). *Which tech trends are rising to the top of the business agenda?* The McKinsey Podcast. https://www.mckinsey.com/capabilities/mckinsey-digital/our-insights/which-tech-trends-are-rising-to-the-top-of-the-business-agenda
Rosenbach v. Six Flags Entertainment Corporation. (2019). *Supreme Court in the State of Illinois*. Document 2019 IL 123186. https://epic.org/wp-content/uploads/amicus/bipa/rosenbach/Rosenbach-v-Six-Flags-Ill-Supreme-Court-Opinion.pdf
Thomas, C., Tayal, A., Stewart, D., Kearns-Manolatos, D., & Parveen, I. (2025). *Is your organization's infrastructure ready for the new hybrid cloud?* https://www.deloitte.com/us/en/insights/topics/digital-transformation/future-ready-ai-infrastructure.html.
Tiwari, S. (2025). *Legacy system modernization with generative AI: A roadmap for enterprise transformation*. https://www.tothenew.com/insights/article/legacy-system-modernization-generative-ai
Topel, R. H., & Ward, M. P. (1992). Job mobility and the careers of young men. *The Quarterly Journal of Economics, 107*(2), 439–479.
Turing, A. M. (1980). Computing machinery and intelligence. *Creative Computing, 6*(1), 44–53.
World Economic Forum (2023). *How 3D printing really helps accelerate sustainable manufacturing*. https://www.weforum.org/stories/2023/06/how-3d-printing-helps-accelerate-sustainable-manufacturing/

Learning Curves: Educational Design Principles for Ethical GAI Adoption in Industry 5.0

Lisa R. Hughes and Kristen Palmer

Abstract Generative Artificial Intelligence (GAI) presents opportunities and challenges for ethical organizational integration. This chapter examines how educational institutions serve as laboratories for responsible GAI implementation, providing insights for Industry 5.0 contexts. We explore key challenges including AI hallucinations, embedded biases, privacy vulnerabilities, and potential erosion of human expertise. The chapter introduces the ADDIE instructional design framework as a systematic approach prioritizing human agency over technological efficiency. Through practical lessons from higher education—including the PCRAFT prompt engineering framework—we demonstrate how organizations can harness GAI capabilities while maintaining critical thinking and professional development. Successful integration requires moving beyond efficiency metrics to embrace human-centered design, continuous learning, and ethical leadership. By applying educational design principles, organizations can navigate GAI adoption to amplify human potential, ensuring artificial intelligence serves as a collaborative partner rather than replacement for essential capabilities.

Keywords Generative artificial intelligence (GAI) · Human-centered design · Instructional design · ADDIE · Prompt engineering · Digital equity · Data privacy · Higher education · Industry 5.0 · Sustainability

1 Introduction

In a neighborhood on the Northwest side of Chicago, there's a small design shop that promotes slow living: steeping morning tea or brewing French-press coffee as a morning ritual, making bread by hand, that kind of thing. They operate from the belief that convenience is not ultimate, rituals are grounding, and struggle can be good for us. There's also an aspect of slow living that is *embodied* and engages the five senses, like kneading bread until it feels like an earlobe, smelling yeast as

L. R. Hughes (✉) · K. Palmer
Butler University, Indianapolis, IN, USA
e-mail: lrhughes@butler.edu

A. M. Waite et al. (eds.), *Leading in Industry 5.0*, Management and Industrial Engineering, https://doi.org/10.1007/978-3-032-19207-3_3

the fermentation process begins, and hearing the squelch and pats of hands forming dough.

Yet "slow living" still uses technology to accomplish these rituals. Cast-iron bread cloches crisp the crusts of bread baked in a home oven, and a pour over dripper slowly drops water infused with coffee beans into a pitcher. The technology is used purposefully to support a goal. The technology connects the person to the world around them.

As humans, creating technological extensions of ourselves seems to be second nature. Writing is one such example. Writing is a technology—a human tool we use to communicate. Writing is not natural but must be taught, learned, and passed down. We use the technology of writing purposefully to communicate our ideas to others and to clarify our own thinking. Similar to baking bread, there's value in the process as well as the product. Otherwise, why are we paying so much for these artisan loaves?

Enter the seismic shift of November 2022. The company OpenAI released ChatGPT 3.5 to the public, and users adopted this tool faster than any other before it. ChatGPT is one example of generative artificial intelligence technology, also called generative AI, genAI, or GAI. What is the purpose of generative AI? Like writing, GAI is a technology that can be used to communicate ideas and clarify thinking; it can be used to amplify the writing process (and now many other processes) as well as the product. However, because of its nature, GAI should be used thoughtfully to amplify, and not undermine or usurp, human skills.

2 The Rise and Promise of GAI

The release of ChatGPT, and its abundant descendants, is causing all industries to reevaluate norms, igniting a parallel evolution of GAI in education and industry. The essential question in both academia and the workplace is, *What knowledge and skills should be uniquely human, and what can be offloaded to technology?* The answer, however, is still quite amorphous and blurry.

Because GAI is a young technology, it's difficult to evaluate potential harms or threats that may result from its use (Wong, 2024). Digital transformation expert Dave Birss (2025) argues that companies using GAI may be trading long-term innovation for short-term efficiency if the technology is not adopted correctly. With this powerful and disruptive technology, we risk not only significant reduction in creativity, resilience, expertise, motivation, curiosity, accountability, emotional intelligence, critical thinking, innovation, and autonomy that so many warn against but also the creation of an insurmountable fissure in the workforce (Birss, 2025; Rasky, 2024; Wong, 2024). As Wong (2024) explains, GAI "has the potential to decimate legions of white-collar jobs," a tangible fear that is quickly spreading across the workforce and dramatically impacting public perception and use of AI systems.

Clearly, this digital revolution requires more than a simple plug-and-play approach. Yet, it also has the potential to amplify efficiencies, upskilling content creation, data analysis, personalization, and engagement (Birss, 2025; Mittal et al., 2024; Rasky, 2024; Terwisch, 2023; Wong, 2024). OpenAI (2023a), in fact, promises AI could "elevate humanity by increasing abundance, turbocharging the global economy, and aiding in the discovery of new scientific knowledge that changes the limits of possibility," even advancing "ingenuity and creativity." So, where do we go from here? How do we take advantage of all that GAI offers while simultaneously considering the very real threats it generates?

3 Higher Education as the GAI Laboratory

Higher education finds itself in a unique position—serving simultaneously as testbed and training ground for GAI integration. Unlike industry, which can often pivot quickly based on market demands, higher education bears the responsibility of credentialing student achievements while preparing future professionals for an AI-augmented workforce (Wong, 2024). This dual role creates both constraints and opportunities that offer valuable insights for Industry 5.0 adoption.

In our work with faculty and students, we've observed that successful GAI integration requires distinguishing between using the technology as a learning tool versus an efficiency tool. For example, when a student uses GAI to generate a paper without engaging in the cognitive processes of research, analysis, and synthesis, they bypass essential learning opportunities. However, when that same student uses GAI to brainstorm initial ideas, check their reasoning, or refine their arguments while maintaining ownership of their intellectual work, the technology serves as a powerful amplifier of human capability.

In higher education, not unlike in industry, this polarization of GAI often emerges when the desire or need for efficiencies take precedence over the weighty and often time-consuming mental work needed to gather, process, and accurately interpret new information. This distinction becomes critical when we consider Industry 5.0's human-centric approach. Just as we ask students to develop domain expertise before leveraging GAI as an efficiency tool, organizations must ensure their workforce maintains the critical thinking and creative problem-solving skills that make human-AI collaboration truly effective.

Yet automating mundane or repetitive tasks *can* increase productivity exponentially, freeing time for more challenging, rewarding, or creative endeavors (Kimmel, 2023; Seo et al., 2021; Straub, 2023; Wong, 2024). This principle holds true in both industry and academic contexts. Faculty members can leverage GAI to streamline various administrative and preparatory tasks: creating exam questions and rubrics, responding to frequently asked student questions, expediting initial grading passes, generating detailed real-world case studies, or drafting course announcements and email communications. These efficiency gains create space for higher-value activities such as complex research, individualized student mentoring, curriculum innovation, or meaningful peer collaboration.

The time savings can be dramatic. Where creating a focused script, robust slide deck, and instructional video might previously require several hours, we can now produce high-quality educational content in under fifteen minutes. Similarly, developing detailed, supportive course announcements that once consumed hours of crafting and revision now takes just minutes with GAI assistance, elevating both the quantity and quality of course communications. Other faculty report similar transformations: literature professors using GAI to generate discussion prompts for diverse texts, statistics instructors creating varied problem sets, and writing teachers developing personalized feedback frameworks. These tools don't replace the intellectual work of teaching, but rather eliminate the friction that often prevents educators from implementing their best pedagogical intentions due to time constraints.

Students, too, benefit from the efficiencies afforded. Tools like Notebook LM and Google Gemini allow learners in my business writing course to generate structured, detailed feedback through custom prompts in a matter of seconds.[1] Without GAI assistance, providing equivalent individualized feedback to all 35 learners would require several hours—creating an insurmountable time barrier in our condensed 8-week online course format. The efficiency of GAI transforms what was previously impossible into routine practice, enabling the kind of precise, personalized, and just-in-time feedback that enhances learning outcomes (Mittal et al., 2024; Seo et al., 2021).

While many fear the lure of efficiency will ultimately replace human capital, industries are also anticipating the potential cost savings and productivity the efficiencies will allow. With the help of GAI, lawyers to laypersons can quickly summarize essential meeting information, craft more succinct and distinct documents, analyze complex data, produce engaging marketing campaigns, and streamline their workflows. Furthermore, GAI can reduce the frequency of mistakes more commonly produced by humans who are working under intense deadlines, such as grammatical errors and structural problems (Choi et al., 2022).

4 The Challenges We Face Together

Both educational and industrial contexts grapple with a complex web of GAI implementation challenges that extend far beyond simple technology adoption. From AI-generated hallucinations and embedded social biases to privacy vulnerabilities and rapidly shifting regulatory landscapes, organizations across sectors face similar obstacles in their quest to harness GAI's benefits while mitigating its risks. These shared challenges require collaborative solutions that acknowledge both GAI's transformative potential and its current constraints.

[1] The personal experiences referenced here are those of Lisa Hughes, co-author, educator, and instructional designer.

4.1 *Confident and Wrong: The Hallucination Dilemma*

Like a 47-year-old who confidently sings the words of a song they have known for the past three decades, GAI can knock out grandiose tomes or even entire databases in mere seconds. Yet also like that same quadragenarian, who boldly bellows, "Chain, chain, chain…chain of *prunes*" instead of the song's original lyrics, GAI can impressively generate completely fictitious content while projecting unwarranted hubris and authority. Such hallucinations—when GAI generates plausible but incorrect information—pose risks whether we're talking about a student's research paper or a company's strategic report (Chakraborty et al., 2023; Ng et al., 2025; Rasky, 2024; Seo et al., 2021; UNESCO, 2023; Wong, 2024; Yee et al., 2024). A core promise of GAI is the speed at which it can produce or analyze, yet without human intervention and oversight, which slows down the process immensely, the results may be riddled with delusional dross. The potential consequences of such erroneous outputs are staggering, from reputational fallout to financial ruin.

In higher education, a blind trust in GAI can lead to notorious results. As a writing instructor,[2] I've seen student work pockmarked with fictitious sources, completely made up "facts," and images that spotlight deformed figures and mutated words that would make even the Jabberwock cringe. When working with faculty in my role as an instructional designer, I've encountered output for data analytics or financial management courses that doesn't quite add up. Sometimes even simple requests can lead to incredibly confusing and wildly inaccurate productions (Terwisch, 2023). I once asked Claude for suggestions on local happenings that evening in the area I was visiting. He[3] produced a beautiful list, customized to my interests—from live music at a winery to an outdoor fair on the boardwalk… for the previous, already past evening. *Thanks, Claude.* Or the time I sought Claude's advice for a fun weekend getaway. He recommended I take a multi-day trip on the Spirit of Peoria to align with my love of water. I was absolutely stoked… until I did my own research and learned the riverboat permanently left Illinois in 2022, 3 years prior. I called Claude out on his misinformation, and he apologized. But my hopes were already crushed. While my personal anecdotes are certainly forgiving and inconsequential, they also punctuate the importance of human intervention. Imagine the repercussions of a manufacturing engineer who accepts GAI's equipment maintenance recommendations without verifying current sensor data or a supply chain manager who implements GAI-suggested

---

[2] The personal experiences referenced here are those of Lisa Hughes, co-author, educator, and instructional designer.

[3] Throughout this chapter, I refer to Claude (and other GAI) using "he/him" pronouns as a form of personification that reflects my engagement with this AI system. While I understand Claude is not human and lacks consciousness, this linguistic choice mirrors how I naturally interact with the technology—as a collaborative partner in ideation and problem-solving, similar to how I might work with a human colleague. Standard practice varies widely, with some users employing "it/its" pronouns to emphasize the non-human nature of AI systems, while others use gendered pronouns or the system's chosen identity markers to facilitate more natural interaction. This personification doesn't suggest I believe Claude possesses human qualities but rather acknowledges the relational dynamic that emerges when humans engage meaningfully with sophisticated AI systems in collaborative work.

vendor changes without considering recent geopolitical factors. The results could be irreversible. We must approach GAI with thoughtful skepticism, maintaining our critical thinking capabilities and applying our professional expertise to evaluate and refine AI-generated outputs (Magesh et al., 2024; Yee et al., 2024).

4.2 *The Boundaries of the Possible: Technical and Usage Limitations*

Beyond the cognitive and ethical challenges of GAI implementation, users in both educational and industrial contexts must navigate a maze of technical constraints that can significantly impact workflow efficiency and output quality. These limitations manifest in various forms, from fundamental capability gaps to usage restrictions that interrupt collaborative processes at critical moments.

Many GAI systems struggle with multimodal tasks, particularly image generation and manipulation (Art of Sol Studio, 2024; Bittla, 2025). While some platforms can create visual content, most cannot incorporate text into images reliably, forcing users to seek additional tools or manual intervention for tasks as simple as creating labeled diagrams or branded presentations. This fragmentation across different AI capabilities means that what should be seamless workflows often become multi-platform processes, reducing the efficiency gains that initially attracted users to GAI solutions.

Usage limitations present another significant hurdle, particularly for organizations trying to balance cost control with productivity enhancement. Free versions of GAI tools typically impose strict limits on the number of queries, conversation length, or processing complexity, often cutting users off mid-task when they're deep in collaborative problem-solving (Pierrés et al., 2024; Southern, 2025). Even paid versions may throttle usage during peak times or impose daily limits that don't align with project deadlines or intensive work periods. Work productivity hits a standstill when you are counting down to when you no longer have to endure the condemning "You are out of free messages" response. These constraints force users, whether learners or employees, to ration their GAI interactions, potentially disrupting the natural flow of iterative refinement that characterizes effective human-AI collaboration.

Perhaps most frustrating is the "memory-fade" phenomenon that occurs in extended conversations. As discussions become more complex and context-rich, many GAI systems begin to lose track of earlier instructions, established parameters, or specific requirements that were carefully outlined at the beginning of the session (Thakker, 2025). Although more advanced systems, like ChatGPT and Gemini, now offer users the optional ability to track previous conversations, the results are still not always accurate and consistent (Morris, 2024). A marketing professional who spends time establishing brand voice guidelines and target audience specifications may find that after twenty exchanges, the AI has forgotten these crucial constraints and reverted to generic responses. This degradation of conversational context forces users to constantly re-establish parameters or start fresh conversations, undermining the cumulative value that should build through extended collaboration. These technical realities require both educational institutions and industry organizations to develop strategies that work within GAI's current boundaries while planning for future capability improvements.

4.3 Coded Assumptions: The Bias Baked into Our Digital Assistants

At a foundational level, the very outputs on which we rely are rife with the social biases inherent in GAI algorithms (Kolisko & Anderson, 2023; Mittal et al., 2024; Ng et al., 2025; Seo et al., 2021; UNESCO, 2023; Wong, 2024). Upon deeper examination, these systems reveal subtle and sometimes overt inequities relating to gender, race, ethnicity, and political leanings. Such discrimination manifests in the cultural assumptions and sociolinguistic biases embedded within GAI output, reflecting the perspectives and prejudices present in their training data (Kolisko & Anderson, 2023). These biases aren't incidental flaws but systemic features that mirror the inequalities and assumptions prevalent in the digital content used to train these models.

The implications extend far beyond individual interactions to shape how entire organizations approach problem-solving and decision-making. When a GAI system consistently generates marketing copy that assumes family structures, or when it provides career advice that subtly reinforces gender stereotypes, these outputs can perpetuate discrimination at an institutional level. In educational settings, biased GAI responses might influence how faculty frame assignments, assess student work, or develop learning materials, potentially reinforcing existing educational inequities. Industry applications face similar risks when GAI-generated hiring criteria, performance evaluations, or strategic recommendations embed cultural assumptions that disadvantage certain groups.

Perhaps most concerning is the invisible nature of many GAI biases, which can appear reasonable and authoritative even when they reflect problematic assumptions. A GAI system might generate seemingly objective analysis that nevertheless incorporates Western-centric perspectives, male-dominated leadership examples, or assumptions about socioeconomic status that users may not immediately recognize as biased. Addressing these subtle biases requires both individual vigilance and systemic change: users must learn to question GAI outputs through equity lenses, while organizations need policies that promote inclusive GAI use, including diverse review teams for content evaluation, bias testing protocols, and ongoing training that helps staff recognize and counter discriminatory outputs. The goal is not perfect neutrality—an impossible standard—but rather conscious awareness of bias and proactive efforts to mitigate its harmful effects while leveraging GAI's benefits responsibly.

4.4 Living in the Past: GAI's Knowledge Freeze

Knowledge cut-off dates also plague GAI outputs (Savicky, 2025; Wong, 2024). With its current limitations, most GAIs cannot retrieve the most up-to-date information or data, producing results that are dated or even incorrect. This temporal blindness creates particular challenges in rapidly evolving fields where currency of information directly impacts decision quality and outcomes. Unlike human experts who can access real-time data, consult recent publications, or draw on the latest industry developments, GAI systems remain frozen at their training data cut-off points, often months or years behind current reality.

The implications vary significantly across different domains and use cases. In academic contexts, students researching contemporary issues may receive GAI responses that ignore recent policy changes, emerging scientific discoveries, or current event developments that fundamentally alter the landscape of their inquiry. A student writing about renewable energy policy might receive information that predates significant legislative changes or technological breakthroughs, leading to analysis based on outdated assumptions. Similarly, faculty members using GAI to develop course materials may inadvertently incorporate superseded theories, deprecated technical standards, or historical data presented as current fact.

Industry applications face even more acute challenges, particularly in sectors where real-time information drives strategic decisions. Marketing professionals seeking consumer trend analysis may receive insights based on pre-pandemic behavior patterns that no longer reflect current market realities. Financial analysts relying on GAI for economic forecasting might work with models that ignore recent regulatory changes, geopolitical developments, or market disruptions. Technology professionals using GAI for coding assistance may receive recommendations for libraries, frameworks, or security protocols that have since been updated, deprecated, or identified as vulnerable.

The knowledge freeze problem becomes particularly insidious because GAI systems typically present outdated information with the same confidence and authority as current data. Users may have no indication that the detailed market analysis, regulatory compliance guidance, or technical specifications they've received are months or years out of date. This temporal disconnect requires users to develop verification habits and maintain awareness of GAI limitations, treating AI-generated information as a starting point for investigation rather than a definitive source of current knowledge. Organizations must establish protocols for cross-referencing GAI outputs with recent sources and ensure that time-sensitive decisions aren't based solely on potentially outdated AI-generated insights.

4.5 Use It or Lose It: The Cognitive and Creative Cost of AI Convenience

Even if (or dare we say when) GAI technology advances to achieve perfect accuracy, objectivity, and relevance, concerns remain about its potential to deteriorate human-driven cognition, creativity, and domain knowledge (Khan, 2025; Ng et al., 2025; Pierrés et al., 2024; Seo et al., 2021). The fundamental question isn't whether GAI can help students meet academic demands or serve as a steppingstone for qualified individuals struggling with time constraints—it's whether the efficiency gains justify the potential long-term costs to human intellectual development. This dilemma requires careful consideration of what we might sacrifice in our rush toward technological convenience.

Many argue the overreliance on GAI can lead to weakened neural pathways (Birss, 2025; Kosmyna et al., 2025). The concept of cognitive "offloading" lies at the heart of these concerns, as GAI use can enable individuals to bypass essential learning processes in favor of immediate results. Consider the well-documented difference between a student using GAI to summarize lecture content versus taking notes by hand. Research consistently demonstrates that the physical and cognitive processes involved in manual notetaking enhance learning retention, comprehension, and critical thinking development (Flanigan et al., 2024). The act of listening, processing, synthesizing, and writing engages multiple cognitive pathways that strengthen understanding in ways that passive consumption of AI-generated summaries cannot replicate. Similarly, when industry professionals delegate report writing, data analysis, or strategic planning entirely to GAI systems, they may forfeit opportunities for skill development and the deeper insights that emerge through intellectual struggle and practice (Birss, 2025).

Educational contexts reveal particularly concerning patterns of cognitive dependency. Students who routinely use GAI for essay writing may never develop the critical thinking skills required to construct coherent arguments, analyze complex texts, or synthesize multiple sources effectively (Kosmyna et al., 2025). Mathematics students who rely on GAI for problem-solving may miss the pattern recognition and logical reasoning development that comes from working through calculations independently. In creative disciplines, students using GAI for brainstorming or concept development may fail to develop the tolerance for ambiguity, iterative thinking, and creative risk-taking that characterize innovative minds.

Industry applications present parallel risks across diverse professional contexts. Marketing professionals who depend on GAI for campaign concepts may lose their ability to understand target audiences intuitively or develop culturally resonant messaging. Engineers who outsource design calculations to AI systems might atrophy in their fundamental problem-solving capabilities, potentially missing critical safety considerations or innovative solutions that emerge from deep technical understanding. Financial analysts relying heavily on GAI for market assessment may lose the nuanced judgment that comes from years of interpreting complex economic indicators and recognizing subtle patterns that automated systems might overlook.

The parallel creativity deficit deserves particular attention, as innovation represents one of humanity's most distinctive capabilities. While newer GAI releases promise enhanced creative features—generating detailed images, processing visual inputs, and producing multimedia content—these capabilities may paradoxically diminish human creative capacity (Mannan, 2025; Truly, 2024). When designers immediately turn to GAI for visual concepts rather than sketching initial ideas, they may bypass the iterative thinking that leads to breakthrough innovations. Writers who rely on AI for plot development or character creation might never develop the imaginative muscle memory that enables truly original storytelling. The concern isn't that GAI lacks creative potential, but that overreliance on AI-generated creativity could atrophy the uniquely human processes of imagination, intuition, and inspired connection-making.

The intelligence testing capabilities of advanced GAI systems further complicate these considerations. Initial studies showed that with heavy reliance on early GAI tools, learners could attain only average results (Choi et al., 2022; Terwisch, 2023). However, the quickly changing landscape has resulted in improved outputs that now often mirror those of top performers. ChatGPT-4's success on high-stakes examinations like the Uniform Bar Examination, SAT, and GRE, for example, demonstrates impressive pattern recognition and information synthesis capabilities (OpenAI 2023b). However, test performance shouldn't be confused with professional competency or cognitive sophistication. While AI can rapidly process vast amounts of information and identify statistical patterns, it remains unable to demonstrate consistent ethical reasoning, provide nuanced legal counsel, interpret complex multi-layered problems, or perform the advanced cognitive abilities that characterize expert human judgment (Choi et al., 2022; Kimmel, 2023; Mittal et al., 2024).

This disconnect between GAI capabilities and human expertise suggests the need for a more nuanced approach—one that views the relationship between human cognition and artificial intelligence as a continuum rather than a binary choice. The critical question becomes: Is there a way to harness GAI's efficiency while preserving the cognitive processes that foster learning, reflection, and growth? The answer likely lies in intentional integration strategies that use GAI as a collaborative tool rather than a replacement for human intellectual engagement, ensuring that the convenience of artificial intelligence enhances rather than undermines our most essential human capabilities (Yee et al., 2024).

4.6 The Upgrade Trap: Forever Playing Catch-Up with AI

At a practical level, the constantly changing GAI landscape challenges institutions and organizations to incessantly evolve, evaluate, and adapt. While novel integrations and new capabilities offer powerful upgrades and should be explored and embraced, with new standalone and integrated tools emerging nearly daily, how can we keep up? The relentless pace of GAI development creates what might be called "innovation fatigue"—a state where the energy required to assess, pilot, and implement new technologies begins to exceed the capacity of even well-resourced organizations (Castrillon, 2025; Fedulov, 2025). Educational institutions and businesses find themselves caught in an exhausting cycle of evaluation, where each promising new tool demands time for testing, training, and integration just as the next wave of innovations arrives.

This acceleration creates practical challenges that extend far beyond simple technology adoption. IT departments struggle to maintain security protocols and compatibility standards when AI tools proliferate faster than governance frameworks can accommodate them. Training programs become obsolete before completion, as the platforms and features they address evolve or disappear entirely. Budget planning becomes nearly impossible when organizations cannot predict which tools will remain viable, which will be superseded, or which will require costly upgrades within months of implementation. The result is often a piecemeal approach to GAI adoption, with different departments using incompatible tools and developing divergent workflows that resist organizational coherence.

Perhaps most problematically, the rapid tool evolution can prevent organizations from achieving meaningful mastery with any single platform. Just as teams begin to develop sophisticated prompting techniques, workflow integrations, or quality control processes for one GAI system, market pressures or feature improvements may push them toward newer alternatives. This constant churn undermines the deep familiarity that enables truly effective human-AI collaboration, leaving users perpetually in learning mode rather than optimization mode. Educational institutions face the additional challenge of maintaining consistency in student experiences when faculty members adopt different tools mid-semester or when institutional GAI policies struggle to keep pace with available options.

The economic implications compound these operational challenges, as organizations find themselves making technology investments with increasingly uncertain return timelines (Fedulov, 2025). The subscription costs, training expenses, and integration efforts required for GAI adoption multiply when tools change frequently, and the sunk costs of abandoned platforms accumulate quickly. This creates a paradox where the organizations most committed to staying current with GAI developments may find themselves most financially strained by the effort, while those who adopt more conservative approaches risk falling behind in an increasingly AI-augmented competitive landscape. Success in this environment requires not just technological sophistication, but strategic wisdom about when to jump on new developments and when to focus on optimizing existing capabilities.

4.7 The Regulatory Race: Writing Rules for a Moving Target

The regulatory landscape surrounding GAI adds another layer of complexity to an already challenging implementation environment. With all the challenges organizations face in GAI's wake, it is not surprising that government and organizational frameworks are rapidly emerging to address potential risks and establish operational boundaries. Fears about AI's long-term societal impact, combined with its potential for complete industry disruption, have spurred government entities, technology giants, sundry industries, and key educational players to call for comprehensive regulations surrounding GAI development and deployment (Clayton, 2023; Ng et al., 2025; Rasky, 2024). This regulatory urgency has created a complex web of compliance requirements, policy development pressures, and governance challenges that both educational institutions and industry organizations must navigate while simultaneously trying to harness GAI's benefits.

Educational institutions find themselves at the epicenter of regulatory complexity, tasked with protecting academic integrity and privacy while enabling innovation in teaching and learning (Choi et al., 2022; Rasky, 2024). From university-wide policies to individual classroom practices, administrators and faculty are continually reframing, reshaping, and reimagining the parameters that define the relationship between GAI and formal education. The challenge extends beyond simple prohibition or permission—institutions must develop nuanced frameworks that distinguish between appropriate research assistance and academic misconduct, between learning enhancement and intellectual shortcuts. Faculty members face the additional burden of crafting detailed syllabi statements that establish expected, encouraged, and prohibited uses of AI within their specific disciplinary contexts (Choi et al., 2022; Hibbert et al., 2024). These policies must be specific enough to provide clear guidance yet flexible enough to accommodate the diverse ways GAI might legitimately support learning across different fields and assignment types. With this, educational leaders face complex questions about consistency and modeling: Is it ethical for faculty to use GAI for creating teaching materials, managing a class, or grading student work while prohibiting learner use entirely? These dilemmas illustrate the need for context-specific policies that consider different roles, learning objectives, and professional responsibilities rather than blanket prohibitions or permissions.

Industry sectors confront their own regulatory Wild West, where professional standards, intellectual property laws, and employment regulations intersect with rapidly evolving AI capabilities in unpredictable ways. The 2023 Hollywood screenwriters' strike exemplifies these tensions, with writers demanding studio restrictions on AI use for script development, arguing that algorithmic content creation diminishes artistic integrity and threatens creative livelihoods (Straub, 2023). Similar concerns ripple across industries where GAI threatens to disrupt traditional roles and creative processes. Legal professionals grapple with questions of AI-generated brief writing and case research, while healthcare organizations navigate the regulatory implications of AI-assisted diagnosis and treatment recommendations (Gerke et al., 2020; Magesh et al., 2024).

The urgency surrounding GAI regulation has produced unprecedented calls for industry-wide intervention and oversight. An influential open letter circulated during GAI's rapid expansion called for an "immediate pause for at least 6 months" on advanced AI training systems, asserting that "advanced AI could represent a profound change in the history of life on Earth, and should be planned for and managed with commensurate care and resources" (The Future of Life Institute, 2023). The letter's high-profile signatories, including technology leaders like Elon Musk, signal widespread concern about the pace of AI development relative to regulatory preparedness. Notably, even OpenAI (2023a), the company behind ChatGPT, has advocated for government oversight of their own systems, acknowledging the substantial risks inherent in advanced AI capabilities.

In the absence of comprehensive legal frameworks, independent organizations and industry groups have stepped forward to provide ethical guidelines and best practice recommendations. The Asilomar AI Principles, established at the 2017 Future of Life Institute's Beneficial AI conference—years before the current GAI surge—offer 23 foundational principles designed to guide responsible AI development and deployment (The Future of Life Institute, 2017). These principles address research ethics, safety considerations, and long-term societal implications, providing a voluntary framework that many organizations have adopted as they develop their own GAI policies. Similarly, The United Nations Educational, Scientific, and Cultural Organization offers a framework to guide the implementation of GAI in educational and research-based organizations (UNESCO, 2023). UNESCO's principles for GAI use embody a "human-centered approach," which values "human agency, inclusion, equity, gender equality, and linguistic and cultural diversities, as well as plural opinions and expressions." The guidelines urge government protection of data privacy and age restrictions, among other restraints designed to elevate inclusivity and honor ethical considerations.

However, the voluntary nature of most current guidelines creates inconsistency across organizations and industries, leaving individual institutions to interpret broad principles within their specific contexts. Educational institutions may find themselves developing contradictory policies based on different interpretations of academic integrity principles, while companies in the same industry may adopt vastly different approaches to GAI integration based on varying risk tolerances and ethical frameworks. This regulatory patchwork creates challenges for students, employees, and organizations trying to maintain compliance across multiple contexts, and it complicates efforts to establish industry-wide standards or professional norms around GAI use.

The fundamental challenge lies in the mismatch between the pace of technological development and the necessarily deliberate process of regulatory development (UNESCO, 2023). By the time comprehensive policies are drafted, debated, and implemented, the GAI landscape may have evolved substantially, potentially rendering specific regulations obsolete or creating unintended consequences. This dynamic tension requires organizations to develop adaptive governance frameworks that can evolve alongside technological capabilities while maintaining core ethical principles and operational standards.

5 Ethical Imperatives Across Contexts

While these challenges might seem daunting, they also reveal the shared ethical terrain that both educational institutions and industry organizations must navigate. The ethical considerations surrounding GAI use in education mirror those in industry, offering transferable insights for responsible implementation.

5.1 The Privacy Imperative: Maintaining Confidentiality in AI-Enhanced Work

Data privacy and security concerns represent some of the most immediate and tangible risks associated with GAI adoption, yet they often receive insufficient attention from users eager to harness the technology's capabilities (Chakraborty et al., 2023; Hamilton, 2024; Mittal et al., 2024; Ng et al., 2025; Seo et al., 2021). Unlike traditional software applications that process data locally or within controlled environments, most GAI systems operate through cloud-based platforms where user inputs become part of the interaction data stream. This architecture creates potential vulnerabilities that both educational institutions and industry organizations must address proactively.

The fundamental challenge lies in the disconnection between user convenience and data security protocols. In our workshops and consultations, we've witnessed participants inadvertently sharing sensitive financial information, confidential research data, student records, and proprietary business strategies while crafting GAI prompts. The conversational nature of these tools can create a false sense of privacy, leading users to treat AI interactions as they would have confidential conversations with trusted colleagues. However, many GAI platforms retain conversation histories, use interactions for model training, or operate under terms of service that grant broad usage rights to submitted content.

Educational institutions face particular vulnerabilities due to the sensitive nature of student data protected under regulations like FERPA. When faculty members use GAI to analyze student work, generate feedback, or create personalized learning materials, they may unknowingly expose protected educational records to third-party systems. Similarly, industry professionals working with proprietary information, client data, or strategic plans risk inadvertent disclosure when they incorporate specific details into their prompts. The challenge is compounded by the fact that effective GAI prompts often require substantial context and specificity—the very details that create privacy and security risks. Recognizing these vulnerabilities, certain educational GAI platforms such as ChatGPT Edu and BoodleBox have implemented enhanced privacy protections and security features, though these improvements remain inconsistent across the broader tool landscape and are frequently inaccessible to industry users (Hamilton, 2024).

Addressing these concerns requires both technical safeguards and comprehensive user education. Organizations must establish clear policies about what information can and cannot be shared with GAI systems, provide secure alternatives for sensitive data processing, and train users to recognize and avoid privacy pitfalls. This might involve teaching techniques for anonymizing data before GAI interaction, using hypothetical examples rather than real cases, or implementing enterprise-level GAI solutions with enhanced security features. The goal is not to eliminate GAI use, but to ensure that its benefits don't come at the cost of compromising sensitive information or violating regulatory requirements.

5.2 Democratizing AI: Addressing Disparities in Tool Quality and Access

Equity and access issues present key challenges in GAI implementation, creating new forms of digital divide that intersect with existing socioeconomic inequalities (Chakraborty et al., 2023; Yee et al., 2024). Premium GAI tools consistently produce superior outputs compared to their free counterparts, offering more sophisticated reasoning capabilities, longer conversation histories, faster processing speeds, priority access during high-demand periods, and advanced features like image generation or document analysis. The result is a stratified GAI ecosystem where users with greater financial resources gain compound advantages in productivity, creativity, and problem-solving capacity, establishing a tiered system where access to high-quality artificial intelligence becomes a function of financial resources rather than merit or need.

In higher education, these disparities create particularly troubling implications for academic equity and integrity. Students from affluent backgrounds may access premium GAI subscriptions that produce more sophisticated analyses, better-structured arguments, and more polished writing, while their peers relying on free versions receive more basic assistance. This technological stratification can exacerbate existing educational inequalities, where socioeconomic status already influences access to tutoring, test preparation, and academic support resources. The irony is that GAI, which holds promise for democratizing access to high-quality educational assistance, may instead deepen the achievement gaps it has the potential to narrow.

The academic integrity implications compound these equity concerns in unexpected ways. Many institutions have adopted AI detection tools to identify potentially AI-generated student work, but these systems demonstrate varying effectiveness across different GAI platforms and subscription tiers (Furze, 2024). Detection algorithms often struggle more with outputs from free or older GAI versions, which may produce text with different linguistic patterns or statistical signatures than premium tools. This creates a perverse situation where students using free GAI tools may face higher rates of false-positive detection, potentially subjecting them to academic misconduct investigations, while their peers using premium services may evade detection entirely. Non-native English speakers face additional vulnerability, as their natural writing patterns may already trigger detection algorithms, compounding the disadvantage when combined with free GAI tool use. The result is an equity gap not just in GAI access but in the consequences of GAI use.

Professional contexts reveal parallel disparities with equally significant implications for career advancement and workplace equity. Employees who can afford premium GAI subscriptions may demonstrate enhanced productivity, creativity, and analytical capabilities that lead to better performance evaluations, promotion opportunities, and salary increases. Marketing professionals with access to advanced GAI tools may be able to produce more sophisticated campaigns, while their colleagues using free versions could struggle to compete for high-visibility projects. Sales representatives with premium AI assistance may outperform peers who lack similar technological support, creating performance disparities that reflect economic privilege rather than inherent capability or work ethic.

Addressing these equity challenges requires intentional intervention at multiple levels. Educational institutions are exploring institutional GAI licenses to provide universal access and redesigning assessments to minimize advantages from premium tools. Employers face similar decisions about providing GAI subscriptions as standard workplace resources. Without broader policy interventions—such as public funding for educational GAI access, regulatory requirements for equitable distribution, or industry initiatives for underserved populations—GAI risks amplifying existing inequalities rather than augmenting human potential across diverse communities.

5.3 Intellectual Property: Redefining Ownership and Originality

Copyright and intellectual property concerns exist whether we're discussing student plagiarism or corporate content creation, creating complex legal and ethical challenges that both educational institutions and industry organizations must navigate carefully (Art of Sol Studio, 2024; Milton and Enright n.d.; Mittal et al., 2024; UNESCO, 2023). In some educational circles, AI has been blacklisted as a tool to facilitate cheating (Furze, 2024; Hamilton, 2024; Ng et al., 2025; Wong, 2024), but the reality involves more nuanced questions about originality, attribution, and intellectual contribution. How do you know if the information GAI provides is copyright compliant? Who owns the copyright of AI-generated content? What counts as "plagiarism"? These questions become particularly pressing as GAI systems demonstrate increasing sophistication in generating text, images, code, and multimedia content that may incorporate or remix existing copyrighted materials in ways that are difficult to detect or trace.

The foundational copyright challenge stems from GAI training processes that involve ingesting vast amounts of copyrighted material—books, articles, images, music, and other creative works—often without explicit permission from copyright holders (Milton and Enright n.d.). When GAI systems generate outputs, they may inadvertently reproduce copyrighted content, paraphrase protected expressions, or create derivative works that build upon copyrighted materials in ways that could constitute infringement. Users have no reliable way to determine whether GAI-generated content incorporates protected intellectual property, creating potential liability for individuals and organizations that use AI-assisted work in commercial or academic contexts. This uncertainty is compounded by the "black box" nature of most GAI systems, which makes it impossible to trace the sources and influences that shaped any particular output.

The ownership question presents equally complex challenges across educational and professional settings (Milton and Enright n.d.). Current U.S. Copyright Office (2025) guidance indicates that works produced solely by artificial intelligence cannot be copyrighted, as copyright requires significant human authorship. However, the boundary between AI-generated and human-authored content remains murky in practice. When a student uses GAI to brainstorm ideas that they then develop into an original essay, when a marketing professional uses AI to generate initial copy that they substantially revise, or when a researcher employs GAI to analyze data and suggest interpretations, determining the extent of human creative contribution becomes crucial for both copyright eligibility and academic integrity assessment.

Industry applications reveal particularly high stakes around intellectual property protection and competitive advantage. Companies investing significant resources in GAI-enhanced product development, marketing campaigns, or strategic analyses need clarity about their ownership rights in AI-assisted work products. If GAI-generated content cannot be copyrighted, organizations may find themselves unable to protect valuable intellectual property that incorporates substantial AI assistance. Conversely, companies may face infringement claims if their GAI-enhanced work inadvertently incorporates competitors' copyrighted materials. The pharmaceutical, entertainment, and technology industries face especially acute challenges, where intellectual property protection directly impacts market value and competitive positioning.

Related to the question of copyrightability, the question of transparency—when and how to acknowledge GAI use—applies equally to academic citations and professional deliverables, but current practices vary widely and often inconsistently (Yee et al., 2024). Academic institutions are developing diverse approaches to GAI attribution requirements, with some demanding detailed disclosure of all AI interactions while others focus on substantial assistance that materially impacts the final work product. The challenge lies in distinguishing between GAI use that requires acknowledgment (such as content generation or significant analytical assistance) and more routine applications (such as grammar checking or basic formatting) that might be considered analogous to spell-check software.

Professional contexts face similar transparency dilemmas with additional considerations around client relationships, competitive advantage, and industry standards. Marketing agencies must decide whether to disclose GAI use in creative campaigns, potentially affecting client perceptions of value and authenticity. Legal firms grapple with whether GAI-assisted research requires disclosure to clients or courts, particularly when AI assistance might affect billing practices or professional responsibility requirements. Technology companies face questions about transparency in GAI-enhanced software development, where AI assistance might be considered proprietary methodology rather than something requiring disclosure.

The ethical dimensions of GAI transparency extend beyond legal compliance to questions of honesty, authenticity, and professional integrity. In educational settings, transparency requirements serve multiple purposes: they help maintain academic honesty standards, enable appropriate assessment of student learning, and prepare students for professional environments where GAI disclosure may be required. However, overly rigid transparency requirements might discourage beneficial GAI use or create administrative burdens that outweigh educational benefits. The goal should be developing frameworks that promote honest engagement with AI tools while recognizing their legitimate role in contemporary work and learning processes.

Looking forward, both educational institutions and industry organizations need adaptive policies that can evolve with changing technology and emerging legal precedents. This includes establishing clear guidelines for GAI attribution, developing protocols for assessing copyright compliance in AI-assisted work, and creating frameworks for evaluating the human creative contribution in collaborative human-AI outputs. As courts begin to address GAI-related copyright disputes and regulatory agencies refine their guidance, organizations must remain prepared to adjust their practices while maintaining ethical standards and protecting intellectual property rights. The ultimate challenge lies in balancing the innovative potential of GAI with respect for existing intellectual property frameworks and the fundamental principles of academic and professional integrity.

5.4 *The Long View: Environmental Sustainability in GAI*

The long-term sustainability of GAI raises critical concerns that extend far beyond immediate implementation challenges, encompassing both environmental impact and the preservation of human expertise across generations. The environmental costs of GAI operation are staggering and largely invisible to end users who experience only the convenience of instant responses (Art of Sol Studio, 2024). Training advanced language models requires enormous computational resources, with some estimates suggesting that training a single large GAI model consumes as much energy as hundreds of homes use in an entire year (Chen, 2025; Institute of Energy and the Environment, 2025). The ongoing operational costs compound these initial expenditures, as millions of daily queries demand constant server processing, cooling systems, and data center infrastructure that contribute significantly to global carbon emissions.

The environmental sustainability question becomes particularly acute when considering the exponential growth in GAI usage across educational and industry contexts. As more institutions integrate AI tools into routine operations, the aggregate energy consumption multiplies rapidly. A university where thousands of students use GAI for daily assignments, combined with faculty utilizing AI for research and course development, creates substantial cumulative environmental impact that may rival the institution's physical infrastructure in terms of carbon footprint. Similarly, when businesses deploy GAI across multiple departments for content creation, data analysis, and customer service, the environmental costs scale dramatically. The promise of more efficient AI models offers hope for reducing per-query energy consumption, but whether these efficiency gains can offset the increasing volume of GAI use remains an open question that will require ongoing monitoring and potentially regulatory intervention.

5.5 *Instructional Design as a Framework for Ethical Integration*

The challenges outlined in the preceding sections—from hallucinations and bias to sustainability concerns and regulatory complexity—might seem insurmountable to organizations seeking to harness GAI's potential while avoiding its pitfalls. However, the field of instructional design offers a tested framework for navigating exactly these types of implementation complexities through systematic, human-centered approaches to technology integration. The ADDIE model (Analyze, Design, Develop, Implement, Evaluate), traditionally used for curriculum and training program development, provides a robust framework for ethical GAI adoption that prioritizes human learning and development over mere technological efficiency.

This instructional design approach transforms GAI integration from a technology deployment challenge into a learning design opportunity, ensuring that artificial intelligence enhances rather than replaces human cognitive processes. By applying ADDIE's systematic methodology, Industry 5.0 organizations can address the multifaceted concerns surrounding GAI adoption—ethical implications, sustainability considerations, and implementation challenges—through a structured process that maintains human agency while leveraging AI capabilities.

5.6 *Analyze: Understanding Context and Needs*

Before implementing GAI, organizations must conduct comprehensive assessments that go beyond efficiency gains to examine specific needs, existing capabilities, and desired outcomes. What problems is GAI solving? What human skills must be preserved or enhanced? This analysis requires honest inventories of current processes, identifying not just pain points that GAI might address, but also valuable human cognitive work that should be protected rather than automated away. A manufacturing company, for example, might discover that while GAI can streamline report generation, the collaborative problem-solving discussions during manual reporting can actually drive innovation and team cohesion.

The analysis must also examine organizational culture, stakeholder readiness, and technological infrastructure to determine how GAI integration might affect human dynamics and workflow patterns. This includes assessing whether team members possess necessary digital literacy and critical thinking skills, identifying where GAI adoption might create unintended dependencies or skill atrophy, and evaluating whether current quality control mechanisms can adapt to AI-augmented outputs. Rather than simply identifying tasks that GAI can perform, effective analysis maps relationships between routine work and expertise development, ensuring that GAI integration preserves learning opportunities and cognitive challenges that build institutional knowledge over time.

5.7 Design: Creating Human-Centered GAI Integration Plans

The design phase focuses on creating workflows that maintain human agency while strategically leveraging AI capabilities to enhance rather than replace human cognitive work. This requires intentional frameworks that treat GAI as a collaborative partner, preserving essential human elements of creativity, critical thinking, and expertise development while utilizing AI for appropriate support functions. Organizations must design assignments and responsibilities that require uniquely human capabilities alongside AI assistance—complex, multifaceted problems that benefit from GAI support but cannot be completed through AI alone.

Context-specific design principles emerge as critical, requiring different GAI parameters across organizational functions. A creative writing course might prohibit GAI use entirely to foster original expression, while a data analysis course might encourage it for certain tasks while requiring human interpretation. Industry 5.0 organizations must similarly develop guidelines recognizing that customer service interactions require different GAI protocols than financial analysis, and entry-level training needs more restrictive policies than senior-level strategic work.

5.8 Develop: Building Ethical Implementation Strategies

Development focuses on building GAI literacy that extends beyond technical proficiency to encompass ethical reasoning, critical evaluation capabilities, and professional judgment skills. Teams need training in prompt engineering, platform navigation, and assessing GAI outputs for accuracy, bias, and appropriateness within their professional contexts. This multifaceted skill development recognizes that effective

GAI use requires the same critical thinking abilities that characterize expertise in any domain.

Programs must address the full spectrum of GAI-related competencies: technical skills like crafting effective prompts, higher-order capabilities like detecting hallucinations and identifying bias, and maintaining appropriate boundaries between human judgment and AI assistance. A financial analysis team learning GAI for investment research needs prompt engineering skills plus training in regulatory compliance, risk assessment methodology, and recognizing when AI-generated projections require independent verification or complete rejection.

5.9 *Implement: Rolling Out GAI with Human Support Systems*

Implementation emphasizes gradual, scaffolded adoption that allows for adjustment and learning while maintaining organizational stability. Phased GAI adoption with adequate support systems provides safety nets during the learning curve, recognizing that successful integration requires cultural adaptation, skill development, and establishment of new collaborative patterns between humans and AI systems.

Pilot programs should begin in low-risk contexts where teams can experiment while maintaining traditional backup processes and quality control mechanisms. Educational institutions might start with faculty development workshops before expanding to student-facing applications; industry organizations might pilot internal communications before client-facing work. Effective implementation requires robust human support networks providing both technical assistance and ethical guidance, including help desk resources, mentorship programs pairing experienced users with newcomers, and clear escalation procedures for quality or ethical concerns.

5.10 *Evaluate: Measuring Success Beyond Efficiency Metrics*

Evaluation extends far beyond traditional efficiency measurements to ensure GAI use aligns with organizational values and learning objectives. This includes monitoring for overreliance patterns, skill degradation, and unintended consequences that may only become apparent over time. While productivity gains represent important metrics, effective evaluation must assess whether GAI use enhances human capability development, maintains professional quality standards, and supports long-term organizational sustainability.

Assessment demands both quantitative measures (accuracy rates, time allocation shifts, output quality comparisons) and qualitative evaluation identifying subtler effects like changes in collaborative dynamics, shifts in professional confidence, and emergence of new dependencies. Longitudinal assessment protocols must track human-AI collaboration evolution over extended periods, recognizing that some impacts may only become apparent months or years after implementation. The evaluation phase should culminate in regular strategy refinement cycles that use findings to enhance GAI integration approaches, ensuring artificial intelligence serves human flourishing rather than merely operational convenience.

6 Lessons from the Academic Trenches

While the ADDIE framework provides a systematic approach to GAI integration, the real-world application of these principles in higher education has yielded specific insights that extend far beyond theoretical models. Through working with diverse faculty and students across multiple implementation cycles, several key lessons have emerged that directly translate to Industry 5.0 contexts and offer concrete guidance for organizations navigating their own GAI adoption journeys.

6.1 The Prompt Matters

Crafting effective GAI prompts requires more than summarizing desired outputs. Effective prompt engineering takes finesse, intentionality, and domain expertise (Birss, 2025; Choi et al., 2022; UNESCO, 2023; Yee et al., 2024). Organizations investing in GAI must also invest in developing their team members' prompting skills and understanding of the technology's capabilities and limitations, recognizing that prompt quality directly impacts output usefulness and reliability.

While numerous frameworks exist for prompt engineering (Birss n.d.; Choi et al., 2022; Fitzpatrick, 2024; Yee et al., 2024), many prove overly complex for widespread organizational adoption or lack the specificity needed for consistent results across diverse user groups. Based on extensive experience with GAI implementation in educational settings, we've developed a streamlined framework that synthesizes the most effective elements from existing models into a memorable, accessible approach. PCRAFT (see Fig. 1) addresses the core components that research identifies as essential for effective prompting while remaining simple enough for routine use by busy professionals and students who need reliable results without extensive technical training. This framework prioritizes practical application over theoretical completeness, providing users with a systematic approach that improves output quality while building transferable skills in human-AI collaboration. Using PCRAFT sets the stage for more controlled, purposeful outputs that align with professional standards and organizational objectives.

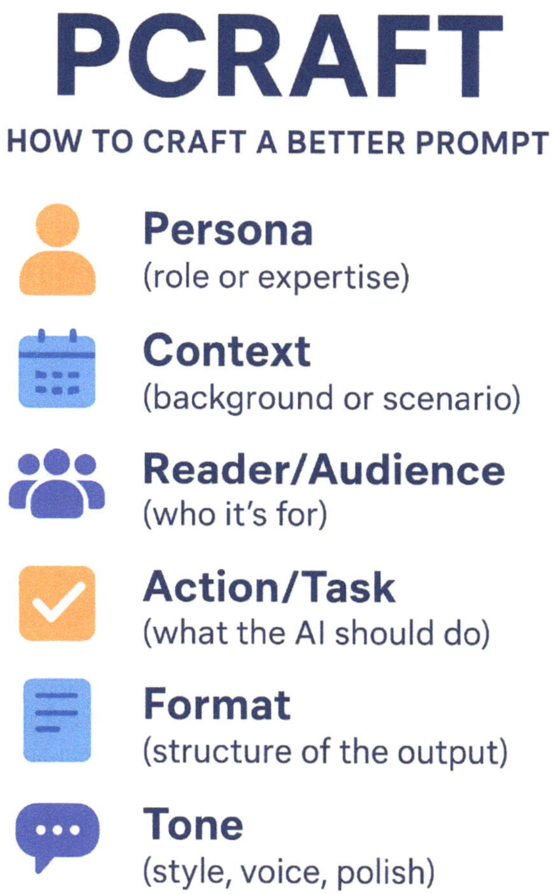

Fig. 1 Prompt engineering with PCRAFT. *Note* The authors created this image with the assistance of ChatGPT

6.2 Persona (Role or Expertise)

Establishing a clear *persona* for the AI helps activate relevant knowledge domains and communication patterns. By specifying a role or area of expertise, you're essentially telling the AI which "lens" to use when approaching your request. For example, asking an AI to respond "as an experienced marketing director" will yield different insights than requesting it respond "as a data scientist" or "as a customer service representative." This element of PCRAFT draws on the AI's training across diverse professional contexts, allowing it to access industry-specific vocabulary, frameworks, and problem-solving approaches.

The persona element works because it provides cognitive scaffolding for the AI's response generation. When you specify, "Write as a project manager with 10 years of experience in software development," you're not only setting expectations for the type of expertise to draw upon, but also establishing credibility markers and communication patterns typical of that role. This becomes particularly powerful when combined with other PCRAFT elements, as the persona informs how the context should be interpreted and what the audience expect from someone in that position.

6.3 Context (Background or Scenario)

Context provides the situational framework that makes AI responses relevant and actionable. Without sufficient context, even the most sophisticated AI can produce generic or misaligned outputs. This element should include relevant background information, current circumstances, constraints, and any specific challenges or goals that inform the request. For instance, rather than asking for "supply chain optimization," providing context such as "a pharmaceutical manufacturer with FDA compliance requirements facing component shortages from overseas suppliers during trade disruptions" yields far more targeted and useful recommendations.

Effective context setting also includes temporal considerations, organizational culture, available resources, and stakeholder dynamics. The more precisely you can paint the picture of the current situation, the better the AI can tailor its response to fit your specific needs. Context acts as the bridge between abstract knowledge and practical application, ensuring that the AI's output acknowledges real-world constraints and opportunities that generic advice might overlook.

6.4 Reader/Audience (Who It's for)

Specifying the intended *reader* or *audience* helps the AI calibrate its language, depth of explanation, and focus areas appropriately. A technical explanation for software engineers will differ significantly from one designed for C-suite executives, even when covering the same topic. The audience element ensures that vocabulary, examples, and level of detail match the recipients' expertise, interests, and decision-making needs. For example, explaining cybersecurity protocols to IT security specialists requires different technical depth than explaining it to hospital administrators who need to understand compliance requirements and budget implications.

Understanding the audience also influences the persuasive elements and priorities within the response. When writing for busy executives, the AI should lead with high-level insights and business impact. When writing for implementers, it should emphasize practical steps and potential challenges. This audience awareness prevents the common pitfall of receiving technically accurate but inappropriately pitched information that fails to serve its intended purpose.

6.5 Action/Task (What the AI Should Do)

The *action* element provides clear direction about what type of cognitive work the AI should perform. Rather than vague requests like "help me with this," effective prompts specify whether you want analysis, creation, comparison, summarization, brainstorming, or problem-solving. Strong action statements use precise verbs: "analyze the competitive landscape," "create an implementation timeline," "compare three strategic options," or "generate five alternative solutions." This specificity helps the AI understand not just what topic to address, but how to approach it.

Well-defined actions also establish success criteria for the output. When you ask the AI to "evaluate" something, you're requesting analytical thinking with pros and cons. When you ask it to "design" something, you're seeking creative construction with consideration for user needs and constraints. The action element transforms broad subject areas into focused deliverables, making it easier for both you and the AI to determine whether the response successfully addresses your needs.

Effective action statements also require defining boundaries and limitations alongside the primary task. Not only must the user define what they want the GAI to produce or do, but they also must indicate the limitations or boundaries (Choi et al., 2022). It may be necessary, for example, to specify what the GAI should *not* do, such as fabricate sources or data. These negative constraints are particularly important given GAI's tendency toward hallucination and its eagerness to fulfill requests even when lacking adequate information. Clear boundary-setting might include instructions like "do not include speculative information," "limit recommendations to evidence-based practices," or "acknowledge when information is unavailable rather than generating placeholder content."

When crafting action statements, it's important to consider GAI's current limitations with complex, multi-step outputs (Choi et al., 2022). Rather than requesting comprehensive deliverables in a single prompt, effective practitioners often break complex tasks into sequential, manageable components. For instance, when developing a business action plan, you might first request a well-organized table of contents, then use that structure to guide subsequent prompts for each section. This iterative approach—asking for an introduction, then providing a detailed outline for the first section before requesting its development—allows you to maintain quality control while leveraging GAI's strengths in focused, bounded tasks. Scaffolded methodologies not only produce higher-quality outputs, but also provide opportunities for human oversight and course correction throughout the development process.

6.6 Format (Structure of the Output)

Format specifications guide how information should be organized and presented, making responses more scannable and actionable. This might include requesting bullet points for quick reference, numbered steps for processes, tables for comparisons, or narrative paragraphs for complex explanations. Format and layout choices should align with how you intend to use the information.

Consider also specifying length constraints, section headers, or visual organization elements that enhance usability. For instance, requesting "a two-page memo with executive summary, three key recommendations, and implementation considerations" provides much clearer direction than asking for "information about this topic." Format specifications help ensure that valuable insights don't get buried in poorly organized responses and that the output integrates smoothly into your existing workflows.

6.7 Tone (Style, Voice, Polish)

Tone definitions influence the personality and energy of the AI's response, ensuring it matches your communication context and organizational culture. In fact, some have found that establishing a clear tone is actually even more important than specifying a role the GAI will assume (Choi et al., 2022). Defining the tone includes establishing a clear formality level (casual vs. professional), confidence level (tentative suggestions vs. definitive recommendations), and emotional resonance (inspiring vs. analytical vs. urgent). For internal team communications, you might request a collaborative, conversational tone, while external client presentations might require more formal, authoritative language.

Tone also encompasses stylistic elements like whether to include humor, how direct to be with criticism or challenges, and what level of enthusiasm is appropriate. This consideration becomes especially important when seeking authentic feedback, since many AI systems are inherently genial and may require intentional prompting to provide more candid and accurate assessments. Different contexts will call for distinctly different tonal approaches to achieve optimal results. A brainstorming session might benefit from an energetic, creative tone that encourages bold thinking, while a risk assessment requires a measured, cautious approach. By explicitly setting tone expectations, you ensure that the AI's communication style supports rather than undermines your intended message and relationship with your audience.

6.8 Human Expertise Remains Essential

GAI outputs cannot stand alone, regardless of prompt quality or technological sophistication. Critical thinking and domain knowledge remain essential for evaluating, refining, and applying GAI-generated content appropriately within professional and educational contexts (Birss, 2025). When professionals develop deep domain knowledge alongside GAI literacy, they become capable of recognizing AI limitations, identifying opportunities for creative application, and maintaining quality standards that reflect human values and professional excellence. This expertise also provides crucial resilience against technological failures, ensuring that organizations can continue functioning effectively when GAI tools encounter novel situations, experience outages, or prove inadequate for specific challenges.

6.9 Context Determines Application

Effective GAI integration requires dramatically different approaches across professional domains based on risk tolerance, regulatory requirements, and quality standards (Ng et al., 2025). Healthcare teams might use GAI for preliminary literature review but maintain strict human control over patient diagnosis, while financial services teams may leverage AI for market analysis but require human verification for investment recommendations. The key distinction lies not in whether to use GAI, but in establishing appropriate boundaries that match the stakes and consequences of each professional context. Entry-level professionals must develop foundational competencies through direct practice before GAI becomes a collaborative tool, ensuring they can recognize when AI assistance is appropriate versus when human expertise must take precedence.

6.10 Process and Product Both Matter

The cognitive processes involved in problem-solving, creating, and learning have inherent developmental value that extends far beyond their outputs. When students work through complex mathematical proofs or engineers iterate through design solutions, these struggles build neural pathways and develop the intuitive expertise that distinguishes professionals from task completers. Organizations must resist optimizing purely for output efficiency when doing so undermines the developmental processes that build long-term capability. The goal is designing workflows that preserve valuable learning opportunities while strategically leveraging AI assistance where it genuinely enhances rather than replaces human cognitive engagement, ensuring that technology serves human intellectual development rather than substituting for the cognitive challenges that make work meaningful.

7 A Shared Learning Curve

As we look toward the future, collaboration between higher education and industry becomes increasingly vital. Universities are developing the GAI-literate workforce that Industry 5.0 organizations need, while industry applications provide real-world laboratories for testing educational approaches. Communities of practice that span sectors can accelerate learning and prevent costly mistakes. When a university discovers an effective approach to GAI ethics training, those insights can inform corporate professional development. When a company develops successful human-AI collaboration models, those frameworks can enhance academic preparation programs.

This collaborative imperative becomes even more urgent as we navigate an uncharted frontier that brings daily innovations, novel tools, and complex ethical quandaries. We venture forward with both excitement and trepidation about the expansive possibilities before us, but we cannot embark on this journey unprepared. Even OpenAI (2023a), the GOAT of GAI, warns us that the infusion of AI includes "serious risk of misuse, drastic accidents, and societal disruption." As we look ahead, then, we must consider what we have learned. Using GAI comes with responsibility to elevate the human essence, to honor the skills and characteristics that make us unique and valuable (Birss, 2025).

The parallel evolution of GAI in education and industry creates unprecedented opportunities for cross-sector learning and growth. Both contexts share the fundamental challenges: maintaining human agency while leveraging technological capability, balancing efficiency gains with skill development, and weighing innovation against responsibility. Both require a fundamental shift in how we envision learning and what it means to prepare for the future. As we continue on this shared learning curve, several principles emerge as essential for ethical GAI adoption.

7.1 Human-Centered Design

Whether designing learning experiences or business processes, humans must remain at the center of GAI integration efforts, with artificial intelligence serving as an amplifier of human capabilities rather than a replacement for human judgment (Wong, 2024). This fundamental principle requires organizations to resist the temptation to automate away human cognitive engagement in favor of efficiency gains, instead designing systems that preserve and enhance the uniquely human elements of creativity, ethical reasoning, and contextual understanding. Students and employees alike need explicit instruction in maintaining their agency within AI-augmented environments, learning to leverage GAI's capabilities while retaining ownership of critical decisions, quality assessments, and creative direction.

The human-centered approach demands comprehensive support systems that develop the critical thinking and decision-making capabilities essential for ethical

GAI use across educational and professional contexts (Choi et al., 2022). This means teaching individuals not just how to craft effective prompts or evaluate AI outputs, but how to maintain professional judgment, recognize the boundaries of AI capability, and preserve the cognitive processes that build expertise over time. Educational institutions and industry organizations must invest in developing these higher-order thinking skills alongside technical GAI literacy, ensuring that human wisdom guides artificial intelligence rather than being supplanted by it. The goal is creating human-AI partnerships where technology enhances human potential while preserving the essential human qualities that drive innovation, ethical behavior, and meaningful work.

7.2 Continuous Learning

GAI technology evolves rapidly, requiring ongoing education and adaptation that extends far beyond initial implementation training to encompass sustained organizational learning commitments. Users recognize and even demand the training necessary to advance their AI literacy for ethical and professional utilization (Ng et al., 2025). Organizations and educational institutions must fully embrace lifelong learning approaches that keep pace with technological change while maintaining focus on human development and ethical considerations (Hamilton, 2024). This continuous learning imperative recognizes that GAI capabilities, limitations, and best practices will shift dramatically over months rather than years, making one-time training programs insufficient for long-term success. The most effective organizations treat GAI literacy as an evolving competency that requires regular updates, refresher training, and opportunities for practitioners to share emerging insights and challenges with their peers.

The continuous learning approach must balance staying current with technological developments while avoiding the exhausting cycle of constant tool-switching that characterizes less strategic GAI adoption. This requires developing learning frameworks that emphasize transferable principles—such as critical evaluation skills, ethical reasoning, curiosity, and human-AI collaboration strategies—that remain valuable even as specific platforms and capabilities evolve (Hibbert et al., 2024; Mittal et al., 2024; Yee et al., 2024). Educational institutions have found success with faculty learning communities that meet regularly to discuss GAI developments, share effective practices, and collaborate on policy refinements. Industry organizations can adopt similar approaches through professional development programs, cross-departmental GAI user groups, and partnerships with educational institutions that provide access to cutting-edge research on AI ethics and implementation. The goal is building organizational capacity for adaptive GAI use that evolves thoughtfully rather than reactively, ensuring that technological advancement serves human flourishing rather than creating perpetual disruption.

7.3 Ethical Leadership

Leaders in both sectors must model responsible GAI use and create cultures that prioritize long-term value over short-term convenience, recognizing that their approach to artificial intelligence will profoundly shape organizational norms and employee behavior. Ethical leadership in the GAI era requires more than policy development or compliance oversight—it demands visible commitment to human-centered principles, transparent communication about AI use and limitations, and willingness to forgo efficiency gains when they compromise human development or ethical standards. When educational administrators demonstrate thoughtful GAI integration in their own work while maintaining emphasis on critical thinking and academic integrity, they establish expectations that cascade throughout their institutions. Similarly, when industry executives use GAI as a collaborative tool rather than a replacement for strategic thinking, they signal that technology should enhance rather than substitute for human expertise.

This leadership approach requires courage to resist the pressure for immediate productivity gains when they conflict with sustainable human development or ethical considerations. Effective leaders establish decision-making frameworks that weigh GAI's short-term benefits against potential long-term costs to workforce development, creativity, and organizational resilience. They invest in comprehensive training programs even when the immediate return on investment remains unclear, maintain quality standards even when GAI shortcuts might accelerate timelines, and preserve opportunities for human skill development even when automation appears more efficient. Most importantly, ethical leaders in the GAI era recognize that their technology adoption decisions today will determine whether their organizations develop sustainable competitive advantages through enhanced human-AI collaboration or create dependencies that ultimately undermine their long-term viability and mission fulfillment.

7.4 Transparent Communication

Clear policies and open dialog about GAI are not only necessary for ethical and sustainable use, but also build trust and ensure alignment with organizational and human values (OpenAI 2023a). Transparent communication goes beyond simply publishing GAI usage guidelines to encompass ongoing conversations about implementation challenges, ethical dilemmas, and evolving best practices that emerge as organizations gain experience with artificial intelligence tools. Educational institutions that engage faculty, students, and staff in collaborative policy development create shared ownership of GAI integration approaches while building institutional knowledge about effective human-AI collaboration. Similarly, industry organizations that maintain open channels for discussing GAI successes, failures, and concerns enable continuous improvement while demonstrating commitment to responsible innovation.

Effective transparent communication requires acknowledging the uncertainties and complexities inherent in GAI adoption rather than presenting oversimplified solutions or definitive answers to evolving challenges. This includes honest discussions about GAI limitations, potential risks, and the ongoing learning process that characterizes responsible implementation. Organizations must create safe spaces for reporting GAI-related problems, sharing concerns about overreliance or skill degradation, and proposing refinements to existing policies without fear of penalty or judgment. When leaders communicate transparently about their own GAI learning experiences—including mistakes, unexpected challenges, and evolving perspectives—they normalize the experimental, iterative approach necessary for ethical AI integration. This transparency builds organizational resilience by ensuring that GAI adoption decisions reflect collective wisdom rather than individual assumptions, creating cultures where technology serves explicitly articulated human values rather than operating according to unstated or poorly understood principles.

8 Conclusion

The question is not whether GAI will transform our work and learning environments—it already has. The question is whether we will approach this transformation with the intentionality of the slow living movement, using technology purposefully to enhance human potential rather than diminish it. By learning from each other's experiences and maintaining focus on human flourishing, we can navigate this technological revolution in ways that honor both efficiency and growth, both innovation and wisdom.

Like the artisan baker who uses modern tools to create something that nourishes both body and soul, we can harness GAI's power while preserving the essentially human processes that make work meaningful and learning transformative. The curve may be steep, but we're climbing it together.

Acknowledgements This chapter was developed through collaborative human-AI partnerships that exemplify the ethical integration principles discussed throughout the work. Claude (Anthropic) served as a primary collaborative partner for ideation, structural refinement, and prose development, helping to organize complex ideas, suggest alternative phrasings, and provide feedback on clarity and flow. ChatGPT was used specifically to process initial ideas for the PCRAFT prompt engineering framework and to generate a related visual illustration. Both AI systems functioned as collaborative tools rather than content generators—all conceptual frameworks, personal experiences, research interpretations, and pedagogical insights originate from human expertise and judgment. The AI assistance enhanced the writing process by enabling more efficient organization and refinement of ideas while preserving human authorship of all substantive content. This collaborative approach demonstrates the human-centered GAI integration that the chapter advocates: leveraging artificial intelligence to amplify human capabilities while maintaining intellectual ownership and critical oversight of all outputs.

References

Art of Sol Studio. (2024, December 11). *5 Reasons why generative AI art could lose its hype in 2025.* Medium. https://medium.com/@artofsolstudio/5-reasons-why-generative-ai-art-could-lose-its-hype-in-2025-8238d60881f1

Birss, D. (2025, May 12). *Adequate work is the new office plague.* LinkedIn. https://www.linkedin.com/pulse/adequate-work-new-office-plague-dave-birss-zdaxe/

Birss, D. (n.d.). *The prompt collection.* https://davebirss.com/documents/the_prompt_guide.pdf

Bittla, S. R. (2025, February 19). *AI vs. reality: Why ChatGPT struggles with accurate image enhancements.* Medium. https://bittla.medium.com/ai-vs-reality-why-chatgpt-struggles-with-accurate-image-enhancements-7351020638c7

Castrillon, C. (2025, June 24). Why 'AI fatigue' is wearing you down and how to beat it. *Forbes.* https://www.forbes.com/sites/carolinecastrillon/2025/06/24/why-ai-fatigue-is-wearing-you-down-and-how-to-beat-it/

Chakraborty, U., Roy, S., & Kumar, S. (2023). *Rise of generative AI and ChatGPT: Understand how generative AI and ChatGPT are transforming and reshaping the business world.* BPB Publications.

Chen, S. (2025, March 5). How much energy will AI really consume? The good, the bad and the unknown. *Nature.* https://www.nature.com/articles/d41586-025-00616-z

Choi, J. H., Hickman, K. E., Monahan, A., & Schwarcz, D. B. (2022). ChatGPT goes to law school. *Journal of Legal Education, 387.* https://doi.org/10.2139/ssrn.4335905

Clayton, J. (2023, September 13). *'Overwhelming consensus' on AI regulation—Musk.* BBC. https://www.bbc.com/news/technology-66804996

Fedulov, I. (2025, April 17). Innovation fatigue: Why most companies get stuck (and how to break free). *Forbes.* https://www.forbes.com/councils/forbestechcouncil/2025/04/17/innovation-fatigue-why-most-companies-get-stuck-and-how-to-break-free/

Fitzpatrick, D. (2024, September 7). *The perfect ChatGPT prompt doesn't exi.* The AI Educator. https://www.theaieducator.io/post/the-perfect-chatgpt-prompt-doesn-t-exi

Flanigan, A. E., Wheeler, J., Colliot, T., Lu, J., & Kiewra, K. A. (2024). Typed versus handwritten lecture notes and college student achievement: A meta-analysis. *Educational Psychology Review, 36*(78). https://doi.org/10.1007/s10648-024-09914-w

Furze, L. (2024, April 9). *AI detection in education is a dead end.* https://leonfurze.com/2024/04/09/ai-detection-in-education-is-a-dead-end/comment-page-1/

The Future of Life Institution. (2017, August 11). *Asilomar AI principles.* Beneficial AI Conference. https://futureoflife.org/open-letter/ai-principles/

The Future of Life Institute. (2023, March 22). *Pause giant AI experiments: An open letter.* https://openai.com/index/planning-for-agi-and-beyond/

Gerke, S., Minssen, T., & Cohen, G. (2020). Ethical and legal challenges of artificial intelligence-driven healthcare. *Artificial Intelligence in Health Care*, 295-336. https://doi.org/10.1016/B978-0-12-818438-7.00012-5

Hamilton, I. (2024, June 6). Artificial intelligence in education: Teachers' opinions on AI in the classroom. *Forbes.* https://www.forbes.com/advisor/education/it-and-tech/artificial-intelligence-in-school/

Hibbert, M., Altman, E., Shippen, T., & Wright, M. (2024, June 3). *A framework for AI literacy.* EDUCAUSE. https://er.educause.edu/articles/2024/6/a-framework-for-ai-literacy

Institute of Energy and the Environment. (2025, September 8). *Why AI uses so much energy–and what we can do about it.* PennState. https://iee.psu.edu/news/blog/why-ai-uses-so-much-energy-and-what-we-can-do-about-it

Khan, K. (2025). *Automated by atrophied? Student over-reliance vs expert augmentation of AI in learning and cybersecurity.* https://doi.org/10.48550/arXiv.2507.21062

Kimmel, L. (2023, April 7). ChatGPT passed the uniform bar examination: Is artificial intelligence smart enough to be a lawyer? *International and Comparative*

Law Review. https://international-and-comparative-law-review.law.miami.edu/chatgpt-passed-the-uniform-bar-examination-is-artificial-intelligence-smart-enough-to-be-a-lawyer/
Kolisko, S., & Anderson, C. J. (2023). Exploring social biases of large language models in a college artificial intelligence course. *Proceedings of the 37th AAAI conference on artificial intelligence, Vol. 37*(13), pp. 15825–15833. https://doi.org/10.1609/aaai.v37i13.26879
Kosmyna, N., Hauptmann, E., Yan, Y. T., Situ, J., Liao, X.-H., Beresnitzky, A. V., Braunstein, I., & Maes, P. (2025). *Your brain on ChatGPT: Accumulation of cognitive debt when using an AI assistant for essay writing task.* MIT. https://doi.org/10.48550/arXiv.2506.08872
Magesh, V., Surani, F., Dahl, M., Suzgun, M., Manning, C. D., & Ho, D. E. (2024, May 23). *AI on trial: legal models hallucinate in 1 out of 6 (or more) benchmarking queries.* Human-Centered Artificial Intelligence, Stanford University. https://hai.stanford.edu/news/ai-trial-legal-models-hallucinate-1-out-6-or-more-benchmarking-queries
Mannan, J. (2025, August 15). *What is GPT-5? OpenAI's latest model explained.* Digital Trends. https://www.digitaltrends.com/computing/gpt-5/
Milton, D. C., & Enright, H. A. (n.d.). *Case tracker: Artificial intelligence, copyrights, and class actions.* BakerHostetler. https://www.bakerlaw.com/services/artificial-intelligence-ai/case-tracker-artificial-intelligence-copyrights-and-class-actions/
Mittal, U., Sai, S., Chamola, V., & Sangwan, D. (2024). A comprehensive review on generative AI for education. *IEEE Access, 12*, 1427330–2142759. https://doi.org/10.1109/ACCESS.2024.3468368
Morris, C. (2024, February 14). *ChatGPT and Google's Gemini will now remember your past conversations.* Fast Company. https://www.fastcompany.com/91029395/chatgpt-google-gemini-remember-past-conversations
Ng, D. T. K., Chan, K. C. C., & Lo, C. K. (2025). Opportunities, challenges and school strategies for integrating generative AI in education. *Computers and Education: Artificial Intelligence, 8*. https://doi.org/10.1016/j.caeai.2025.100373
OpenAI. (2023a, February 24). *Planning for AGI and beyond.* https://openai.com/index/planning-for-agi-and-beyond/
OpenAI. (2023b, March 14). *GPT-4.* https://openai.com/index/gpt-4-research/
Pierrés, O., Darvishy, A., & Christen, M. (2024). Exploring the role of generative AI in higher education: Semi-structured interviews with students with disabilities. *Education and Information Technologies, 30*, 8923–8952. https://doi.org/10.1007/s10639-024-13134-8
Rasky, E. (2024). *Generative AI policy in higher education: A preliminary survey* [Working Paper]. Center for International Governance, Digital Policy Hub. https://www.cigionline.org/publications/generative-ai-policy-in-higher-education-a-preliminary-survey/
Savicky, R. (2025, February 5). *LLM data cutoff dates: Why the matter for AI content.* Medium. https://medium.com/writing-for-humans/llm-data-cutoff-dates-why-they-matter-for-ai-content-9709afba73e3
Seo, K., Tang, J., Roll, I., Fels, S., & Yoon, D. (2021). The impact of artificial intelligence on learner-instructor interaction in online learning. *International Journal of Educational Technology in Higher Education, 18*(54). https://doi.org/10.1186/s41239-021-00292-9
Southern, M. G. (2025, September 8). Google publishes exact Gemini usage limits across all tiers. *Search Engine Journal.* https://www.searchenginejournal.com/google-publishes-exact-gemini-usage-limits-across-all-tiers/555433/
Straub, J. (2023, May 10). Can AI write 'Ted Lasso'? Writers strike may open door to ChatGPT-written scripts. *USA Today.* https://www.usatoday.com/story/opinion/2023/05/10/wga-strike-pave-way-ai-generated-tv-movie-scripts/70198801007/
Thakker, A., (2025, May 8). *Why your AI assistant sometimes forgets what you just said.* Medium. https://medium.com/@aasthathakker/why-your-ai-assistant-sometimes-forgets-what-you-just-said-7bd969de885a
Terwisch, C. (2023). *Would ChatGPT3 get a Wharton MBA? A prediction based on its performance in the operations management course.* Mack Institute for Innovation Management at the

Wharton School, University of Pennsylvania. https://mackinstitute.wharton.upenn.edu/wp-content/uploads/2023/01/Would-ChatGPT-get-a-Wharton-MBA.pdf

Truly, A. (2024, December 6). *GPT-4: Everything you need to know about ChatGPT's standard AI model.* Digital Trends. https://www.digitaltrends.com/computing/chatgpt-4-everything-we-know-so-far/

UNESCO. (2023). *Guidance for generative AI in education and research.* https://doi.org/10.54675/EWZM9535

U.S. Copyright Office. (2025, January). *Copyright and artificial intelligence: Part 2: Copyrightability* [Report]. https://www.copyright.gov/ai/Copyright-and-Artificial-Intelligence-Part-2-Copyrightability-Report.pdf

Wong, W. K. O. (2024). The sudden disruptive rise of generative artificial intelligence? An evaluation of their impact on higher education and the global workplace. *Journal of Open Innovation: Technology, Market, and Complexity.* https://doi.org/10.1016/j.joitmc.2024.100278

Yee, K., Uttish, L., Main, E., & Giltner, E. (2024, July 17). *AI hacks for educators: 50+ practical tips for faculty to save time by using GenAI.* FCTL Press. https://stars.library.ucf.edu/oer/9/?utm_source=Iterable&utm_medium=email&utm_campaign=campaign_10587896_nl_Teaching_date_20240801

Preparing to Lead in Industry 5.0

Leaders' Roles and Responsibilities in Industry 5.0

Heesun Kim and Walid El Mansour

Abstract This chapter examines the evolving landscape of leadership in the Industry 5.0 era, where reoriented emphasis on human-centricity, sustainability, and resilience is critical. By outlining global macrotrends identified by the World Economic Forum and highlighting the challenges introduced by increased human–machine collaboration, it reframes the leader's roles as one of adaptive facilitation rather than rigid authority. Through a narrative review of contemporary research, the chapter identifies foundational, implementation, and advanced themes of Industry 5.0 leadership, including the cultivation of supportive cultures, workforce development through continuous learning, ethical technology integration, and ecosystem-based partnerships. The analysis emphasizes that leaders must not strive to adopt every responsibility simultaneously but instead prioritize, delegate, and collaborate in response to organizational context, maturity, and evolving demands. Ultimately, the chapter provides insights to guide leaders in aligning technological innovation with human values to achieve sustainable, purpose-driven organizational outcomes.

Keywords Leadership · Reorientation · Industry 5.0 · Organization success

1 Introduction

Industry 5.0 is not a radical break from Industry 4.0 but rather its logical and necessary evolution. While Industry 4.0 was characterized by automation, the Internet of Things (IoT) and data-driven manufacturing systems that prioritized efficiency and productivity, Industry 5.0 shifts the emphasis toward the integration of technological advancement with broader societal objectives (Hexagon, 2024; Nafukho et al., 2024; Raiche, 2022). Academic discussions further highlight that the shift is not necessarily a linear replacement of one paradigm by another. Instead, Industry 4.0 and 5.0 may coexist or form hybrid configurations (Golovianko et al., 2023).

H. Kim (✉) · W. El Mansour
Department of Educational Administration and Human Resource Development, Texas A&M University, College Station, TX, USA
e-mail: heesun@tamu.edu

A. M. Waite et al. (eds.), *Leading in Industry 5.0*, Management and Industrial Engineering, https://doi.org/10.1007/978-3-032-19207-3_4

For example, cognitive digital twins (CDTs) are increasingly being used to offer data-driven insights that supplement human decision-making rather than replace it (Anwer et al., 2025; Zheng et al. 2022). Leng et al. (2022) describe this as collaborative intelligence, where humans and machines work symbiotically toward shared goals. Unlike the fully automated vision of Industry 4.0, this approach recognizes the irreplaceable value of human intuition, empathy, and creative problem-solving, particularly in complex and uncertain environments.

The transition from Industry 4.0 to 5.0 can be understood through the lens of Steve Jobs' quote: "*Creativity is just connecting things*." Jobs' observation captures the essence of how transformative change often arises not from inventing entirely new elements, but from strategically combining existing components into configurations that are both novel and meaningful. The iPhone, launched in 2007, illustrates this principle. Rather than being invented from scratch, it integrated established technologies (e.g., telephone, portable music, internet connectivity, and touchscreen interaction) into a single, human-centered product that redefined communication and digital interaction. Similarly, Industry 5.0 does not emerge from nothing. It builds upon the technological achievements of Industry 4.0 while aligning them into urgent priorities such as human well-being, environmental stewardship, and organizational adaptability.

Although there is no single, universally accepted definition, Breque et al. (2021) from the European Commission (i.e., Directorate-General for Research and Innovation) suggested three core elements that describe Industry 5.0:

1. **Human-Centricity**: Prioritizing human needs and interest and adapting new technology to guide and train individuals.
2. **Sustainability**: Aligning industrial processes with environmental responsibility and long-term societal well-being.
3. **Resilience**: Enhancing the capacity of systems to anticipate, withstand, and adapt to disruptions.

These defining features distinguish Industry 5.0 from its predecessor. Industry 5.0 seeks to optimize the relationship between humans and machines, situating technology as an enabler rather than the central actor. The paradigm underscores that innovation is most impactful when it connects technological sophistication with human creativity and ecological responsibility, transforming industrial systems into engines not just of productivity, but also of sustainable and resilient prosperity.

This progression reflects a growing consensus that industrial development must be both technologically advanced and socially responsible. This evolution acknowledges that the industrial future cannot be defined solely by technical capability; it must also be guided by values that place people, the planet, and resilience at the forefront of decision-making.

Drawing from a wide range of scholarly discourse, this chapter examines the varied and evolving leadership paradigms through a narrative literature review, highlighting the increasing importance of leadership that is responsive to dynamic technological and social changes. Central to this analysis is the principle of human-centricity, an emerging core value for fostering sustainable and resilient organizations. Rather than presenting Industry 5.0 as a radical break from its predecessor, the chapter situates it as a logical evolution: one that builds upon automation and digital transformation by prioritizing human well-being, environmental stewardship, and organizational adaptability. Through this integrated lens, the chapter underscores how leaders are called to guide organizations not only toward increased productivity but also toward a future that values people, purpose, and societal impact.

1.1 Global Trends Driving the Industry 5.0

The future of Jobs Report 2025 by the World Economic Forum highlights several converging trends that are fundamentally transforming labor markets. First, artificial intelligence (AI), automation, and information processing technologies are accelerating change worldwide, with more than 58% of employers anticipating significant impacts on their operations by 2030 (World Economic Forum, 2025). This transformation is amplifying the demand for technology-oriented skills, including generative AI, big data analysis, networks, cybersecurity, and technological literacy (World Economic Forum, 2025). Second, organizations are facing inflation and the increasing cost of living, alongside a general economic slowdown projected to displace 1.6 million jobs globally (World Economic Forum, 2025). These economic disruptions emphasize the importance of resilience, flexibility, and agility skills closely aligned with the resilience elements of Industry 5.0. Third, "climate-change mitigation" has emerged as a decisive factor in workforce planning, with nearly half of companies expecting environmental policies to shape their strategies (World Economic Forum, 2025). This shift is fostering rapid growth in fields such as renewable energy, sustainability management, and environmental services, while environmental stewardship has, for the first time, entered the list of the ten fastest growing skills (World Economic Forum, 2025). This trend directly reflects the sustainability element of Industry 5.0. Finally, demographic shifts, aging populations in higher-income countries and expanding working-age populations in lower-income regions are reshaping labor markets (World Economic Forum, 2025). In addition, "geoeconomic fragmentation and geopolitical tensions" are expected to transform 34% of organizations in the next five years (World Economic Forum, 2025, p. 5). Restrictions on trade and industrial policies are further fueling demand for cybersecurity, resilience, leadership, and human-centered skills (World Economic Forum, 2025). To respond effectively to dynamic change, leaders who "recognize that sustainable transformation is critical for long-term success, and that it requires more than operational and structural changes—it demands a deep understanding of how different intelligence domains (e.g., intellectual, emotional, spiritual, cultural, and social) shape behavior, interactions, and decision-making" are more essential than ever (Tessema, 2025, p. 2).

1.2 Challenges of Human–Machine Integration

A recent report from MIT and Project Networked Agents and Decentralized Architecture (NANDA) which interviewed 52 enterprise stakeholders, systematically analyzed over 300 public AI announcements and initiatives, and surveyed 153 leaders revealed that despite enterprise investments of $30–40 billion into generative AI (GenAI), 95% of organizations are seeing no return on investment (i.e., GenAI Divide). While generic tools like ChatGPT are widely adopted (over 80% explored, nearly 40% deployed), these applications primarily enhance individual productivity rather than impacting profit-and-loss performance.

The report highlights a critical learning gap: enterprise tools often lack memory, adaptability, and seamless integration into existing workflows, which hinders scaling. In addition, barriers include users' unwillingness to adopt new tools, concerns about model output quality, poor experience, limited executive sponsorship, and challenges in change management. This suggests that adopting AI in organizational systems is not a "magic wand"; rather, it requires strategic integration within a human–machine learning loop and a balance between advanced technology and a clear understanding of workers' needs and expectations, current workflows, and leadership support.

Niederhoffer et al. (2025), writing in Harvard Business Review, further critique AI adoption, noting that it can actually drain productivity instead of enhancing it. They describe the "workslop" phenomenon, which is defined as "AI generated work content that masquerades as good work, but lacks the substance to meaningfully advance a given task." Workslop erodes trust and collaboration, leading colleagues to question each other's creativity, capability, reliability, trustworthiness, and intelligence. This raises a critical question: Is success defined by the speed of AI adoption, or by the ability to use AI effectively and ethically in work?

As economics and organizations pursue multiple paths of transformation in response to global trends and AI usage, leaders' ability to recalibrate strategies becomes critical. In the following sections, we examine the roles and responsibilities that leaders must assume to not only survive but also to thrive in the emerging paradigm of Industry 5.0. Importantly, just as Industry 5.0 builds on and complements Industry 4.0 rather than replacing it, leadership in this new paradigm does not require entirely novel approaches. Instead, it calls for a reorientation of emphasis: greater attention to human-centricity, sustainability, and resilience than ever before.

2 Leader's Roles and Responsibilities in Industry 5.0 Era

As the Industry 5.0 paradigm and the definition of three core elements are evolving, Hermawati et al. (2025) identified several themes defining human-centricity in manufacturing sector across the literatures: ensuring both mental and physical health at work, taking into account social, autonomous, and ethical factors, workplaces that are inclusive, using employee participatory design method, humanizing technology to create meaningful sociotechnical system, investing in human potential, and synergetic and symbiotic relationship between humans and machines while humans focus on creative and meaningful tasks and automated systems work on mundane and repetitive tasks. Industry 5.0 principle of human-centricity acknowledge that technology is designed to augment rather than replace human capabilities. Thus, the emphasis is on well-being, inclusivity, continuous learning, and meaningful human–machine collaboration. Then what is leaders' roles and responsibilities to support human-centricity?

We conducted a narrative review using a keyword such as *leader, leadership, human-centric*, sustainability, resilience,* and *Industry 5.0*. To fully reflect the context of Industry 5.0, we focused on the articles published after 2020, considering 2021 was the first appearance of human-centricity as a core element of Industry 5.0. Although it is not a systematic literature review, we identified several recurring leaders' roles and responsibilities across the industry sectors.

Through this narrative review of recent leadership research spanning digital transformation, human-centric technology implementation, and organizational change management, ten interconnected themes emerge from the literature that define the scope of leadership responsibility in Industry 5.0. Scholars across multiple disciplines converge on recognizing that these themes form a sophisticated hierarchy progressing from foundational principles through practical implementation to advanced integration capabilities (Abbu et al., 2025; McCoy, 2025; Schiuma et al., 2024). The literature demonstrates that while these leadership dimensions are deeply interconnected, they represent different levels of organizational maturity and leadership sophistication, requiring leaders to build capabilities systematically while maintaining focus on human flourishing as the ultimate measure of success (Olsson et al., 2025; Rhodes et al., 2025).

It is important to note that leaders are not expected to embody all listed roles and responsibilities at once. Leadership is not about being a superhero but about adopting to what the organization and sector require in a given situation. The relevance of certain responsibilities depends on the industry context, organizational maturity, and evolving challenges. Effective leadership involves prioritizing, delegating, and cultivating collaborative ecosystems rather than individually mastering every possible role. Thus, it is critical for leaders to first ask fundamental questions: What do we need? What can we do? How can we achieve what we need? Who should be involved? What resources and partnerships are required? What risks and opportunities might arise? How will we measure success, adapt to feedback, and continue to learn?

2.1 Foundation Themes: Establishing the Prerequisites for Human-Centric Leadership

The scholarly literature identifies three foundational themes that represent the essential mindsets and principles leaders must establish before any Industry 5.0 transformation can succeed (Castillo et al., 2025; Fenwick et al., 2024; Rhodes et al., 2025). These themes create the philosophical and cultural context that enables all subsequent leadership activities, serving as what multiple scholars describe as the bedrock upon which human-centric organizations are built (Bauer & Grosse, 2024; Olsson et al., 2025).

Strategic Vision and Cultural Transformation. Recent research shows that Industry 5.0 leaders must design organizational transformations by creating compelling visions that merge technology with human-centered value creation while driving cultural shifts to embed human-centric principles into organizational DNA (Castillo et al., 2025). Leaders are responsible for developing long-term Industry 5.0 strategies and integrating these into corporate vision and policies. This extends beyond traditional strategic planning to include comprehensive AI frameworks that balance innovation with ethical considerations and workforce development (Fenwick et al., 2024).

Cultural transformation represents the most complex leadership challenge. Multiple studies confirm that leaders must facilitate cultural evolution to overcome resistance while aligning organizational culture with Industry 5.0 values, which requires both systematic approaches and authentic commitment (e.g., Castillo et al., 2025; McCoy, 2025; Rhodes et al., 2025). Bauer and Grosse's (2024) case study of Weinmann, a German mechanical engineering company, demonstrates how leaders can successfully reshape corporate culture using coaching as the central tool to establish values and influence employee attitudes. Their documented success achieving stronger growth through strategic transformation illustrates the power of authentic leadership that builds trust and team stability.

McCoy's (2025) healthcare cybersecurity research extends these insights by advocating for embedding new paradigms into core organizational strategy through systems thinking. Aligning technology initiatives with organizational values using agile, data-informed approaches that incorporate multiple perspectives (McCoy, 2025). The article mentioned Kotter's Eight-Step Change Management Model that guides organizations through transformation: create urgency, build cross-disciplinary coalitions, share the transformation vision, enable organization-wide participation, achieve early victories, build on success without premature celebration, and embed changes in the culture (Kotter et al., 2021).

Ethical Leadership and Responsible Technology Implementation. Research consistently identifies leaders as ethical stewards of technological advancement, emphasizing that AI and digital technologies must promote fairness, transparency, and alignment with human values (Abbu et al., 2025; Albannai & Raziq, 2025; Altassan, 2025). This goes beyond mere compliance to what Abbu et al. (2025)

validate as "ethical AI use"; a core digital leadership dimension requiring fair and explainable AI outcomes. Altassan's (2025) findings demonstrate that ethical leadership characterized by integrity and transparency is essential for responsible AI implementation, particularly in addressing employee concerns about AI adoption.

Sustainability awareness forms another crucial component of Industry 5.0 ethical leadership. Schiuma et al. (2024) show through their study of small-medium enterprises that leaders must consider their actions' impact on stakeholders and the environment, prioritizing long-term objectives and ethical decision-making. This aligns with Shvetsova's (2025) "double bottom line" approach, which balances financial success with societal impact by treating global talent as strategic assets whose values must be integrated into organizational goals.

Albannai and Raziq (2025) introduce trust and responsibility orchestration as a defining aspect of ethical leadership, reframing the role as managing trust between humans and AI systems while ensuring innovation aligns with human values. Anwar et al. (2025) extend this perspective by positioning leaders as drivers of comprehensive cybersecurity strategies that go beyond technical solutions to create security-conscious cultures and maintain human-centered organizational defense against cyber threats.

Employee Empowerment and Inclusive Participation. Another consensus about leader's responsibility is that they should transform traditional top-down control to facilitating inclusive participation, making employees active partners in technological transformation rather than passive change recipients (Abbu et al., 2025; Hermawati et al., 2025; Rhodes et al., 2025). Abbu et al.'s (2025) digital leadership scale identifies "participatory style" as essential, requiring collaborative approaches that move beyond consultation to genuine shared decision-making. Participative leadership that involves workers in digital tool design and rollout positively affects workforce perception of change and technology adoption (Castillo et al., 2025; Rhodes et al., 2025).

Effective employee involvement requires sophisticated methodological approaches. Hermawati et al. (2025) identify employee participatory design approach as essential for human-centricity, advocating for active involvement in new technology design and implementation that consider social, autonomous, and ethical workforce factors. This aligns with Rhodes et al.'s (2025) recommendation for a "whole person view" of workers, including them in decision-making at all levels while moving beyond hierarchical structures toward collaborative governance.

Transparency and information sharing form the communication foundation for employee empowerment. Leaders need to openly share business strategy information to enable collective decision-making (Abbu et al., 2025). In addition, reducing employee resistance requires transparent communication, with leaders demonstrating system trust while involving employees in implementation decisions, which creates a paradigm shift from AI implementation to human-centric adoption through comprehensive engagement strategies (Fenwick et al., 2024).

2.2 Implementation Themes: Translating Principles into Practice

Implementation themes represent how leaders translate foundational principles into specific programs, systems, and practices. These themes require the ethical frameworks, cultural readiness, and participatory principles from the foundation phase, converting them into concrete organizational capabilities (Krywalski-Santiago, 2024; Panteli et al., 2025).

Human-Centric Technology Integration. Research establishes that leaders must ensure technology augments rather than replaces human capabilities while maintaining meaningful connections and leveraging technological potential to create "meaningful sociotechnical systems" (Hermawati et al., 2025). Transformational leadership highlights the need for leaders to maintain balance between the adoption of new technologies and the cultivation of human engagement, ensuring that employees remain motivated and fulfilled while organizations undergo digital transformation through personalized support and ongoing development (Krywalski-Santiago, 2024). Synergistic human–machine relationships come from where humans focus on creative tasks while automated systems handle routine work (Hermawati et al., 2025). This requires leaders to demonstrate that technology empowers human potential through careful design of collaboration models that optimize both efficiency and satisfaction. However, it is important to note that sequential division of labor can work, but only when employees have the requisite skills to capitalize on the opportunity (Jia et al., 2024). In other words, for AI to successfully augment creativity, leaders must consider the skill level of the employees involved. Creativity and innovative idea rarely appears instantly; rather, it unfolds through iterative experimentation and sustained effort. Consequently, it is important for leaders to resist removing challenging or time-constrained tasks that offer struggling employees valuable opportunities for learning and growth.

Contextual implementation ensures technological solutions align with employees' daily realities. Leaders must ensure security measures are usable within employees' contexts, considering human factors in implementation (Panteli et al., 2025). In addition, leaders need to consider employee realities when implementing technologies, developing collaboration frameworks that define complementary roles and enhance rather than complicate existing work processes (Albannai & Raziq, 2025).

Workforce Development and Continuous Learning. While AI systems continuously learn from human-provided data and interactions, it is critical to recognize that these innovations remain fundamentally human constructs, shaped and advanced through purposeful human input and supervision. Therefore, leaders must create learning organizations that invest in human potential and continuously develop capabilities for Industry 5.0 demands, fostering environments where knowledge sharing and skill development become integral to organizational culture (Schiuma et al., 2024; Shvetsova, 2025). Leaders are the facilitators who create environments nurturing digital knowledge generation and empowering cross-departmental collaboration to address complex digital challenges (Schiuma et al., 2024).

While learning is a shared responsibility, leaders create the conditions for it to flourish by promoting environments that broaden knowledge bases and continuously upskill employees (Olsson et al., 2025). Skills development also requires leaders to shape both digital talent acquisition and the development of the existing workforce (Abbu et al., 2025). Importantly, leaders must upskill themselves to lead digital transformation discussions and communicate technology's value, reflecting the recursive nature of organizational learning (Rhodes et al., 2025). Finally, comprehensive development programs that recognize individual capabilities and supporting personal and professional goals constitute investments in human potential (Hermawati et al., 2025).

Well-Being and Inclusive Workplace Design. Leaders must prioritize comprehensive employee well-being, recognizing that human flourishing forms the foundation for sustainable organizational success in Industry 5.0 (Hermawati et al., 2025; Panteli et al., 2025; Shvetsova, 2025). Leaders can achieve competitive advantage while prioritizing human flourishing by prioritizing employee well-being, happiness, health, and safety through "double bottom line" strategies that balance financial success with societal impact (Shvetsova, 2025). Caring employees' mental health and work-life balance impact cybersecurity as it prevents burnout and mental health challenges which is significant risks for teams managing technological transformation (Panteli et al., 2025). Both mental and physical health at work can be achieved through social, autonomous, and ethical workplace design factors, requiring leaders to ensure employee autonomy over work and data while maintaining respect and dignity (Hermawati et al., 2025). Leaders need to ensure environments accommodate diverse needs and backgrounds while creating comfortable, creative atmospheres that unlock employee potential (Hermawati et al., 2025). This comes from a viewpoint where leaders see employees as whole person rather than resources (Rhodes et al., 2025).

Trust Building and Transparent Communication. Leaders must establish and maintain trust as the foundation for successful Industry 5.0 transformation, using transparent communication and authentic leadership behaviors to build confidence in both leadership and technological systems (Bauer & Grosse, 2024; John et al., 2025; Olsson et al., 2025). The establishment and maintenance of trust require leaders to model vulnerability and authentic behavior, which build trust and restore team equilibrium, proving critical for group growth during digital transformation processes (Bauer & Grosse, 2024). Olsson et al.'s (2025) co-workership approach demonstrates how leaders can create environments where trust enables innovation and collaboration. Albannai and Raziq (2025) extend this concept through their thematic literature review, establishing that trust must be built between humans and AI systems, not merely between humans, requiring sophisticated orchestration of trust across human-technology interfaces. Successful AI adoption is not just about having good technology or even a high-trust company culture (Vuori et al., 2025).

It depends on critically on understanding and managing the diverse combinations of cognitive and emotional trust among employees, because these combinations predict behaviors that can actively sabotage the very data the AI needs to function effectively (Vuori et al., 2025).

Transparent communication excellence requires leaders to exhibit robust communication competencies that inspire and motivate through effective discourse. Schiuma et al.'s (2024) research on transformative leadership competencies identifies "communicator" competencies as essential, requiring leaders to craft compelling visions, adapt communication to diverse audiences, and inspire motivation through effective discourse. In addition, leaders need to foster strong organizational identity while internalizing empathy, purpose, and societal responsibility (John et al., 2025).

The facilitation of continuous dialogue creates ongoing communication channels and feedback mechanisms that support sustained engagement and continuous improvement. Olsson et al. (2025) stress fostering continuous dialogue as critical alongside providing clear vision, requiring leaders to be visible, humane, and clear in their communication approaches while ensuring unified and decisive management. In digital-intensive environments, leaders can strategically upskill areas such as developing digital presence and engagement, managing asynchronous communication, mastering virtual platforms, and enhancing cybersecurity awareness (Kim et al., 2025). Cybersecurity is framed as a core managerial competency in this era, embedded in strategy and guided by systems thinking, agility, and data-informed decision-making, to manage uncertainty and align protective practices with organizational values (McCoy, 2025). These competencies are essential for building trust and facilitating seamless, transparent organizational communication (Kim et al., 2025).

2.3 Advanced Leadership Themes: Sophisticated Integration and Systems Mastery

The advanced leadership themes represent higher-order leadership capabilities that integrate and optimize across all foundational and implementation themes. These competencies require mastery of foundational principles and implementation experience, enabling leaders to facilitate complex, multi-stakeholder collaborations while maintaining strategic perspective on broader ecosystem impacts (John et al., 2025; McCoy, 2025).

Collaborative Leadership and Facilitation. Leaders must transition from traditional command-and-control models to facilitate approaches that enable collaboration across boundaries while empowering distributed leadership throughout organizations (Hermawati et al., 2025; Rhodes et al., 2025; Schiuma et al., 2024). Leader serves as a facilitator who empower knowledge sharing and collaborations across disparate companies, departments, and functions, addressing multifaceted digital challenges through cross-functional cooperation (Schiuma et al., 2024). The empowerment and distributed authority aspects of collaborative leadership involve creating governance structures that distribute decision-making authority effectively while maintaining accountability and strategic alignment (Olsson et al., 2025; Rhodes et al., 2025). Rhodes et al. (2025) emphasize through their research on human-centric manufacturing culture the importance of ensuring that multidisciplinary teams are involved in design and testing processes, requiring leaders to break down organizational silos and create collaborative structures that leverage diverse expertise.

Expert integration and partnership represent sophisticated capabilities that enable leaders to orchestrate expertise from multiple sources to address complex challenges in Industry 5.0. Experts can provide strategic guidance across entire project lifecycles and internal and external collaboration and partnerships can strengthen organizational capabilities (Hermawati et al., 2025; McCoy, 2025). This requires leaders to create networks of expertise that support ongoing innovation and problem-solving while maintaining strategic coherence across multiple partnerships and collaborations.

Systems Thinking and Stakeholder Engagement. Leaders are urged to adopt ecosystem perspectives that account for interdependencies among stakeholders, technologies, and institutions, balancing multiple objectives across organizational boundaries (John et al., 2025; McCoy, 2025; Shvetsova, 2025). Stakeholder engagement similarly shifts from self- to system-orientation, with corporate social responsibility effectiveness hinging on leaders' sensemaking: how they interpret and respond to stakeholder expectations to design socially responsive, contextually grounded initiatives (John et al., 2025). Multi-stakeholder balance extends this capability beyond the firm, requiring collaboration with policymakers, educators, and other external actors to build sustainable talent ecosystems (Shvetsova, 2025).

A long-term orientation complements these practices. Strategic choices are evaluated against generational impacts and sustainable development goals, aligning human-centricity with business strategy to deepen trust and competitiveness while advancing well-being and sustainability (Castillo et al., 2025). Transformative leadership emphasizes ethical stewardship and durability, attending to cross-time-horizon effects and positioning human flourishing as a central success criterion (Schiuma et al., 2024). Realizing these aims depends on collaborative ecosystem development by sustaining cross-sector partnerships, shared learning networks, and industry collaborations that codify and diffuse human-centric best practices (Hermawati et al., 2025; Rhodes et al., 2025).

Adaptive Leadership and Change Management. Leaders are expected to cultivate adaptive competencies that enable them to steer technological transitions without sacrificing organizational performance or human connection (Abbu et al., 2025; Albannai & Raziq, 2025; McCoy, 2025). A recent thematic review positions adaptability and structured change practices as core to leadership, reframing the role from merely directing people to orchestrating trust and responsibility across human-AI systems (Albannai & Raziq, 2025). Complementing this view, a validated digital leadership scale emphasizes a growth mindset (e.g., continuous skill development, digital literacy, and readiness for disruption) as a central determinant of effectiveness in evolving contexts (Abbu et al., 2025).

Leaders must act as change agents to create urgency, build coalitions, communicate vision, and institutionalize new practices, including shifting cybersecurity from compliance to culture (McCoy, 2025). As mentioned above, technology adoption succeeds when leaders pair integration efforts with transformational behaviors (e.g., personalized support, motivation, and meaning-making) that enhance employees' emotional commitment, while transparent communication and a clear, shared vision help mitigate resistance and facilitate the uptake of human-centric technologies (Castillo et al., 2025; Krywalski-Santiago, 2024; Olsson et al., 2025).

3 Conclusion

The evolution toward Industry 5.0 represents not merely an extension of technological advancement but a paradigm shift placing human-centricity, sustainability, and resilience at the heart of industrial and organizational leadership. As this chapter has discussed, effective leaders in the Industry 5.0 era must move beyond operational efficiency and embrace multidimensional responsibilities. They are not expected to fulfill all roles at once, but rather to adapt responsively, prioritize wisely, and orchestrate collaboration across complex ecosystems. The future of leadership lies in the ability to harmonize technological innovation with human values such as empowering workers, reinforcing well-being, and aligning business success with societal and environmental impact. Leaders and organizations that proactively develop these capabilities will be best positioned to navigate complexity, address future disruptions, and achieve sustainable prosperity in an increasingly digital yet rooted in human values.

References

Abbu, H., Khan, S., Mugge, P., & Gudergan, G. (2025). Building digital-ready leaders: Development and validation of the human-centric digital leadership scale. *Digital, 5*(1), 1–18. https://doi.org/10.3390/digital5010007

Albannai, N. A. A., & Raziq, M. M. (2025). Navigating ethical, human-centric leadership in AI-driven organizations: A thematic literature review. *The Service Industries Journal*, 1–28. https://doi.org/10.1080/02642069.2025.2534360

Altassan, M. A. (2025). AI-powered HRM and ethical leadership: Balancing efficiency and human-centricity. *International Journal of Innovative Research and Scientific Studies, 8*(4), 2219–2228. https://doi.org/10.53894/ijirss.v8i4.8358

Anwar, R. W., Souissi, Y., & Ali, S. (2025). Cyber threats and vulnerabilities in Industry 5.0: A review. *Arab Journal of Administrative Sciences, 32*(2), 435–454. https://doi.org/10.34120/ajas.v32i2.1289

Anwer, N., Stark, R., Tao, F., & Erkoyuncu, J. A. (2025). Developing and leveraging digital twins in engineering design. *CIRP Annals, 74*, 843–868. https://doi.org/10.1016/j.cirp.2025.05.002

Bauer, M. C., & Grosse, E. H. (2024). Enhancing human-centricity for strategic alignment: The value of coaching in digital transformation. *IFAC-PapersOnLine, 58*(19), 551–556. https://doi.org/10.1016/j.ifacol.2024.09.270

Breque, M., De Nul, L., & Petridis, A. (2021). *Industry 5.0: Towards a sustainable, human-centric and resilient European industry* (p. 46). European Commission, Directorate-General for Research and Innovation.

Castillo, C., Otero-Romero, T., & Alvarez-Palau, E. J. (2025). Navigating the transition to industry 5.0: Advancing sustainability, resilience, and human-centricity in Spanish supply chain management. *Discover Sustainability, 6*(1), 1–20. https://doi.org/10.1007/s43621-025-01190-0

Fenwick, A., Molnar, G., & Frangos, P. (2024). The critical role of HRM in AI-driven digital transformation: A paradigm shift to enable firms to move from AI implementation to human-centric adoption. *Discover Artificial Intelligence, 4*(1), 1–16. https://doi.org/10.1007/s44163-024-00125-4

Golovianko, M., Terziyan, V., Branytskyi, V., & Malyk, D. (2023). Industry 4.0 vs. Industry 5.0: Co-existence, transition, or a hybrid. *Procedia Computer Science, 217*, 102–113. https://doi.org/10.1016/j.procs.2022.12.206

Hermawati, S., Correa, R., Mohan, M., Lawson, G., & Houghton, R. (2025). Defining human-centricity in Industry 5.0 and assessing the readiness of ergonomics/human factors communities in UK. *Ergonomics, 68*(6), 857–876. https://doi.org/10.1080/00140139.2024.2343947

Hexagon. (2024, November). *Industry 5.0: A comprehensive introduction.* AliResources. https://aliresources.hexagon.com/operations-maintenance/industry-5-0-a-comprehensive-introduction

Jia, N., Luo, X., Fang, Z., & Liao, C. (2024). When and how artificial intelligence augments employee creativity. *Academy of Management Journal, 67*(1), 5–32. https://doi.org/10.5465/amj.2022.0426

John, J., Smiju, I. S., & Ahmed, U. (2025). CSR and stakeholder engagement in the age of industry 5.0: A framework for developing nations. *Corporate Social Responsibility and Environmental Management*, 1–17. https://doi.org/10.1002/csr.70159

Kim, H. S., Waight, J., Xu, X., & Yoon, S. W. (2025). Navigating leadership in hybrid or remote workplaces: A systematic review of employee engagement strategies. *Human Resource Development Review*, 1–29. https://doi.org/10.1177/15344843251381386

Kotter, J. P., Akhtar, V., & Gupta, G. (2021). *Change: How organizations achieve hard-to-imagine results in uncertain and volatile times.* Wiley.

Krywalski-Santiago, J. (2024). Rethinking transformational leadership and workforce dynamics for the human-centric tech era: Insights into employee engagement, satisfaction and job performance. *Journal of Intercultural Management, 16*(4), 42–94. https://doi.org/10.2478/joim-2024-0014

Leng, J., Sha, W., Wang, B., Zheng, P., Zhuang, C., Liu, Q., Wuest, T., Mourtzis, D., & Wang, L. (2022). Industry 5.0: Prospect and retrospect. *Journal of Manufacturing Systems, 65*, 279–295. https://doi.org/10.1016/j.jmsy.2022.09.017

McCoy, E. (2025). How cybersecurity leadership became the new critical managerial competency in healthcare administration. *Health Economics and Management Review, 6*(2), 50–59. https://doi.org/10.61093/hem.2025.2-04

Nafukho, F. M., El Mansour, W., & Van, H. T. M. (2024). The fourth and fifth industrial revolutions: HRD'S role in preparing graduates for future work. In T. S. Rocco, M. L. Morris, & R. F. Poell (Eds.), *The Sage handbook of human resource development* (pp. 248–263). Sage Publications Ltd.

Niederhoffer, K., Kellerman, G. R., Lee, A., Liebscher, A., Rapuano, K., & Hancock, J. T. (2025, September 25). *AI-generated "workslop" is destroying productivity.* Harvard Business Review. https://hbr.org/2025/09/ai-generated-workslop-is-destroying-productivity

Olsson, A. K., Eriksson, K. M., & Carlsson, L. (2025). Management toward industry 5.0: A co-workership approach on digital transformation for future innovative manufacturing. *European Journal of Innovation Management, 28*(1), 65–84. https://doi.org/10.1108/EJIM-09-2023-083

Panteli, N., Nthubu, B. R., & Mersinas, K. (2025). Being responsible in cybersecurity: A multi-layered perspective. *Information Systems Frontiers*, 1–19. https://doi.org/10.1007/s10796-025-10588-0

Raiche, J. P. (2022, November 14). *Industry 5.0: Revolutionizing work by putting people first.* Proaction International Blog. https://blog.proactioninternational.com/en/industry-5.0-the-next-industrial-revolution-is-people-centric

Rhodes, D. H., Cuddy, S., Jeffers, M., & O'Rourke, F. (2025). Human-centric manufacturing culture: A research study of MedTech manufacturers in Ireland. *Production & Manufacturing Research, 13*(1), 2445623. https://doi.org/10.1080/21693277.2024.2445623

Schiuma, G., Santarsiero, F., Carlucci, D., & Jarrar, Y. (2024). Transformative leadership competencies for organizational digital transformation. *Business Horizons, 67*(4), 425–437. https://doi.org/10.1016/j.bushor.2024.04.004

Shvetsova, O. (2025). Managing global talent: Innovative solutions and a sustainable strategy using a human-centric approach. *Administrative Sciences, 15*(5), 1–29. https://doi.org/10.3390/admsci15050190

Tessema, D. B. (2025). From personal to social intelligence: A pathway to sustainable organizational transformation-analysis based on Tessema's multiple intelligence framework. *Land Forces Academy Review, 30*(1), 83–98. https://doi.org/10.2478/raft-2025-0009

Vuori, N., Burkhard, B., & Pitkäranta, L. (2025). It's Amazing–But Terrifying!: Unveiling the Combined Effect of Emotional and Cognitive Trust on Organizational Member' Behaviours, AI Performance, and Adoption. *Journal of Management Studies*, 1–42. https://doi.org/10.1111/joms.13177

World Economic Forum. (2025). *The future of jobs report 2025.* World Economic Forum. https://www.weforum.org/reports/future-of-jobs-report-2025/.

Zheng, X., Lu, J., & Kiritsis, D. (2022). The emergence of cognitive digital twin: Vision, challenges and opportunities. *International Journal of Production Research, 60*(24), 7610–7632. https://doi.org/10.1080/00207543.2021.2014591

Human-Centered Leadership in Industry 5.0: Fostering Inclusivity, Virtual Collaboration, and Emotional Intelligence

Violeta Vidaček Hainš, Antonela Devčić, Karolina Kokot, and Marina Klačmer Čalopa

Abstract Leadership in Industry 5.0 is based on a balance between technological advances and a human approach, with digital leaders playing a key role in empowering employees, developing their skills and ensuring their well-being. The focus is on emotional and cultural intelligence, ethics and sustainability to foster collaboration, and resilience and innovation in a dynamic digital environment. Digital leaders in Industry 5.0 are not just managers of change—they are human enablers. Their job is to guide employees through uncertainty, encourage creativity, and ensure that technological solutions serve people and not the other way round. Industry 5.0 brings a shift in how we think about the role of technology in society and the economy. Unlike previous industrial revolutions that emphasized efficiency, automation, and technological advancement, the new paradigm puts the human back at the center. For this reason, technology and advanced information systems are no longer an end in themselves, but a tool for achieving sustainable development. In this effort, strategic human resource management should play an important role as a driver of change and adaptation. The role of human resource management is becoming increasingly important in shaping an organizational culture that not only recognizes the importance of human capital and the cultural backgrounds of new employees but also fosters a new intercultural environment. Industry 5.0 places people at the center of technological progress, emphasizing the importance of inclusion, employee well-being, and the balance between digital transformation and human needs. Successful leadership in this context requires inclusive management, the development of diverse skills, respect for diversity, and the creation of a work environment in which every individual feels valued and equal. However, despite the emphasis on people, and flexible working in the era of artificial intelligence, leadership in virtual collaboration faces challenges where it is critical to make a balance between technology and human needs, such as workplace well-being and employee satisfaction. The successful leadership of virtual teams requires the effective management of communication challenges

V. V. Hainš · A. Devčić · K. Kokot (✉) · M. K. Čalopa
Department of Organization, University of Zagreb Faculty of Organization and Informatics, Varaždin, Croatia
e-mail: kkokot@foi.hr

A. M. Waite et al. (eds.), *Leading in Industry 5.0*, Management and Industrial Engineering, https://doi.org/10.1007/978-3-032-19207-3_5

related to flexibility, an ethical approach to the use of artificial intelligence in organizations, positive organizational climate toward artificial intelligence and employee awareness and training to create the conditions for the successful adoption of artificial intelligence technologies and the use of their benefits to achieve competitive advantage.

Keywords Industry 5.0 · Digital leadership · Strategic human resource management · Human-centered approach · Emotional intelligence · Employee well-being · Artificial intelligence

1 Introduction

Industry 5.0 represents a significant shift in relation to the values of Industry 4.0, which emphasized technological efficiency and automation. Industry 4.0 has brought the latest technological advances that have fundamentally changed manufacturing, logistics, and supply chains. In Industry 5.0, the emphasis is on harmonizing the relationship between technology and people. Human is once again at the center, and technology does not replace him, but supports him in his work—for example, as a tool for encouraging creativity and innovation. Key values of Industry 5.0 include inclusion, sustainability, resilience, and employee well-being.

In this context, the role of digital leadership is also important, given the increasing number of processes that take place digitally, as well as new communication practices. Leaders are no longer just managers of organizational change but become facilitators and empower key individuals within organizations. That is why it is crucial that they possess skills such as emotional and intercultural intelligence, which are necessary for the leadership of the future. On the other hand, human resource management is also undergoing a transformation—from a supportive and administrative role, it is moving into a strategically important area that shapes organizational culture, promotes values, and focuses on talent development, career development, and employee retention.

However, in a new era marked by artificial intelligence, virtual collaboration, and digital communication, a key question arises: how to harmonize technology and human needs and achieve a balance that will be sustainable for both the individual and the organization? In order to answer this question, this chapter discusses the following topics: the role of digital leadership, emotional and cultural intelligence, strategic HRM and intercultural communication, inclusive management and diversity, virtual collaboration, and the challenges of artificial intelligence.

All of the above provides a theoretical and practical framework that helps theorists, practitioners, leaders, and HR enthusiasts to better understand the issue and find answers and solutions to new challenges.

2 Leadership in the Industry 5.0

The existing environment is defined by fast technological development, volatile conditions, and a focus on a human approach. Companies are aware of the importance of technologies, but alignment with company strategies and human resources is needed to achieve sustainable company performance and competitive advantages. After the hunter-gatherer society, agricultural society, industrial society, and information society, Industry 5.0 aspire to create a human-centered society where products and services are easily provided to meet customer's needs, with focus on reducing economic and social inequality, enabling that all people can live comfortable and active lives (Turan & Doğan, 2025).

2.1 *Industry 5.0*

In the digital age, the environment of a company is changing faster, becoming increasingly unstable and complex, which results in rapid changes in competition, demand, technological, and legislative requirements (Kokot et al., 2021). The current business environment can be described by several major phenomena that are causing the reshaping of the strategic and organizational landscape. Those phenomena include the development of digital competitors that renew industries (e.g., Uber), changes in customer behaviors (large amounts of information and increased demands), and disruptive technologies that create new opportunities for digital businesses. These phenomena are enhanced, especially after the coronavirus (Leso et al., 2023). Scientists define this as Industry 5.0. Industry 5.0, according to Alshaibani et al. (2024), "*creates a personalized and sustainable work environment by using collaborative AI alongside human creativity and well-being, seeking to create a balance between technological advancements and human skills, leading to smarter human-friendly industrial practices.*"

Transformation of Industry 4.0 to Industry 5.0 have following critical success factors: (Sarkar et al., 2024)

- Aligned organizational strategy in transition to I5.0.
- Top management support toward smart manufacturing adoption.
- Employee acceptance and alignment for digital transformation.
- Security and privacy to ensure effective digital transformation.
- Scalability design to ensure digital transformation of smart manufacturing system.
- Legal and ethical concerns in transition to I5.0.
- Upskilling, reskilling, and talent retention.
- Cost and funding in transition from I4.0 to I5.0.
- Cultural collaboration to formalize economic advantages.
- High-powered analytical ability for smart manufacturing.
- Environmental concerns.
- Adoption of advanced technology to create value in I5.0 paradigm.

- Process automation for digital transformation in smart manufacturing.
- Sensitization toward sustainability in transition to I5.0.

Those critical factors indicate that the focus in Industry 5.0 is on a well-defined organizational strategy and strong management support, while also emphasizing a human-centered approach where employees play a crucial role in the digital transformation process. Additionally, security, privacy, legal, and ethical concerns are also important. Continuous learning and cultural concerns are essential. On the other hand, environmental sustainability is a key aspect of Industry 5.0.

Industry 5.0 combines the technological progress of Industry 4.0 with a stronger emphasis on human skills, sustainability and well-being (Rejeb et al., 2025). In line with digital transformation, companies are transforming their workplaces into digital forms, regardless of industry or size (Cortellazzo et al., 2019). Digital transformation presents an opportunity for organizations to innovate (Zulu & Khosrowshahi, 2021) and transforms business processes, business models, customer relationships, and business tasks, leading to significant changes in business structures (Klein, 2020). In the digital transformation process, leadership is crucial for recognizing and defining the positive impact of digital transformation on companies' performance (Zulu & Khosrowshahi, 2021). Digital transformation starts with digital leadership, and a strategic vision enables all activities and processes (Mihardjo & Rukmana, 2018). The process of digital transformation must be entirely supported by top-level management (Kokot et al., 2021). The role of digital leadership is crucial in Industry 5.0, and leaders must prioritize people, sustainability, and technology.

Industry 5.0 focuses on production improvement by interconnecting modern technology and human creativity to achieve a high level of productivity, sustainability, and workplace dynamism (Rejeb et al., 2025). The rapid progress of Industry 4.0 presents challenges such as environmental sustainability, social inequity, and technological complexity (Rejeb et al., 2025). Industry 5.0 is characterized by mass personalization, cognitive cyber-physical systems, design focused on humans, digital health, and equity (Basulo-Ribeiro & Teixeira, 2024), and in line with those challenges, Industry 5.0 emphasizes the resilience, personalized and sustainable production environment (Ghobakhloo et al., 2022). Industry 5.0 enhances people's roles in business processes and aligns human well-being with technological improvement within the framework of sustainability.

To enhance creativity, build employee trust and avoid uncertainty, managers need to understand and relate to employee needs and behaviors during organizational changes, such as digital transformation (Olsson et al., 2024).

Olsson et al. (2024) define four organization prerequisites for digital transformation management from human-centric perspective:

- **Holistic understanding**: transparent vision of digital transformation, internal prioritization of digital transformation, change of culture required for digital transformation, fostering collaboration at all levels.
- **Inclusive organizational change**: change of culture, transparent communication, maintaining and rebuilding trust, handling of disruptive re-structuring, fostering the desire for digitalization, avoidance of ad hoc digital initiatives.

- **Leadership practice**: avoidance of scattered management, recognition of desired leadership, maintaining and rebuilding trust for leadership, clear ownership of responsibilities.
- **Learning and innovation processes**: broadening the knowledge base, continuous up-skilling of employees, avoidance of scatted competence initiatives.

The placement of humans at the center is crucial for highlighting and understanding the interaction between humans and technology in the innovative paradigm of I5.0. A major challenge for organizations in Industry 5.0 is finding qualified employees who can work effectively with advanced robotics (Moreno-Marcial et al., 2024). Therefore, in the manufacturing context, for organizations, one of the challenges is managing workers' diversity in terms of experience, productivity, and physical capacity, especially those characterized by high staff turnover, to increase the well-being of workers (Battini et al., 2022).

2.2 *Digital Leadership*

In this digital environment, leaders must understand and align with the trends of digital transformation and their impact on business, as well as utilize new technologies to maintain a competitive advantage (Chang et al., 2022). Organizations are operating in a disrupted environment, and those conditions lead to changes in traditional leadership and to defining a new leadership recognized by contemporary literature as digital leadership (Kokot et al., 2023). Leading people in the digital era requires adapting to and transitioning to new habits for all employees. These can frequently result in discomfort, anxiety, and a decline in employees' self-confidence due to their insufficient ability to keep up with technological changes (Hadi et al., 2024). In this situation, leaders must not only be able to manage with employees but also serve and motivate their subordinates (Hadi et al., 2024). Leaders must possess the skills to implement the process of digital transformation and stay updated about new technological possibilities, while also coping with all the challenges that digital transformation creates for all employees to achieve a higher level of company performance.

Digital leaders distinguish themselves through a combination of skills, attitudes, and knowledge, as well as professional and personal experiences. This leadership model posits that the transformation of an organization's culture is a result of changes in individuals' attitudes and perceptions, beginning with the leaders themselves. Digital leaders must be humble and collaborative, yet also ambitious, creative, and possess a visionary perspective. Additionally, they put focus on coworkers and building trust, and learn from the teams and experts around them. Digital leaders, who are focused on the process of change, can help sustain the job motivation of new-generation employees. (Turan & Doğan, 2025).

The process of digital transformation must be balanced with reinforcing HR competency within the company, ensuring that every employee can adapt to new technological requirements and provide support for skills improvement (Theng et al., 2024). Empirical studies also show that digital leadership has a positive effect on organizational performance, and digital culture and employees' digital capabilities can have an impact on sustainable organizational performance (Mollah et al., 2023).

Digital leadership is essential for businesses to remain resilient in the digital era, as it enables them to adapt their company strategies (Mollah et al., 2023). Organizations use a digital mindset as a strategic opportunity for investment and profit, and in this process, leaders play an essential role in an organization because of the responsibility they have in managing with followers (Mollah et al., 2023).

Organizations that implement digital leadership tend to demonstrate higher business performance, and digital leaders possess the knowledge necessary to transform their organizations toward a digital direction. Also, it can have a positive effect on employees' performance (Theng et al., 2024). Digital leadership can be defined as a leader's ability to manage a company's digital transformation process, adding value for the company, and being oriented to innovation and creativity. Additionally, it is a strategic approach that leverages the company's digital assets to achieve its organizational goals. (Theng et al., 2024).

2.3 Definition of Digital Leadership

Motivation, control, and team autonomy, as well as digital organizational culture and innovation, are critical pillars for leadership in a digital environment (López-Figueroa et al., 2025). The concept of digital leadership can be examined as a result of this necessity (Oktaysoy et al., 2022). Changes in the environment related to digital technologies led to classical leadership practices becoming ineffective and inadequate under changing environmental conditions. (Oktaysoy et al., 2022) According to the traditional understanding of managers and leaders, digital transformation requires much more than the qualifications attributed to these roles. The new leader type has been defined as the digital leader, and these leaders must have knowledge about technologies as an organizational necessity (Oktaysoy et al., 2022).

"Digital leadership can be understood as a leadership strategy used by a leader in utilizing digital assets to achieve organizational goals" (Nurabadi et al., 2022). Digital leadership can be described as a type of leadership that possesses knowledge of digital technologies that facilitate digital transformation, both for employees and businesses. It is also grounded in modern leadership theories and has a positive impact on organizational culture, enabling the development of the organization (Topcuoglu et al., 2023). In line with definitions of digital leadership, it is clear that digital leadership has two dimensions: technological and impact on people, as described in classical leadership theory.

Promsri (2019) defines six characteristics of a digital leader for digital transformation success:

1. Digital knowledge and literacy. A digital leader needs to understand the voluntary environment and engage with the influence of digital technologies.
2. Vision. A digital leader needs to have a clear vision and purpose relating to digital transformation and communicate that vision to all employees in the organization.
3. Understanding of customers. A digital leader must be aware of customers' needs and changes, incorporating them into the digital transformation process.
4. Agility. A digital leader must possess a high level of agility, flexibility, and adaptability to navigate a rapidly changing digital environment.
5. Risk-taking. A digital leader needs to seek new opportunities and embrace the mistakes that occur during new experiments by employees.
6. Collaboration. A digital leader needs to encourage collaboration across boundaries among employees and teams to ensure that everyone works together toward digital transformation.

Collaboration and focus on employees are crucial segments of digital leadership, aligning with the human-centered characteristics of Industry 5.0, which brings digital leadership into focus in today's world. Digital leaders have to serve as facilitators "in a way that encourages and motivates their followers, helps to reduce their stress levels and promotes innovative behavior during their adaptation to the dynamic requirements of the digital age" (Büyükbeşe et al., 2022).

2.4 *Digital Leadership and Other Leadership Style*

Digital leadership is a combination of several classical leadership theories when it comes to describing the behaviors of leaders. Furthermore, digital leadership has a technological aspect of leading and managing companies.

According to Alshaibani et al. (2024), leaders in Industry 5.0 must demonstrate transformational, adaptive, and servant leadership behaviors. Transformational leadership behaviors are demonstrated by inspiring and motivating employees through a clear vision and empowering them to contribute to the company's goals, while encouraging innovation, creativity, and continuous improvement. Adaptive behaviors are demonstrated through the ability to navigate uncertainty and respond flexibly to changes in the environment. Moreover, servant leadership behaviors are demonstrated through prioritizing the well-being of employees and their development, serving as mentors, listening to employees' concerns, and providing them with support and resources. Furthermore, they must be ethically oriented, prioritizing sustainable practices, job security, and effective communication strategies to foster a positive learning and innovation environment and promote a culture of integrity and ethical behavior. (Alshaibani et al., 2024).

For successfully managing digital organizations, transformational leadership theory is extended in digital leadership, which can provide a company with the possibility of achieving sustainable goals (Mollah et al., 2023). The roles of employees, digital workplace culture, and technological issues are some of the organizational aspects that must be developed, and digital leadership is expected to have a stronger positive impact on sustainable performance than transformational leadership (Mollah et al., 2023). A digital leader must plan and execute a company's strategies, while also shaping and evolving its culture, including the development of digital cultures, to enhance the company's competitiveness (Mollah et al., 2023). These cultures will empower employees to employ digital technologies.

Nevermore, digital leadership has an impact on human emotion and social structure and can also decrease social loafing and other negative organizational behaviors, while also increasing organizational performance, which is similar to transformational leadership (Topcuoglu et al., 2023). Digital leadership is often compared to the leadership approach stated as "transformational leadership" in literature and can be described as a combination of transformational leadership understanding and digital technology competence (Oktaysoy et al., 2022). Digital leadership shares some common characteristics with transformational leadership, such as vision, innovation, and motivation, but digital leaders have a broader scope. (Büyükbeşe et al., 2022) Digital leaders have knowledge and skills about digital technologies, continuously improve these skills, and provide support for the team through encouraging, motivating and guiding their employees in the digital transformation process (Büyükbeşe et al., 2022). Furthermore, digital leaders create a supportive work environment, put the focus on characteristics and behaviors that can facilitate the process of digitalization and move both the organization and the employees forward (Büyükbeşe et al., 2022). Requirements that are put in front of managers in industry 5.0 go beyond every existing leadership theory, so defining a new leadership style, digital leadership, was needed. A link exists between classical leadership theories and digital leadership, which extends to the technical aspects of leading.

2.5 *Digital Leadership and Importance of Emotional and Cultural Intelligence*

Leaders in the Industry 5.0 environment have to prioritize collaboration, adaptability, and continuous learning; promote AI automation and stimulate employees to focus on higher-value tasks. Therefore, leaders must invest in employee development and create opportunities for collaboration to enhance overall performance and learning processes, ensuring transparency and open communication that enables the free flow of information throughout all level of company. (Alshaibani et al., 2024). In line with this, in those environments, psychological safety becomes crucial, as employees must feel comfortable taking risks and expressing concerns without fear of negative consequences, which fosters a culture of trust and collaboration (Alshaibani et al., 2024).

The implementation of new technology results in major changes in human life, which can bring significant benefits to companies, but can also have a negative impact on human resources who think they are incapable of adapting to those changes (Bujang et al., 2023). In line with this, emotional intelligence became crucial for improving employees performance and organizational image in the long term (Bujang et al., 2023). Emotional intelligence is integrated into every aspect of life (work, school, and private), and it is crucial to understanding its role in the digital world because it is related to numerous positive outcomes (Audrin & Audrin, 2023).

Digital leaders must contend with challenges and trends arising from technological development, and technostress is a result of digital transformation. In line with this, emotional intelligence is a fundamental characteristic that enables leaders to cope with technostress (Ertiö et al., 2024). Technostress results increased pace of technological changes and exposure to technology at the workplace. Therefore, constant implementation of digital products and services, requires from employees to update their digital skills on daily bases (Ertiö et al., 2024). Ertiö et al. (2024) define three essential attributes for a digital leader to developed well relationships with their employees in the frame of digital transformation: communication, transparency, and trust. To developed trust, digital leaders have to establish transparency communication (Ertiö et al., 2024). Emotional intelligence enables digital leaders to effectively maintain interpersonal relationships by following skills: (1) self-awareness, (2) self-reflection, (3) self-regulation, (4) social awareness, and (5) empathy (Ertiö et al., 2024).

Anwar & Saraih (2024) claim that emotional intelligence "*involves comprehending and effectively regulating interpersonal relations*". Emotional intelligence is interconnected with different aspects of a job, such as job satisfaction, loyalty, and turnover rate, because optimistic leadership can improve the psychological resources of employees (Anwar & Saraih, 2024).

"*Emotional intelligence defined as the ability to recognize, understand, and manage one's own emotions and the emotions of others serves as a foundational skill that enhances interpersonal relationships and fosters effective communication among team members*" (Yasmeen et al., 2024, pp. 969). Therefore, previous research indicated that teams with a higher level of emotional intelligence are better equipped to manage the challenges of remote work, including feelings of isolation, miscommunication, and reduced morale. On the other hand, existing studies indicate that leaders with high emotional intelligence tend to manage more qualitative communication and collaboration among team members (Yasmeen et al., 2024).

Enhancing emotional intelligence and fostering collaboration can significantly improve leadership effectiveness, ultimately contributing to better team outcomes (Yasmeen et al., 2024). On the other hand, emotional intelligence can enhance team dynamics, and findings suggest that teams led by emotionally intelligent leaders are characterized by greater cohesion and lower conflict levels (Yasmeen et al., 2024).

Emotional intelligence is omnipresent in every aspect of life—at work, school, or home. With the growing number of digital interactions, understanding emotional intelligence in the digital world is fundamental (Audrin & Audrin, 2023). Emotional intelligence can have a significant impact on digital competencies, particularly in adopting appropriate online behaviors and navigating ethical issues (Audrin &

Audrin, 2023). Emotional intelligence enables individuals to work together effectively, manage stress, lead others effectively, and create innovative work behaviors (Hadi et al., 2024).

Leaders must continuously adapt and ensure the optimal use of technology, as well as improve their digital skills, to encourage and motivate their employees to generate and develop creative ideas. Furthermore, leaders must foster a competitive atmosphere and promote collaboration among employees through organizational activities (Hadi et al., 2024). To create this atmosphere, a high level of leaders' emotional intelligence is vital because it provides leaders with the ability to manage employee satisfaction and stress levels in a volatile digital environment.

In the dynamic economic environment, effective leadership requires an excellent understanding of artificial intelligence capabilities, along with emotional intelligence, as the integration of these two can enhance decision-making processes for leaders and lead to excellence in leadership (Dwivedi, 2025). Artificial intelligence provides outstanding capabilities to get analytical insight and correct ideas for the decision-making process, and emotional intelligence plays a vital role in motivating the team, managing complex issues, and fostering innovation (Dwivedi, 2025).

Leaders with strong skills in emotional intelligence can establish a trusting environment, facilitate the team's effective achievement of the company's goals and promote transparent communication. These leaders can provide constructive feedback to employees to enable professional and personal growth, address employees' concerns, and listen to their opinions. Nevertheless, leaders with a high level of emotional intelligence have outstanding skills in maintaining employee relationships and establishing a well-being environment. In other words, emotional intelligence is crucial for properly managing employees' stress, improving the mental well-being of employees, and developing resilience. (Dwivedi, 2025).

In today's business environment, the well-being of employees is essential to achieving the company's goals. Employees with a high level of job satisfaction tend to be more productive, which can ultimately improve the company's overall performance. Suppose leaders are efficient in managing the stress levels of employees, especially technostress, which is common to all companies undergoing digital transformation. In that case, the company will improve its performance at all levels. A high level of emotional intelligence is crucial for leaders to enhance all aspects of leadership.

Additionally, today's leaders must apply cultural intelligence in their leadership approach. The high level of economy globalization, required necessarily an international approach, not just to be enable to serve customer from all over the world but also to cope with different cultures of the team members (Rüth & Netzer, 2020) The concept of cultural intelligence "*provides the toolset to manage a diverse workforce from various cultural backgrounds within one organizational framework*" (Rüth & Netzer, 2020, pp. 3). Digital leadership with cultural intelligence enables providing an effective way of collaboration, decreasing transactional costs to improve company resilience and in the digital condition, cultural intelligence is crucial for planning, communicating, understanding, and leading across cultural boundaries (Rüth & Netzer, 2020). Picture 1 shows the model of digital leadership. Industry 5.0 defines

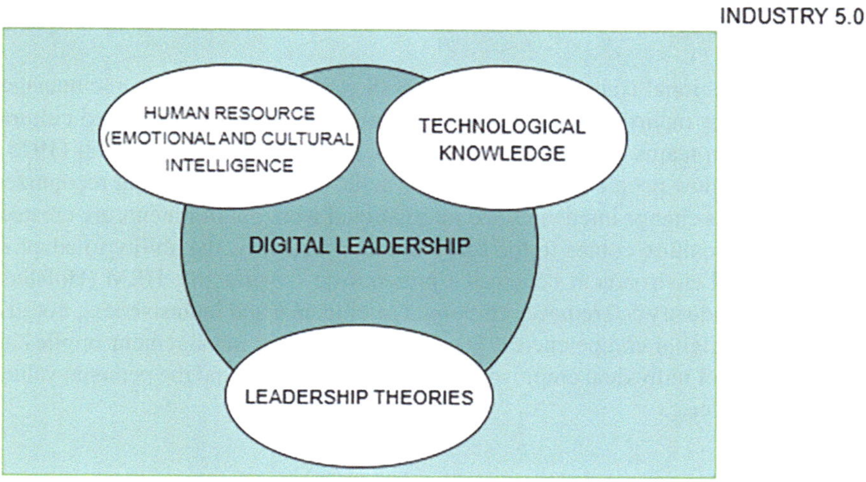

Picture 1 The model of digital leadership (Made by authors)

the environment in which digital leaders operate. Digital leadership encompasses characteristics of leaders defined within the framework of traditional leadership theories, such as transformational leadership. Furthermore, digital leadership comprises technological knowledge. For the success of digital leaders, emotional and cultural intelligence is crucial because humans are the central point of leadership in Industry 5.0.

3 Strategic Human Resource Management and Cross-Cultural Communication in International Human Resource Management

3.1 Introduction

The emergence of Industry 5.0 marks a qualitative shift in how organizations think about the role of technology, labor, and humans in the knowledge economy and digital society. In contrast to the previous phase, Industry 4.0 which was based on automation, the Internet of Things (IoT), big data, and artificial intelligence, Industry 5.0 puts humans back at the center, thereby promoting resilience, sustainability, and social responsibility (Demir et al., 2019; Breque et a., 2021). In this context, the human resources (HR) function is undergoing a transformation. Human resource management should never be merely an administrative function but should evolve into a strategic domain of organizational management. Strategic human resource management (SHRM) is defined as "a systematic approach to developing and implementing

HR strategies aligned with the long-term goals of the organization" (Wright & McMahan, 2011).

In an international context, the challenges of strategic human resource management are further intensified. Globalization, labor migration, and increased cultural diversity within teams demand different competencies. According to Hall (1976), culture affects how people communicate and make decisions, an insight recognized decades ago. As change intensifies and international work environments are created, organizational culture comes to the forefront. Consequently, the ability to adapt in an intercultural environment becomes a prerequisite for effective HRM (Hofstede et al., 2010). Industry 5.0 requires emotional intelligence and inclusiveness, not just technical and digital competencies. Such human resource management implies an understanding of individual employee needs, cultural norms, and the personal values of each employee.

4 Strategic Human Resource Management and Intercultural Communication

4.1 Evolution of the Human Resource Management Role

Traditional human resource management (HRM) was primarily focused on recruitment, record-keeping, and payroll processing, representing a purely administrative function. Today, in the era of Industry 5.0, the role of HR has significantly evolved. HR professionals are now expected to contribute strategically to organizational development, particularly through talent management, support for digital transformation, and the cultivation of an employee well-being culture.

Strategic human resource management (SHRM) involves actively aligning human capital with the organization's long-term goals, a connection that is especially important in an increasingly complex, fast paced, and globally connected digital society (Boxall & Purcell, 2017; Jackson et al., 2014). In this context, HR plays a vital role not only in productivity but also in fostering innovation, resilience, and sustainability within the organization.

4.2 Human-Centered Leadership

Industry 5.0 places the human at the center, with technology serving people, not the other way around. In such an environment, managers are increasingly valued for qualities that are not necessarily technical. Instead, they are expected to possess a range of general leadership traits, making emotional intelligence, empathy, inclusiveness, and the ability to foster intercultural understanding essential leadership skills (Goleman, 1998; Yukl, 2013; Northouse, 2021).

Human-centered leadership means leading an organization through understanding and creating a space where everyone feels seen and motivated. In this kind of environment, HR plays a crucial role from hiring individuals with potential for this leadership approach to introducing mentorship programs and, above all, enabling continuous development.

4.3 Sustainable Development of Human Resource Strategy

Sustainable development is becoming an integral part of HR strategy. Organizations that aim to grow must create jobs that encourage both personal and professional development of employees. For most organizations, especially those operating in an international environment, this means introducing flexible work arrangements, strengthening employee participation in decision-making, and building a culture that respects diversity and promotes inclusion.

Additionally, SHRM increasingly recognizes the importance of so-called green competencies skills that enable employees to contribute to environmental sustainability and innovations related to the circular economy (Renwick et al., 2012; Dumont et al., 2016).

4.4 Intercultural Communication in International Human Resource Management

In an international environment, effective intercultural communication becomes the foundation of a successful organization. It is no longer sufficient to simply know the foreign language spoken by employees nor is it necessarily English, which has become the dominant language of communication in most organizations. Rather, it is essential to understand the values, behavior patterns, communication styles, and nonverbal signals of different cultures. Precisely because organizations are becoming increasingly international, strategic HR management must recognize and actively apply intercultural competencies.

There is a growing body of literature discussing the role of leadership and national culture in business practices. For example, authors Bulog et al. (2024) explain the dimensions of national culture and highlight correlations between cultural value dimensions and leadership behavior preferences.

4.5 The Concept and Importance of Cultural Intelligence

Cultural Intelligence (CQ) refers to an individual's ability to effectively adapt and learn to communicate in different cultural situations. Earley and Ang (2003) define CQ as a set of interconnected components:

- Cognitive: knowledge of cultural norms, values, and belief systems.
- Motivation: willingness to engage in intercultural interactions.
- Behavioral: the ability to act flexibly in accordance with cultural expectations.

These three components distinguish between superficial tolerance and true intercultural understanding within an organization. Today, more organizations operating in international environments with well-developed HR functions use CQ evaluations in recruitment, as well as in identifying and developing managers within the organization (Livermore, 2015; Rockstuhl & Van Dyne, 2018).

4.6 Theoretical Models of Intercultural Communication

In intercultural business environments, challenges can be numerous ranging from unconscious biases and communication noise to deeply rooted stereotypes and cultural misunderstandings. These challenges are often not the result of ill intent but rather a lack of intercultural awareness.

In the development of international HRM, two well-recognized theoretical frameworks stand out for understanding cultural differences and shaping strategies:

- GLOBE 2020 Study (House et al., 2004): adds the dimension of institutional and in-group collectivism within organizations, as well as a dimension to identify the level of participative leadership accepted in different cultures.
- Cultural Agility Model (Caligiuri, 2021) emphasizes the importance of adaptability and the ability to operate in less familiar cultural environments. Culturally agile employees not only adapt but also contribute to innovation and intercultural synergy.

5 Integration of Strategic Human Resource Management and Intercultural Communication

5.1 Human Resource Management Processes in an International Context

In the context of globalization and business internationalization, strategic human resource management (SHRM) increasingly requires the integration of intercultural competencies across all key HR functions. This integration enables organizations to manage diversity and adapt to the cultural context in which they operate. The following outlines intercultural dimensions within core HR functions:

Selection is the first step in the HRM cycle and involves identifying candidates who possess not only technical skills but also intercultural competencies. To this end, organizations are increasingly using models that include Cultural Intelligence (CQ) as a key competency. Selection interviews are often localized to align with specific cultural norms of the labor market. According to Ang and Van Dyne (2008), a high level of CQ in employees positively correlates with successful adaptation in multicultural teams.

Onboarding processes become more effective when they include corporate strategy alongside cross-cultural training. Such training helps new employees understand cultural norms, expectations, and communication styles within the organization. Cross-cultural training reduces intercultural misunderstandings and encourages faster integration of employees into the team (Black & Mendenhall, 1990).

Employee development in an international environment requires an individualized approach. Individual development plans often include intercultural coaching, a professional development process that helps employees develop the ability to work in diverse teams. This development process, focused on intercultural competence, has shown a positive impact on team cohesion (Molinsky, 2013).

Performance evaluation must consider different cultural contexts when defining goals and key performance indicators. For example, in cultures that value collectivism (e.g., Japan, South Korea), evaluation is often based on team performance, whereas in individualistic cultures (e.g., the USA, UK), assessment is primarily based on individual contribution. Recognizing these differences is essential to ensure the evaluation system is fair and motivating (Hofstede Insights, 2020).

5.2 Technological Support for Human-Oriented Leadership: Intercultural Sensitivity and Strategic Approach

The use of modern technologies such as artificial intelligence, virtual reality, augmented reality, and digital platforms for monitoring employee well-being enhances the way HR and management approach human-oriented leadership. These technologies allow for the collection and real-time processing of large amounts of data, enabling personalized decision-making. Radonjić et al. (2024), in their study Artificial Intelligence and HRM: HR Managers' Perspective on Decisiveness and Challenges, show that AI improves decision-making in HR by speeding up data processing. The authors emphasize that HR managers are increasingly open to adopting AI as a data management tool but also point out the challenges related to strategic alignment and organizational structure.

A particularly important aspect of digital transformation is its role in improving the intercultural sensitivity of leadership. Intercultural differences in communication, expectations, and work styles require leaders to be adaptable. Results of moderate mediation analysis show that CQ motivation, the attention and energy individuals invest in intercultural encounters significantly affect the relationship between communication norm alignment and role clarity in global project teams. In this way, it indirectly contributes to project satisfaction and effectiveness (Henderson et al., 2018).

However, what proves to be crucial is that digital solutions must not become a substitute for human judgment. Excessive automation can lead to dehumanization of processes and the loss of an empathetic approach that is essential for HRM. As Vargas and Barcaui (2024) point out, emotional intelligence cannot be fully supported by artificial intelligence; rather, generic human elements of managers must prevail, which significantly influences the shaping of the workplace and personal relationships within the organization.

Successful leaders of the future will be those who know how to balance technology with human values especially in the context of global and culturally diverse teams. This very synergy enables the development of organizations that are not only efficient but also adaptable to the constant changes of the modern business environment.

6 Human-Centered Aspect of Industry 5.0

6.1 Industry 5.0 and Human-Centricity

According to the definition of the European Commission (n.d), the economic and societal transition that is part of Industry 5.0 comprises the elements of human-centricity, sustainability, and resilience alongside technological advancements.

The "human-centricity" concept covers the professional and soft skills which future Industry 5.0 workers and leaders will need. Employees' soft skills shall, inter alia, involve sensitivity related to engaging socially under-represented groups of employees in the work process. Socially less represented groups may be under-represented due to their gender, disability or foreign origin, in particular regarding their nationalities and ethnic minorities and employees from lower socio-economic backgrounds. Even the system of education itself defines under-represented groups of currently enrolled students who will become future employees within Industry 5.0. One example of an under-represented and vulnerable group of students who will be included into Industry 5.0 work environment has been provided by the Ministry of Science, Education and Youth's National Group for the Improvement of the Social Dimension of Higher Education (2024). Under-represented and Vulnerable Groups in Higher Education in the Republic of Croatia—Updated List—(2024). https://mzom.gov.hr./vijesti/objavljena-nova-verzija-popisa-podzastupljenih-i-ranjivih-skupina-u-visokom-obrazovanju/6477, which states the following categories of under-represented and vulnerable groups of future Industry 5.0 employees: students whose parents have a lower level of education, students from families of lower economic status, students with additional family obligations, students from single-parent families or without parents, students who are children of Croatian veterans, students from alternative care, students from the Roma minority, students with disabilities, mature students, students who work and study at the same time, studies, students who have completed vocational school, homeless students and those at risk of homelessness, students who travel to study on day basis, students from rural, traffic-isolated and/or deprived areas, refugees and asylum seekers, LGBT students, female students in technical fields, students in humanities fields, students with part-time status and students of professional studies. The human-centricity is key term which puts human individual in the center of the industry development. This approach includes the development of different skills for future workers (Padovano et al., 2024).

Parallel to the development of Industry 5.0 arises the need for human-centered scientific research. Contrary to the previous iteration, i.e., Industry 4.0, which focused on automation, robotics, and digitalization, Industry 5.0 tries to find a balance between technology and human values. In the product planning process mistakes and cognitive biases may occur. The application of a human-centricity approach and the evaluation of a business organization's human-centered maturity level may additionally decrease a distorted perception and cognitive biases (Rannertshauser et al., 2022).

Sustainability in Industry 5.0 includes green technology and circular economy, which are, amongst other things, directed to reducing waste through responsible waste management and recycling, reducing CO2 emissions and environmentally responsible business. Contemporary production connects technological advances with human values, with one of the focuses being on sustainable production. Sustainable production comprises personalized production to reduce waste, use of material and circular processes, which enable recycling and higher energy efficiency. Industry 5.0 shows high potential and challenges within sustainable development (Rame et al., 2024) depending on the actual legislation and state administration bodies, the leading industries and base-level practitioners.

Industry 5.0 resilience research has been conducted within technologies and concepts of systemics, which include the interdisciplinarity of social and technical sciences (Valette et al., 2023).

Resilience and human centricity in Industry 5.0 are interconnected in the way that resilience refers to raising the levels of human creativity, offering employees the option of decision-making, problem solving and innovation. Thus, AI is used to support people, not to replace humans. The priority is to ensure employees' well-being, mental health and safety. Active participation of employees in decision-making based on the actual needs and experience of employees contributes to system adjustment and system resilience to crises and, eventually, to a more humane industrial environment.

The co-existence of Society 5.0 and Industry 5.0 is an integral part of changes brought forward by progress and globalization within society. Industry 5.0 is human-centered and is followed by Society 5.0, which was proposed by the Japanese government to ensure balance between economic advancement and the need to address social issues (Huang et al., 2022).

Industries can face difficulties related to a disbalance between AI and human intelligence due to various reasons, e.g., vaguely defined roles of technology and humans, which may cause frustration and decrease employee motivation. Furthermore, there is a question of business stability and personal data protection which may result in lower transparency and a refusal by employees to use their personal details. Due to these potential difficulties, Industry 5.0 needs to be a combination of the technological and digital skills of all employees. One potential difficulty may result from the management and employees who do not accept human-centricity. Individual approaches based on the needs of employees may be in discrepant with standards and norms. Employees may evaluate Industry 5.0 as a potential "stressor," i.e., a source of stress, to which they react to by a deterioration in their mental health, fluctuation, absenteeism, and a distrust in technology. With reference to reviews, Pacheco and Iwaszczenko (2004) have created a human-centric model in Industry 5.0 and stated the following dimensions which may cause problems and tensions: automation, well-being, safety, education and creating value, and have emphasized the key role of workers in production. According to authors Tót et al., 2023), the following strategies, which may mitigate problems in Industry 5.0, have been proposed: intelligent automation of production systems in order to reduce employee workloads, adjustments of production, digital transformation aimed at improving production, noise reduction and improved flow of information, implementation of new tools, training of employees which aims to developing new skills and the development of new decision-making models.

Implementation of digital technologies into production processes in the way which makes them human-centered requires adequate human–AI interaction, the Internet of Things and robots, using generic semantic definition, utilizing a time event-driven process. Collaboration architecture, a term used within the human-centricity context, defines structures, principles, and technologies that enable harmonization of employees and technology within a business organization.

6.2 Diversity

Diversity is one of the key elements in Industry 5.0, which facilitates the development of business organization by attracting creative and talented individuals, their innovative and creative approach and collaborative learning, as well as by understanding social inequality through ethics, responsible business, and higher employee satisfaction.

Authors (Kaur and Arora, 2020, pp. 125) define the term diversity as the "human differences (including their intersectionality) on the grounds of color, race, ethnicity, gender, identity, age, physical attributes, ethical values, nationality, education, experiences, knowledge base."

Authors Gardenswartz and Rowe (2003) state the following sources and dimensions of diversity: personality traits being the basic and central dimension of individual differences. The second dimension comprises internal dimensions of diversity acquired by an individual at birth, such as age, gender, sexual orientation, ethnicity. Subsequent dimensions encompass external dimensions of diversity which are changeable, such as choice of residence, profession, education, hobbies, interests, etc. The final organizational dimension of diversity includes the type and period of employment, the workplace, the function, and the field of work.

The 5-factor dimensional model contains individual personality dimensions, which define individual differences amongst individuals: extroversion, content, diligence, neuroticism (emotional stability), and openness to experience (Costa and McCrae, 1992).

Personality dimensions, being personality traits of an individual, impact on work behavior: in the Costa and McCrae (1992) 5-factor dimensional model personality traits are associated with more relevant work behavior referring to career, choice of profession, training and education, and individual performance, as well as the counter-productive work behavior and occupational incidents, group processes, and team performance (Krapić, 2005).

The demographic characteristics of employees, such as gender, age, level of education, work experience, followed by social and cultural characteristics like ethnicity and race, religion, social and economic status, organizational climate in a business organization and attitudes and their system of values and beliefs contribute to diversity in a business environment. Recognizing diversity at workplace includes different socio-cultural, socio-economic, and demographic characteristics of employees and working environment Roberson (2019). Wah (1999) explained diversity as the employee base that is representative of the differences apparent in the broader society.

For the team to be successful, apart from inclusive team leaders, the members themselves have to demonstrate traits of inclusive employees. Inclusive employees notice and recognize individual and social and cultural differences and thus contribute to innovative and inclusive business environment. HR management is an integral part of attracting, selecting, and retaining employees who come from different social and cultural environments, which requires an interdisciplinary, holistic approach of various sciences and scientific disciplines: psychology, medicine, technical sciences, management, economy, law, etc. Whether and to what extent a business organization will recognize diversity shall, inter alia, depend on business organization management and their soft skills.

High quality human resources management shall contribute to diversity management in order to enhance productivity through inclusion, and the results of their research demonstrate the importance of high-quality and inclusive human resources management (Yadav and Lenka, 2020). Using a selective procedure and deployment to adequate workplaces contributes to choosing the best candidates for certain workplaces, their feeling of satisfaction with the workplace and team belonginess and cohesion.

Scientists Ayoko and Fujimoto (2023) emphasize the importance of interdisciplinary approach to human resources practice based on equality, inclusion, belonginess, and intersectionality in management and leadership.

6.3 Diverse Management

Whether and to what extent a business organization will appreciate diversity shall, among other things, depend on a business organization's management and their soft skills. Inclusive leaders improve team identification, reduce perception of status differences which, in return, improves the team's impact (Mitchell et al., 2015). These results indicate the importance of inclusive leadership in Industry 5.0. Inclusive leaders transfer inclusive organizational climate onto teams which they manage. Loden and Rosener (1991) define diversities as primary, secondary, and tertiary dimensions differentiating one group from another. The primary dimensions of diversity are those that produce a primary impact on the identity of an individual, and include gender, ethnicity, sexual orientation, age, and mental and physical abilities. Secondary dimensions of diversity are less visible and include education, residence, religion, native language, family status, work style, work experience, organizational role, income, and style of communication.

A characteristic of good leaders is that they appreciate diversity and provide opportunities to every individual to achieve their full potential. Leaders who appreciate diversity are focused on employee involvement in creative work (Carmeli et al., 2010).

Rijamampinina and Carmichael (2005) state the following differences that determine employment equality and diversity management: employment equality focuses on race, gender, and disability differences, whether it was legally prescribed with social and moral justification, and in what way did change the image of an organization; diversity management focuses on all dimensions of differences and similarities, changes the method of organization, affects organization transformation and in such a way, changes an organization's method of operation. Furthermore, according to Rijamampinina and Carmichael (2005) diversity management encourages creation of an open and equal learning environment and composition of human resources development strategic plans that facilitate this.

6.4 Fostering Inclusivity: Gender, Disability and Adjustments, Cultural Differences, and Socio-Economic Background

Fostering inclusivity is a process which is closely connected to Industry 5.0, because it provides equal employment conditions to under-represented groups of employees and employees whose social exclusion status is borderline. Daya (2013) defines inclusion as a process which provides the feeling of belonginess, empowerment, integration and respect toward all employees, regardless of their differences.

Modern business involves communication within international virtual teams whose leaders and members respect diversity. Inclusion is a process which recognizes individual differences in a virtual working environment aimed at enhancing work performance of every individual as much as possible (April et al., 2009). The process of inclusion within working teams depends on various factors being, inter alia, its team members, organizational culture, type of tasks, team leadership, etc. The attitudes of team leaders and their model of behavior will impact on the inclusivity of the entire team. If the leaders support inclusivity and cherish quality social relationships with all employees, not only with certain members of diverse teams, but employee fluctuations are also less common Nishii and Mayer (2009).

Research conducted by Cillo et al. (2022) revealed that a combination of human resources and knowledge management methods enables companies to appreciate their employees' needs, implement human-centered production and, at the same time, increase production. The analysis made over four years of at more than 24 thousand companies showed that inclusion, diversity and people employment are efficient factors of Industry 5.0. In business organizations, organizational culture, i.e., the way employees think, communicate, make decisions, and work together, shall be connected to knowledge management as well as the sharing of information and knowledge important for successful business operations. Process rethinking requires rethinking of current business processes, their modernization and adjustment.

6.5 Gender Issue

In Industry 5.0. the gender issue follows up on the issues of personal data monitoring, of ensuring ethical procedures in organizational process digitalization and reflection of digitalization on gender equality. Research results (Aydin et al. 2023) confirm the presence of algorithms' gender nature in machine learning which uses AI systems that produce gender-based results and reflect gender prejudices. It is necessary that business organizations, together with interdisciplinary teams' experts, develop software that considers gender differences and protects the dignity of employees, regardless of their gender, ethnicity or nationality. Kaur and Arora (2020) analyzed 60 sources of literature and find that gender diversity and inclusion are becoming an element of organizational growth. Inclusion requires employee creativity and innovation, contributes to the reputation of a business organization and boosts productivity. Teams' heterogeneity helps with problem solving and decision-making.

Gender-acceptable and inclusive workplaces facilitate work to working mothers, contribute to their employment and help eliminate unconscious biases. Flexible hours is one element of gender-acceptable and inclusive workplaces. Including diversity in business organizations is a moral and social issue. At the moment, the status of gender diversity has slightly improved, but the development of policies and strategies on workplace inclusion is crucial.

Research (Aydin et al., 2023) in which experts in HR management participated posed a question on the importance of human resources in software development teams when it comes to the gender issue. Special attention was paid to business ethics in human–machine interaction in Industry 5.0. The results show that the Modified Grounded Theory (Kinoshita, 2003 in Aydin et al., 2023) explains communication processes and team members' interactions when using digital technology, when it comes to theory generation, coding of data, comparative method, and categorization evidence (Aydin et al., 2023). The results indicate the importance of the right to protection of privacy in processing personal data, of ensuring ethical norms in the digitalization process and the link between digitalization and gender inequality. Authors state that in Industry 5.0 the patriarchal gender role form is applied in connection to, e.g., digital games. They have provided gendered nature of machine learning algorithms in AI systems which can produce gender biases and discrimination in male-dominated patterns. In order to establish gender equality, cooperation of experts in the areas of human resources with a computer professionalist is necessary (Aydin et al., 2023). Authors suggest creation of team values through cooperation and technology design in accordance with digital culture, workplace standards, codes of ethics, and the principles of appreciating gender role diversity.

6.6 Disability and Adjustments

People with disabilities can significantly contribute to organizational success. Diversity may be one of key elements encouraging team motivation and innovations. By including disabled employees, a business organization shows social responsibility, provides a positive impact on organizational culture and can benefit from legal and financial incentives. Diversity contributes to a positive organizational culture. Application of the System Dynamics and Discrete Event Simulation Method (Litwin et al., 2022) determined that effective production participated in by disabled employees depends on the number of disabled employees, job requirements, duration of work, number of produced units, product price and level of service. The results of longitudinal research conducted by Jurado-Caraballo and Quintana-García (2024) show that hiring people with disabilities positively contributes to a company's success and reputation and increases competitiveness and visibility of a company's brand. Schwitter et al. (2025) performed an analysis of 69 experimental research studies on discrimination of the disabled conducted from 1972 to 2025. An imbalance was determined regarding the hiring procedure of people with disabilities, i.e., certain chronic conditions and disabilities were more focused on during research than others. Hiring people with disabilities should be approached in an interdisciplinary manner, taking into consideration individual differences and various causal mechanisms linked with unfavorable outcomes of hiring among different sub-groups of people with disabilities.

To enable inclusion of people with disabilities into labor market, key prerequisites are essential at all levels of education—from pre-school to tertiary education and lifelong learning. To this end, all levels of educational system provide adjustments for students regarding studying and attending educational facilities. One example is the adjustment of courses and exams at institutions of tertiary education. During their studies, students can adjust attendance of courses and taking exams by, e.g., extending the duration of a written exam, exchanging a written exam with an oral one (wherever applicable), peer support for students with disabilities during and outside classes, digitalization of teaching material in a format adjusted to the blind and visually impaired students, engaging a sign language interpreter and using various digital tools tailored to persons suffering from dyslexia, the blind and visually impaired and adding subtitles to videos of study material using assistive technology etc. (Vidacek Hains, 2024). Below is an example of possible adjustments for under-represented and vulnerable students attending classes and taking exams at one University of Zagreb Faculty which was analyzed for the 2007 to 2023 academic years. The analysis of adjustments was conducted based on the University's Central Office recommendation for students with disabilities and then summarized in 140 documents for one faculty for approx. 20 years. The recommendations refer to 1.5% of students who used one or more adjustments. Content analysis of these documents demonstrated that students use, on average, of 2 or more adjustments referring to support for students with learning difficulties (dyslexia, dysgraphia, dyscalculia), ADHD who those using digital technologies (Spell Checker options, online learning materials, different fonts) and personal support (extended time, counselling, tutoring). Students with motor difficulties used assistive technology and online learning materials, together with forms of support like counselling and medical external treatment. During their studies, students with mental health issue(s) have been provided with various forms of counselling, e.g., psychological counselling and medical/external treatment. Students, whose chronic illnesses and frequent hospitalizations prevent them from regularly attending classes, more commonly use online teaching materials and medical external treatments. Sensory impaired (visually or hearing) students have ICT digital tools at their disposal: video materials with subtitles, a sign language interpreter, digital library with textbooks and learning materials. In addition they have access to counselling, tutoring, and assistive technology (Vidacek Hains, 2024).

6.7 Cultural Differences

Employees from various cultures bring to their business environment various behavioral patterns, habits, customs, ways of communication etc. Popescua et al. (2014) conducted research aimed at identifying cultural diversity in international project teams. They evaluated intercultural competence, communication style, the emotional intelligence level and the potential role in international project teams. The results have shown that teams which work in a multi-cultural environment are characterized by cross-cultural competences expressed as cultural empathy, open-mindedness, social initiative, and emotional stability. The style of communication in such teams is focused both on tasks and on processes, and team members are flexible at work. The goal was set for the employees and the management to raise the level of communication skills in an international environment, resulting in the need for further strengthening of and education on improving cross-cultural competences. One example of a program that encourages cross-cultural competences is the Global Leadership and Organizational Behavior Effectiveness (GLOBE) program focused on including cross-cultural differences in management and organization of business processes. The results have shown that teams which effectively complete their tasks appreciate cultural differences of their team members more and demonstrate mutual trust among team members. Their team leaders encourage their team members to cooperate and communicate openly (Karlsen et al. 2024).

The feelings of satisfaction at work and recognition of personal integrity are an integral part of business environmental inclusion. Itam and Bagali (2018) emphasize that employees can feel excluded and not feel that their social identity has been recognized if their business organization does not appreciate diversity and shows no need for inclusion. In such cases employee motivation decreases and fluctuation and absenteeism increase when they are not satisfied with their job and dissipate negative "words of mouth."

The impact of cultural differences on the issue of performance has been explained using the divergence and convergence models of cultural diversity (Canney, & Ekelund, 2004). Employee mindset and creative problem solving are important factors when it comes to the performance of team tasks. While a convergent mindset focuses on finding one possible correct solution, a divergent mindset looks for several possible solutions to the same problem.

Divergent processes within a team are associated with cultural diversities of the team; multi-cultural teams usually seek creative ideas and divergent thinking (Canney & Ekelund, 2004). Cultural diversity exists in teams of employees in various business activities. Multi-culturalism in a business organization appears when employees of different cultural origin, with numerous different values, life experience and ways of expressing themselves work together, but consider appreciation of diversity and creation of a positive organizational climate in which all individuals feel accepted (Rice, 2010).

Work performance is associated with team effectiveness. Teams can be mono-cultural, i.e., all team members belong to the same culture, or multi-cultural, in which team members come from diverse cultural backgrounds.

The results of research conducted by Misoc (2017) on a sample comprising 15 mono-cultural and 15 multi-cultural teams demonstrate that multi-cultural teams, compared to the mono-cultural teams, achieve higher scores, are more innovative, provide new ideas, are more creative, their assisting others is influenced by individual differences and they communicate more intensively within their teams.

6.8 Socio-Economic Background

The main characteristic of an employee's social and economic background is their family income, which is often associated with their level of education and employment status. Today's global business is characterized not only by growing awareness of appreciating diversity based on gender, sex, ethnicity, race, social and economic, and other differences, but also by the positive impact of diversity to business. Positive impacts of appreciating diversity through inclusion are visible in creativity, innovation, employee inclusion and higher productivity (Chugh, 2024).

6.9 The Model of Inclusive Organization

The model of inclusive organizations (Shore et al., 2018) is applicable in a global environment and comprises elements which refer to inclusive work groups, inclusive leaders, perception of inclusion in an organization, inclusive organizational climate and practical implementation based on science. Inclusiveness involves understanding challenges associated with diversity, a unity of science and practice, conducting research, existence of business organizations which enable scientists to conduct their research in an inclusive work environment, and an inclusive management recognizes and provides support to an implosive business environment.

The concept of "workspace inclusion" refers to inclusive environment which enables workers to feel respected and recognized and to receive support, considering their origin and identity.

On the other hand, exclusion and discrimination at work may cause feelings of not-belonging and adversely impact physical and mental health. Exclusion can manifest itself through open or more subtle forms of discrimination.

Open discrimination is easily recognizable, whilst subtle discrimination is harder to notice but is equally detrimental. The forms of open discrimination differ—rejecting to employ a worker, inappropriate comments, hate speech, preventing promotion at work, etc. Subtle discrimination is more difficult to notice and prove, as it refers to non-information about important business decisions, exclusion from business meetings, social isolation, etc.

The model of inclusive business organization management (Shore et al., 2018) consists of two main orientations: management prevention orientation and management promotion orientation.

The conditions required for management prevention orientation are top management commitment to compliance and management focus to prevention and inclusion. If the conditions of top management commitment to compliance and management focus to prevention of exclusion have been complied with, a business organization shall develop compliance practices and policies for management of harassment and discrimination, recruitment of protective social categories and diversity training, as well as those for the management of microinequities and subtle discrimination.

Compliance to these prerequisites ensures creation of an inclusive climate, retention of employees, development of their talents, and perception of inclusive environment as part of management promotion orientation. Management promotion orientation unites inclusive processes and practices, starting with lower ones like honoring and advancing of diversity, and authentically recognizing, toward higher levels of inclusive organization like psychological safety, involvement in the work group, feeling respected and valued, influence in decision-making, feeling respected and valuated, involvement in the work group and psychological safety. To achieve successful management promotion orientation, the necessary prerequisites include top management commitment to diversity and inclusion and management focus on enhancement of inclusion.

A gap was identified between Industry 5.0 and employee-centered production processes. Authors Ghobakhloo et al. (2023) used the content-centric review of the literature method to identify 12 functions which facilitate inclusive sustainable environment which puts the employee in the center of production processes:

Sustainable business model innovation (SBI) involves innovative production methods; Sustainable employment (SEP) provides a stable working environment for employees, where they are satisfied with their working conditions; Sustainable skill development (SKD) requires employees to have the knowledge and skills necessary for sustainable production and individual training in order to achieve sustainable production; Sustainable technology governance (STG) covers the creation of laws and policies that govern sustainable management of technology; Value network integration (VNI) is associated with employee networking and cooperation values, and Work environment smartification (WES) facilitates the use of smart technologies.

In order to minimize the gap of Industry 5.0 in relation to the human factor and production processes leading institutions have become involved and confirmed that technological progress and business digitalization are integral parts of modern business and have emphasized the need for the integration of social and environmental needs into Industry 5.0. In order to ensure sustainability in Industry 5.0, modifications shall be implemented gradually. According to Roberson (2020) the process of inclusion enables harmonization of individual differences to provide better cooperation at work, whilst respecting human rights and organizational entities. Corporate leaders shall take measures to ensure diversity and inclusion.

6.10 Inclusive Management

The leaders of inclusive business organizations recognize diversities and create work environment which accepts diversities and, together with employees, create new opportunities for all employees. They are authentic and original in finding solutions for individuals which are based on individual differences (Boekhorst, 2015). Authentic leaders have built a relationship of trust with their co-workers, they are consistent, transparent in communication, and have built a system of stable ethical codes which they strictly respect. They accept criticism well and are also self-critical.

Different communication channels and informing employees in a timely manner are an integral part of Industry 4.0 and have been transferred to Industry 5.0 to focus on homo-centricity. Clearly visible presentations of business results and the success of employees proved to be a factor in this productivity boost. As a part of Digital Shop Floor Management activity, displays on visualization boards provide clear feedback to employees and have a motivational function.

Daya (2013) differentiates three levels of inclusion elements: organizational, inter-personal relations and individual. The elements of the organizational level of inclusion are leadership at higher hierarchy level, organizational climate, organizational affiliation, communication, transparent hiring, progress and development. The elements of inter-personal relations level of inclusion are respect and acceptance, equality of collaborators at different levels of hierarchy by respecting the dignity of each employee, participation in decision-making and access to information. The elements of individual level of inclusion are personality traits of employees and locus of control. Locus of control may be mostly external or internal. Individuals who possess internal locus of control believe they can control events, whilst individuals with external locus of control believe that events are beyond their control.

Inclusive management has a significant role in emergency situations and in high-risk business operations. In assessing risk management the new methods, functional resonance analysis method (FRAM), the system-theoretic accident model and processes (STAMP, system theoretic process analysis (STPA)), and the global risk-informed decision-making approach (RIDM) are used, which contribute to assessment of management performance focused on human-centricity (Diop et al., 2022). Leaders who create a positive work environment and psychological climate contribute to inclusion of members of different race and female employees. The role of leaders is huge, they can contribute to positive inclusive environment and in cases when the environment disregarded diversity (Randel et al., 2016). By respecting diversity and considering individuality of every employee, Industry 5.0 may result in content and motivated employees who will actively participate in production processes.

7 Fostering Virtual Collaboration and Leadership in Era of Artificial Intelligence

7.1 Introduction to Virtual Collaboration in Industry 5.0

The COVID-19 era has had a significant impact on leadership and team management practices, both during and after the pandemic (Kniffin et al., 2021). In addition to increased employee's competences in the use of digital tools, particularly communication and collaboration platforms, the pandemic has led to changes in leadership concepts, especially in terms of working conditions: flexible working times and locations. Such new adjustability for leaders across industries made huge changes for the future of the organizations in the era of Leadership 5.0.

Accordingly, virtual teams, remote working, hybrid working, and working from home have become an integral part of modern leadership and human resource management practices. During the COVID-19 pandemic significant scientific and professional papers on the topic digital leadership and different styles have emerged, as organizations and scholars recognized the need for new leadership models to address the unique challenges of this era (see for example: McCarthy et al. (2022); Eberl & Drews, 2021)). Although digital leadership emphasizes the application of various technologies in everyday work (Tigre et al., 2023), it is essential to place a double emphasis on the human element—both on the abilities and qualities of the leader, and on the leader's orientation toward the needs of their team members. In virtual environments, it is crucial to build trust in team and put focus on the well-being of employees, as those factors significantly contributes to the efficiency of virtual work (Badrinarayanan, 2024).

The remainder of this chapter discusses different forms of work in terms of the degree of virtuality and the use of technology and digital leadership challenges in designing the virtual work environment. The final part is dedicated to the balance between humans and technology in the age of AI and Industry 5.0, focusing on leadership skills, ethical challenges, and the future of leadership in the complex, AI-driven times.

7.2 Types and Modalities of Virtual Collaboration

In the literature discussing different forms of virtual work and collaboration, there are numerous definitions that usually take into account the degree of virtuality of team members and their dependence on the use of technology for daily tasks. Table 1 defines the main forms of virtual work that are important for organization of leadership practices and managing people.

Virtual forms of work have a two-way effect on employee well-being, bringing both positive and negative consequences (Hill et al., 2024). Positive outcomes are related to employee well-being and refer to virtual collaboration and forms of remote

Table 1 Types of virtual work

Work form	Description
Virtual work	Virtual teams are work structures that focus on the fulfilment of tasks that need to be completed by the team members. Their work largely depends on communication and collaboration via digital tools (e.g., document management tools, video conferencing tools) in a virtual environment. They usually work without a shared physical workspace, and members share information and knowledge and collaborate regardless of their location and despite different time zones. Virtual teams are often made up of people who have specific expertise and are required for international projects. They focus on developing intercultural competence and digital skills and have a high degree of independence in their work.
Remote work	Remote working, sometimes referred to in the literature as working from home, allows employees to work from a location that is not the company's office, but is slightly further away. The benefits for the employee are time off work, work-life balance, time savings and reduced travelling costs. The contracts can be concluded full-time, part-time, or according to project requirements or expertise.
Hybrid work	Hybrid working is a flexible model in which employees work a certain number of days in the office, while on the remaining days they work from a location outside the office, usually from home. This form of work has become common, especially in the wake of the COVID-19 pandemic. To preserve interpersonal contact and strengthen organizational culture, many organizations have introduced so-called *return-to-office* (RTO) policies, which usually require 3 days of work in the office and 2 days of flexible work. However, despite the obligation to be physically present on certain days, employees retain a significant level of flexibility, if they respect availability rules and agreed work commitments.

Source Made by authors

work. On the one hand, they enable work flexibility—both in terms of place and time—which contributes to a better balance between work and private life.

On the other hand, the dependency and extensive use of digital tools can cause so-called "tool fatigue," technical complexity and information overload. These outcomes, regardless of intensity, affect employee well-being and as such present challenges for leaders in the framework of Leadership 5.0, in which the emphasis is on the human aspects of and the balanced use of technology.

7.3 Communication Challenges and Technology Use

Despite numerous advantages of communication and collaboration technologies offer to leaders in digital era, downsides are also present in virtual forms of work organization. Negative effects for leaders related to virtual team communication include difficulties in developing trust among team members, managing conflicts, task delegation and monitoring, avoidance of take responsibility, social loafing, and increased possibility of employee procrastination in task execution (Hill et al., 2024). Likewise, it is difficult to establish a supportive climate from management, which also presents a challenge for leaders in achieving team effectiveness. Asynchronous communication is often challenging because different perceptions of task priority and message importance can cause communication breakdowns among team members. Also, information overload and message fatigue also occur due to the expected constant availability of team members, and the combination of different tools and communication channels that are used (Arnold et al., 2023) for team communication. Communication, that takes place mostly online, reduces the sense of belonging and social connection among team members which negatively affects well-being of team members (Hill et al., 2024) in comparison to face-to-face communication.

Although modern tools enable rich message transmission (through video, audio, images, and real-time communication), the increasing dependence on digital platforms, as well as their complexity and numerous functionalities, can present a challenge and be a source of dissatisfaction in the workplace (Hill et al., 2024). Balancing the way and timing of technology use, especially video conferencing tools, has proven to be extremely important during the COVID-19 pandemic, when the term *"Zoom fatigue* "emerged. This phenomenon caused employees to feel overwhelmed and stressed by using technology, and one of the main reasons of it was the high visual focus and intense attention required to participate in video meetings from team members (Riedl, 2022).

Therefore, leaders, when choosing organizational tools for everyday communication, it is important for management to consider optimal number the specific characteristics of each tool, such as quality of user experience, ease of use, stability, support for different virtual tasks and the ability to adapt to different working requirements (Hurst et al., 2022).

In this way, it is possible to shape an inclusive digital environment and workspace, but also to recognize the need for additional education and training for employees who are not sufficiently skilled in using certain technologies. This can contribute to greater employee satisfaction, more effective collaboration, and a reduction in feelings of overload or frustration caused by technical challenges, which is in the line of the leadership 5.0 values.

7.4 Managing Flexibility in Virtual Teams

Flexibility is one of the advantages for employee's well-being in various forms of virtual work. They can work from anywhere and at any time, which makes it easier to balance work and private life as it enables better time management and reduces the need for business trips or coming to the office. In addition, virtual team members can decide for themselves when to complete their tasks, which increases their autonomy (Hill et al., 2024). On the other hand, with fully or partially virtual work on a daily, weekly or monthly basis, this type of flexibility can lead to a loss of boundaries between business and private life, making working conditions at home (e.g. due to inadequate workspaces) difficult and extended regular working hours. All of this can lead to increased stress, lower job satisfaction, and difficulties in recovering from work tasks (Hill et al., 2024, pp. 770).

Flexibility is a challenge for leaders for many reasons. Firstly, it is more difficult to delegate tasks and monitor their progress, and it is more complex to maintain team performance due to the different work habits and preferences of employees. Flexibility and working in a fully or partially virtual environment for leaders requires additional communication about the availability of team members. To overcome this challenge, use of software for task or project management is a good supporting solution which allows leaders to more easily manage resources and timely monitor the progress of team goals.

Moreover, leaders should develop and communicate with team members clear guidelines for virtual communication, especially in multicultural working environments (Swart et al., 2022), (e.g., deadlines for responding to emails, regular weekly meetings at specific times), availability requirements, and prioritize tasks and expectations for employees.

However, in terms of technology, instead of strict monitoring the performance of employees, it is recommended to rather monitor their behavior with the aim of ensuring productivity. At the same time, good communication is important, and with new employees, the onboarding process for the virtual environment is important (Mitchell, 2023).

In this way, in line with values of Leadership 5.0, leaders will be able to strike a balance between the use of technology that enables flexibility and compliance with organizational policies, while preserving the freedom and privacy of employees and monitoring their efficiency without putting them in stressful situations.

7.5 AI-Driven Leadership in Industry 5.0

In Leadership 5.0, people are in the focus, but their connection to technology plays a significant role in the terms of establishing equilibrium and leveraging technological progress and human capabilities. Artificial intelligence (AI) plays a huge role in organizational complex environments, even before appearance and strong use of Generative AI (GenAI) tools since November of 2022, when ChatGPT was introduced. In this matter, the role of Leadership 5.0 for Industry 5.0 is to ensure digital transformation and practices for sustainable, ethically aligned and inclusive working practices that will reduce the negative impact of production processes on the environment. In the context of Industry 5.0, artificial intelligence is being developed on five levels: monitoring for greater accuracy, optimization for flexibility, automation with autonomy, collaboration between humans and machines and resilience, which ensures the reliability and adaptability of systems (Bécue et al., 2024, pp. 11).

In terms of capabilities in organizational settings, AI excels beyond human intelligence in areas such as speed of processing, memory capacity and accessibility; while human intelligence has an advantage in situations or tasks that are seeking creativity, empathy, and human awareness. This difference requires the development of new leadership models that can effectively combine the advantages of both artificial and human intelligence in the context of Industry 5.0 (Bécue et al., 2024).

In digital leadership, leaders who are early adopters and of AI tools and AI supported processes are better prepared to later use its full potential in organizations where AI is a key enabler (Hossain et al., 2025). According to Bevilacqua et al. (2025), an important aspect of effective digital leadership, is the ability to manage complex digital environments, which goes in the line with the softening, or complete disappearance of traditional hierarchical structures and fostering the digital culture in the organization. Therefore, strategic innovative thinking is crucial if leaders want to recognize and take advantage of the benefits of artificial intelligence as a means of differentiating organizations from the competition.

Additionally, leading an organization through technological change in AI times requires the leaders' capacity to deal with employe's resistance, as well as to educate them and to promote the potential uses of AI technologies (Bevilacqua et al., 2025). Nowadays, many employees across industries are concerned that artificial intelligence will replace them in workplaces, affect job losses or take away their decision-making autonomy. The human factor, including managing people's anxiety, concerns, and resistance to AI technology is the key factor of the successful integration of AI in digital leadership (Aftab et al., 2025).

In the literature, there are two central theories about the impact of AI on management and leadership: augmentation theories, which emphasize how AI can complement human capabilities, and replacement theories, which suggest that AI can replace certain human functions and thus change traditional leadership patterns (Bécue et al., 2024, pp. 12). Regarding the distrust in AI as focal problem in human minds, digital leaders must overcome this obstacle through development of training and educations related to the use of AI to create the conditions for AI change readiness. Therefore, digital leaders need to focus on both groups of employees in organizations: digitally trained and non-digital employees, to successfully shape the digital transformation in the context of AI (Aftab et al., 2025).

In the concept of Leadership 5.0, leaders who do not trust artificial intelligence and is skeptical about it, have challenges to make full use of its capabilities and influence their subordinates effectively. Therefore, the question arises: whether a leader who has no trust in artificial intelligence can even play the role of a leader in modern reality? (Balcerzyk & Czainska, 2024). Finding the right approach and choosing a suitable leadership style is crucial to overcoming these challenges. Transformational leaders, compared to other types of leaders, are better at encouraging creativity, adaptability and innovation among employees. They successfully address employees' concerns about job security regarding AI usage or possible replacements and effectively communicate the vision and make sense of AI technology, which is in line with the humanistic and participative approach of Leadership 5.0 (Hossain et al., 2025).

Still, management representatives often express their distrust in artificial intelligence or took a neutral or moderate position, stressing that they use AI mainly for private purposes (Balcerzyk & Czainska, 2024), as reported from research conducted in Poland. Some of the challenges could be related to the cognitive limitations on a personal level resulting from the fear of the unknown and a lack of understanding of technology, as well as psychosocial limitations such as the fear of losing one's job and possible influence of conspiracy theories. In addition, there are normative restrictions related to the lack of clear rules or organizational guidelines for the use of artificial intelligence, as well as the lack of legal regulations that would govern responsibility for the decisions suggested by AI (Balcerzyk & Czainska, 2024).

Readiness for AI transformation strengthens the link between digital leadership capabilities and big data analytics capabilities in the organization. This enables leaders to utilize advanced analytics tools to make data-driven decisions, both internally and externally. At the same time, such an approach supports the development of green innovation, facilitates change monitoring and strategic decision-making, and enables more optimal utilization of resources (Aftab et al., 2025). All of this is in line with the sustainability dimension, which is one of the most important components of Leadership 5.0 which is relevant for Industry 5.0. Nevertheless, leaders who support AI transformation, technology adoption in general, and foster a culture of innovation, can positively influence employee attitudes toward AI adoption and increase employee trust and acceptance of AI in later phases of digital organizational transformation (Passalacqua et al., 2025).

7.6 *Key Leadership Skills in the Age of Artificial Intelligence and Ethics Challenges*

In the time of artificial intelligence prevalence, leadership skills must evolve to keep up with significant technological developments but also changes in interpersonal relationships and work habits and perceptions. Although this field of study is relatively new, recent research has identified several key areas in which leaders and managers need to develop their skills to effectively respond to the challenges of the complex work environment and deliver positive outcomes. Thus Hossain et al. (2025) mention following three key areas crucial for leaders:

- **Technical ability**—Leaders need to have the ability to use and understand AI technologies and at the same time continuously expand their technological knowledge. This includes the active use of AI tools and training on their application so that they can make informed decisions based on the data generated by AI. These skills also enable them to encourage and promote the use of AI technologies to other employees.
- **Adaptive ability**—AI requires quick and proactive responses from leaders, so decision-making based on data analyses is critical. Continuously assessing the impact of AI on productivity and encouraging creativity in solving unique problems are also important. Coordination and integration of AI initiatives through cross-functional teams is also necessary for the successful and efficient implementation of AI projects.

- **Transformational ability**—These skills focus on the human aspect of AI deployment. Leaders need a visionary approach to AI transformation that is clearly communicated within the organization. Managing the uncertainties and unpredictability that AI brings involves mentoring employees, training them and adapting to new working conditions. Leaders with transformational skills drive innovation and the development of new business models based on AI technologies.

In addition to these three general areas mentioned above, Bevilacqua et al. (2025) suggest that, for top managers or leaders, **agility** is a key skill in the age of artificial intelligence. This skill enables leaders to respond quickly and effectively to change requirements by adapting resources and adopting new strategies to achieve more effective results in different organizational areas. For example, agility as such could be demonstrated in the areas of operational efficiency, human resource management, and personalization of the user experience. Flexibility is essential for AI-driven leadership to continuously improve business processes and maintain an organization's competitiveness.

Next, **data-driven decision-making** with the support and analysis of AI tools is necessary for leaders to understand information, extract it correctly and transform it effectively into strategic decisions (Bevilacqua et al., 2025).

Nevertheless, **emotional and social intelligence** are crucial in managing individuals—not only recognizing their emotions but also understanding their mindset (Bevilacqua et al., 2025). In the age of AI, where such tools support decision-making, social skills become crucial to use AI ethically and for human benefits only.

Moreover, **ethical awareness i**s key for digital leadership in the AI age, promoting transparency while protecting user privacy and enabling personalized services (Hossain et al., 2025). As Uddin (2023, pp. 412) stressed: *"The primary objective of ethical leadership in the era of artificial intelligence is to effectively utilize the capabilities of AI technology while simultaneously protecting and upholding social values and the overall welfare of humanity."* Therefore, nowadays leaders, if they want to be prepared for the future, must consider a wide range of challenges in the mix of human, technological use, and environment.

The ethical issues of artificial intelligence can be systematically categorized into three levels Huang et al. (2023). At the individual level, it is about security, privacy and data protection, freedom and autonomy and the preservation of human dignity. At the societal level, it is about questions of fairness and justice, responsibility and accountability, transparency, surveillance and datafication, the manageability of artificial intelligence, democracy and civil rights, job replacement and interpersonal relationships. Finally, at the environmental level, the challenges related to the use of natural resources, energy consumption, pollution, and sustainability are emphasized.

All the above ethical issues related to AI can be considered depending on the industry and work context, as the role of AI in career development can also be seen as ambiguous, especially in the field of human resource management. As the (Pandya & Wang (2024) point out, the focus should be on understanding employees' perceptions of their work and the potential disruption caused by the integration of

AI into daily work routines. Such integration can increase the operational efficiency of organizations and reduce labor and production costs.

Finally, ethical AI leadership starts from the integrity of the leader, transparency in decision-making and accountability, while fostering an organizational culture of ethical use of AI in which concerns and positive examples of application are openly communicated (Uddin, 2023). At the organizational level, it is important to establish ethical standards and policies, as well as AI ethics committees composed of experts from different profiles (ethics, technology, law, social sciences) who analyze and evaluate AI projects, with a human-centered approach including continuous learning and development of employees with an emphasis on AI literacy and responsibility (Uddin, 2023).

Such a human-centric approach, which is also the focal concept of Leadership 5.0, but balanced with technology advancements and strong ethical dedication, strengthens the skills of managers for the ethical transition on AI on the one hand, but also on the other hand, prepares them to successfully lead people in the phases of uncertainty from technology adoption to fully exploiting the potential of AI in organizations.

7.7 *Future Directions of AI-Driven Leadership*

One of the most important directions for future research, as suggested by Bevilacqua et al. (2025), includes looking more closely at what soft and hard skills leaders need to possess in order to effectively lead organizations in AI dominant contexts. In addition, there is a need to explore how different leadership styles, e.g., transformational, transactional, and service-oriented, influence the AI adoption and AI driven innovation and how they can promote a more successful digital transformation (Bevilacqua et al. (2025); Hossain et al., (2025).

Research suggests that AI can promote increased organizational complexity, market asymmetry, and uncertainty in decision-making, which as a consequence can lead to the need for strengthening of hierarchical structures and a concentration of industrial actors (Bécue et al., 2024). For example, in systems where there is a clear vertical division of power and responsibility, with higher levels supervising lower ones how AI is used. This could be because the conditions of complexity, resistance, and uncertainty that AI brings to organization means finding ways for stronger control, structure and clearer rules which can help leaders to form a manageable framework that could involve hierarchy flows.

This formatted way of leadership for AI implementation is partly in contrast to leadership 5.0 values which emphasize humanistic approach, collaboration between people and technology, flexibility, and decentralization of decision-making, teamwork and autonomy. In the future, adapting leadership styles that combine the advantages of AI and human capabilities is key to realizing the full potential of Industry 5.0, while avoiding unwanted concentration of power and rigidity of hierarchy.

Leadership 5.0 requires leaders' competencies that include not only technical knowledge, but also emotional intelligence and the ethics of using AI. Because of that, Leadership 5.0 should not focus on technology, but on employee well-being and retention (Uddin, 2023). While AI represents the transformation and automation of workplaces, it can also pose a threat to jobs and raise concerns about their technical competence. Regarding this, education and training are needed in organization or team level, depending of AI need of proficient use.

Regarding human focus but also to better integrate technology into day-to-day operations, leaders in AI era can also improve the recruitment system within an organization and improve employee engagement by analyzing survey results and communication patterns (Hossain et al., 2025). This is an example of balanced cooperation between man and technology, where technology acts as a supporting pillar to the organization in improving the quality of the workplace. Implementing AI strategies that priorities mental health and work-life balance will therefore create a supportive work environment that encourages high performance (Passalacqua et al., 2025). While at an operational level, open communication about the pros and cons of AI is encouraged among employees, involving teams from different departments such as IT, HR, and operations to ensure a balanced and comprehensive approach, at a strategic level, it is important to plan AI implementation that includes a thorough assessment of the work environment and the specific needs of the workforce (Passalacqua et al., 2025).

Results of research conducted by Hu et al. (2025), showed that technophilia, which refers to positive attitude, preference, or enthusiasm toward new technologies and their use, in combination with paradoxical leadership, are key to the acceptance of artificial intelligence in organizations. This flexible approach enables leaders to motivate employees and take advantage of the learning opportunities that AI provides, while technophilia encourages a positive attitude toward technology, which together increases employee engagement and encourages effective knowledge sharing within the organization (Hu et al., 2025).

Therefore, one of the possible approaches to leadership in the context of artificial intelligence is the so-called **paradoxical leadership**. In this leadership style, the leader tries to successfully balance the often-contradictory attitudes, values, goals and needs of the employees. A leader often finds himself in contradictory situations, but offers flexibility and discipline, control and autonomy, proximity, and distance, and harmonizes them to achieve optimal results. This is a representation of the situation in which leaders find themselves nowadays—in a constant dance of position between a human factor that is skeptical, but often abuses the power of technology, which, on the other hand, if used ethically, brings numerous benefits.

8 Conclusion

The concluding reflections of this chapter bring us back to the original and fundamental value of Industry 5.0—people and their well-being, both in and out of the workplace. The philosophy that technology is a tool and not an end in itself, whether it is edge computing, blockchain, robotics, the Internet of Things (IoT), digital twins, augmented (AR) and virtual reality (VR), or artificial intelligence, clearly points to considerations and efforts that while technologies are evolving rapidly, people in different organizations and industries should also be systematically taken care of.

There are many challenges along the way, some of which this chapter highlights:

Balancing technology and human needs, managing virtual teams, overcoming resistance to artificial intelligence and maintaining employee well-being.

Successful digital leadership, as the key to solving these challenges, requires a combination of technical and "soft" skills—emotional, cultural, and ethical competences. It is particularly emphasized that inclusivity is the foundation for sustainable and more resilient organizations.

Although this chapter covers a wide range of topics related to digital leadership, the space for future research remains open. Possible directions include:

- The integration of emotional intelligence and artificial intelligence;
- Measuring the impact of inclusive and human-centered leadership;
- The development of innovative HR and communication practices in global teams;
- The role of virtual and AI assistants in virtual teams.

At first glance, it may seem that humans are losing the race against technological progress. However, the values of Industry 5.0 mean that this "race" is not a competition, but a collaboration between humans and technology, side by side. Leaders in Industry 5.0 play a crucial role in this process: they build bridges of trust, promote innovation, and shape a sustainable future in which technology serves people and society.

References

Aftab, J., Stan, M. R., Srivastava, M., Wei, F., & Abid, N. (2025). The impact of digital leadership on performance: Examining the roles of big data analytical capabilities, green innovation, and AI change readiness in Italian SMEs. *Business Strategy and the Environment*. https://doi.org/10.1002/bse.4322

Alshaibani, E., Bakir, A., & Al-Atwi, A. (2024). The impact of leadership behaviors on organizational innovative performance and learning in AI-driven industry 5.0 environments. *Development and Learning in Organizations*. https://doi.org/10.1108/DLO-06-2024-0159

Ang, S., & Van Dyne, L. (Eds.). (2008). *Handbook of cultural intelligence: Theory, measurement, and applications*. M.E. Sharpe. https://books.google.hr/books?id=FSTzFksL0IkC&printsec=frontcover&hl=hr&source=gbs_ge_summary_r&cad=0#v=onepage&q&f=false

Anwar, S., & Saraih, U. N. (2024). Exploring the abilities of emotional intelligence in psychological empowerment: Digital leadership as mediator. *Journal of Applied Research in Higher Education., 16*(4), 1246–1257. https://doi.org/10.1108/JARHE-06-2023-0245

April, K., Katoma, V., & Peters, K. (2009). Critical effort and leadership in specialised virtual networks. *Annual Review of High-Performance Coaching & Consulting, 1*(1), 187–215.

Arnold, M., Goldschmitt, M., & Rigotti, T. (2023). Dealing with information overload: A comprehensive review. In *Frontiers in psychology* (Vol. 14). https://doi.org/10.3389/fpsyg.2023.1122200

Audrin, C., & Audrin, B. (2023). More than just emotional intelligence online: Introducing digital emotional intelligence. *Frontiers.* InPsychology.

Aydin, E., Rahman, M., & Ozeren, E. (2023). Does industry 5.0 reproduce gender (in)equalities at organisations? Understanding the interaction of human resources and software development teams in supplying human capitals. Information system Frontiers. https://doi.org/10.1007/s10796-023-10450-1

Ayoko, O. B., & Fujimoto, Y. (2023). Diversity, inclusion, and human resource management: A call for more belongingness and intersectionality research. *Journal of Management & Organization, 29*(6), 983–990.

Badrinarayanan, V. (2024). Trust building strategies for virtual leaders in the post pandemic era. *Project Leadership and Society, 5.* https://doi.org/10.1016/j.plas.2024.100126

Balcerzyk, D., & Czainska, K. (2024). Chapter 10 Trust in Artificial Intelligence and the 5.0 leadership. In *Digital Synergy* (pp. 94–100). CRC Press. https://doi.org/10.1201/9781003440406-10

Basulo-Ribeiro, J., & Teixeira, L. (2024). The future of healthcare with industry 5.0: Preliminary interview-based qualitative analysis. *Future Internet.*

Battini, D., Berti, N., Finco, S., Zennaro, I., & Das, A. (2022). Towards industry 5.0: A multi-objective job rotation model for an inclusive workforce. *International Journal of Production Economics., 250.* https://doi.org/10.1016/j.ijpe.2022.108619

Bécue, A., Gama, J., & Brito, P. Q. (2024). AI's effect on innovation capacity in the context of industry 5.0: A scoping review. *Artificial Intelligence Review, 57*(8). https://doi.org/10.1007/s10462-024-10864-6

Bevilacqua, S., Masárová, J., Perotti, F. A., & Ferraris, A. (2025). Enhancing top managers' leadership with artificial intelligence: Insights from a systematic literature review. *Review of Managerial Science.* https://doi.org/10.1007/s11846-025-00836-7

Black, J. S., & Mendenhall, M. (1990). Cross-cultural training effectiveness: A review and a theoretical framework for future research. *Academy of management review, 15*(1), 113–136. https://doi.org/10.5465/AMR.1990.11591834

Boekhorst, J. A. (2015). The role of authentic leadership in fostering workplace inclusion: A social information processing perspective. *Human Resource Management., 54*(2), 241–264.

Boxall, P., & Purcell, J. (2017). *Strategy and human resource management (management, work and organisations, 4th ed.).* Red Globe Press.

Breque, M., De Nul, L., & Petridis, A. (2021). *Industry 5—Towards a sustainable.* The Publications Office of the European Union.

Bujang, S., Aetis Heromi, N., Capel, D. M., & Hadil, H. (2023). Conceptual framework on emotional intelligence., Digital Competency and Organizational Readiness for Change. https://ssrn.com/abstract=4307041

Bulog, I.; Malbašić, I.; Klačmer Čalopa, M. (2024). The interplay of cultural value dimensions and preferred leadership Behaviors: Intersecting pathways in the case of CroatiaCross-Cultural Research, 58 (2024), 4; 300–310. https://doi.org/10.1177/10693971241261420, Q1.

Büyükbeşe, T., Dikbaş, T., Klein, M., & Batuk Ünlü, S. (2022). A study on digital leadership scale (Dls) development. *Kahramanmaraş Sütçü İmam Üniversitesi Sosyal Bilimler Dergisi., 19*(2), 740–760. https://doi.org/10.33437/ksusbd.1135540

Caligiuri, P. (2021). *Build your cultural agility: The nine competencies of successful global professionals.* Kogan Page.

Canney, D. S., & Ekelund, B. Z. (2004). Effective team processes for global teams. *Handbook of global management.*, 227–249. https://doi.org/10.1002/9781405166355.ch12

Carmeli, A., Reiter-Palmon, R., & Ziv, E. (2010). Inclusive leadership and employee involvement in creative tasks in the workplace: The mediating role of psychological safety. *Creativity Research Journal., 22*(3), 250–260.

Chang, C. L., Octoyuda, E., & Arisanti, I. (2022). The Role of Digital Transformation on Strategic Leader: A Systematic Literature Review. ICBIR 2022–2022 7th international conference on business and industrial research, Proceedings. 289–294. https://doi.org/10.1109/ICBIR54589.2022.9786422

Chugh, U. (2024). Diversity and inclusion in human resource management. *Journal of Advanced Management Studies., 1*(1), 16–20.

Cillo, G. L., Gregori, L. M., Daniele, F. C., & Bitbol-Saba, N. (2022). Rethinking companies' culture through knowledge management lens during industry 5.0 transition. *Journal of Knowledge Management., 26*(10), 2485–2498.

Cortellazzo, L., Bruni, E., & Zampieri, R. (2019). The role of leadership in a digitalized world: A review. *Frontiers in Psychology, 10*(AUG), 1–21. https://doi.org/10.3389/fpsyg.2019.01938

Costa, P. T., & McCrae, R. R. (1992). *The revised NEO personality inventory (NEO-PI-R) and NEO five factor inventory (NEO-FFI) professional manual.* Psychological Assessment Resources.

Daya, P. (2013). Diversity and inclusion in an emerging market context. *Equality, Diversity and Inclusion: An International Journal., 33*(3), 293–308.

Demir, K. A., Döven, G., & Sezen, B. (2019). Industry 5.0 and human-robot co-working. *Procedia Computer Science, 158*, 688–695.

Diop, I., & s Abdul-Nour, G.G., & Komljenovic, D. (2022). A high-level risk management framework as part of an overall asset management process for the assessment of industry 4.0 and its corollary industry 5.0 related new emerging technological risks in socio-technical systems. *American Journal of Industrial and Business Management., 12*, 7. https://doi.org/10.4236/ajibm.2022.127071

Dumont, J., Shen, J., & Deng, X. (2016). Effects of green HRM practices on employee workplace green behavior: The role of psychological green climate and employee green values. *Human Resource Management, 56*(4). https://doi.org/10.1002/hrm.21792

Dwivedi, D. (2025). Emotional intelligence and artificial intelligence integration strategies for leadership excellence. *Advances in Research., 26*(1), 84–94. https://doi.org/10.9734/air/2025/v26i11235

Earley, P. C., & Ang, S. (2003). *Cultural intelligence: Individual interactions across cultures.* Stanford University Press. Globe2020. https://globeproject.com/about.html

Eberl, J. K., & Drews, P. (2021). Digital Leadership-Mountain or molehill? A literature review digital Leadership Mountain or molehill? A literature review. *Digital Leadership—Mountain or Molehill? A Literature Review*, 223–237. https://aisel.aisnet.org/wi2021

Ertiö, T., Eriksson, T., Rowan, W., & McCarthy, S. (2024). The role of digital leaders' emotional intelligence in mitigating employee technostress. *Business Horizons., 67*(4), 399–409. https://doi.org/10.1016/j.bushor.2024.03.004

European Commission: Industry 5.0 https://research-and-innovation.ec.europa.eu/research-area/industrial-research-and-innovation/industry-50_en. Accessed 15 June 2025.

Gardenswartz, L., & Rowe, A. (2003). Diverse team at work. 2nd edition. Alexandria: Society of Alexandria, VA: Society for Human Resources Management.

Ghobakhloo, M., Iranmanesh, M., Mubarak, M. F., Mubarik, M., Rejeb, A., & Nilashi, M. (2022). Identifying industry 5.0 contributions to sustainable development: A strategy roadmap for delivering sustainability values. *Sustainable Production and Consumption, 33*, 716–737.

Ghobakhloo, M., Iranmanesh, M., Foroughi, B., Tirkolaee, E. B., Asadi, S., & Amran, A. (2023). Industry 5.0 implications for inclusive sustainable manufacturing: An evidence-knowledge-based strategic roadmap. *Journal of Cleaner Production, 417*. https://doi.org/10.1016/j.jclepro.2023.138023

Goleman, D. (1998). *Working with emotional intelligence.* Bantam Books.

Hadi, S., Setiawati, L., Kirana, K. C., Lada, S., & bin, & Rahmawati, C. H. T. (2024). The effect of digital leadership and organizational support on innovative work behavior: The mediating role of emotional intelligence. *Quality - Access to Success., 25*(199), 74–83. https://doi.org/10.47750/QAS/25.199.09

Hall, E. T. (1976). *Beyond culture.* Anchor Books. https://doi.org/10.1016/j.procs.2019.09.104

Henderson, L., Stackman, R., & Lindekilde, R. (2018). Why cultural intelligence matters on global project teams, October 2018. *International Journal of Project Management, 36*(7), 954–967. https://doi.org/10.1016/j.ijproman.2018.06.001

Hill, N. S., Axtell, C., Raghuram, S., & Nurmi, N. (2024). Unpacking virtual Work's dual effects on employee Well-being: An integrative review and future research agenda. *Journal of Management, 50*(2), 752–792. https://doi.org/10.1177/01492063221131535

Hofstede Insights. (2020). National culture. https://www.hofstede-insights.com/models/national-culture/.

Hofstede, G., Hofstede, G. J., & Minkov, M. (2010). *Cultures and organizations: Software of the mind* (3rd ed.). McGraw-Hill.

Hossain, S., Fernando, M., & Akter, S. (2025). Digital leadership: Towards a dynamic managerial capability perspective of artificial intelligence-driven leader capabilities. *Journal of Leadership and Organizational Studies, 32*(2), 189–208. https://doi.org/10.1177/15480518251319624

House, R. J., Hanges, P. J., Javidan, M., Dorfman, P. W., & Gupta, V. (2004). *Culture, leadership, and organizations: The GLOBE study of 62 societies.* Sage.

Hu, X., Gao, H., Agafari, T., Zhang, M. Q., & Cong, R. (2025). How and when artificial intelligence adoption promotes employee knowledge sharing? The role of paradoxical leadership and technophilia. *Frontiers in Psychology, 16*. https://doi.org/10.3389/fpsyg.2025.1573587

Huang, C., Zhang, Z., Mao, B., & Yao, X. (2023). An overview of artificial intelligence ethics. *IEEE Transactions on Artificial Intelligence, 4*(4), 799–819. https://doi.org/10.1109/TAI.2022.3194503

Huang, S., Wang, B., Li, X., Zheng, P., Mourtzis, D., & Wang, I. (2022). Industry 5.0 and society 5.0— Comparison, complementation and co-evolution. *Journal of Manufacturing Systems., 64*, 424–428. https://doi.org/10.1016/j.jmsy.2022.07.010

Hurst, W., Withington, A., & Kolivand, H. (2022). Virtual conference design: features and obstacles. *Multimedia Tools and Applications, 81*(12), 16901–16919. https://doi.org/10.1007/s11042-022-12402-4

Itam, U., & Bagali, M. M. (2018). Diversity and inclusion management: A focus on employee engagement (chapter 9). In N. Sharma, V. K. Singh, & S. Pathak (Eds.), *Management techniques for a diverse and cross-cultural workforce* (pp. 149–171). IGI Global.

Jackson, S. E., Schuler, R. S., & Jiang, K. (2014). An aspirational framework for strategic human resource management. *The Academy of Management Annals, 8*(1), 1–56. https://doi.org/10.1080/19416520.2014.872335

Jurado-Caraballo, M. A., & Quintana-García, C. (2024). Disability inclusion in workplaces, firm performance, and reputation. *European Management Journal.* https://doi.org/10.1016/j.emj.2024.08.001

Karlsen, H. E., & Nazar, M. (2024). How cultural diversity affects communication and collaboration within global high-performance project teams? *Procedia Computer Science, 239*, 491–497. https://doi.org/10.1016/j.procs.2024.06.197

Kaur, N., & Arora, P. (2020). Acknowledging gender diversity and inclusion as key to organizational growth: A review and trends. *Journal of Critical Reviews, 7*(6), 125–131.

Klein, M. (2020). Leadership characteristics in the era of digital transformation. *Business & Management Studies: An International Journal., 8*(1), 883–902.

Kniffin, K. M., Narayanan, J., Anseel, F., Antonakis, J., Ashford, S. P., Bakker, A. B., Bamberger, P., Bapuji, H., Bhave, D. P., Choi, V. K., Creary, S. J., Demerouti, E., Flynn, F. J., Gelfand, M. J., Greer, L. L., Johns, G., Kesebir, S., Klein, P. G., Lee, S. Y., … Vugt, M. van. (2021). COVID-19 and the workplace: Implications, issues, and insights for future research and action. American Psychologist, 76(1), 63–77. https://doi.org/10.1037/amp0000716

Kokot, K., Đunđek Kokotec, I., & Klačmer Čalopa, M. (2021). Impact of leadership on digital transformation. *IEEE Technology & Engineering Management Conference - Europe*. (TEMSCON-EUR).

Kokot, K., Kokotec, I. Đ., & Čalopa, M. K. (2023). Digital leadership and maturity as a key to successful digital transformation: Country case study of Croatia. *TEM Journal., 12*(1), 192–199. https://doi.org/10.18421/TEM121-25

Krapić, N. (2005). Five-factor personality dimensions and work behavior. *Psychological Topics, 14*(1), 39–56.

Leso, B. H., Cortimiglia, M. N., & Ghezzi, A. (2023). The contribution of organizational culture, structure, and leadership factors in the digital transformation of SMEs: A mixed-methods approach. *Cognition, Technology and Work., 25*(1), 151–179. https://doi.org/10.1007/s10111-022-00714-2

Litwin, P., Antonelli, D., & Stadnicka, D. (2022). Disabled employees on the manufacturing line: Simulations of impact on performance and benefits for companies. *IFAC-PapersOnLine, 55*(10), 848–853.

Livermore, D. (2015). *Leading with cultural intelligence: The real secret to success* (2nd ed.). AMACOM.

Loden, M., & Rosener, J. B. (1991). *Workforce America! Managing employee diversity as a vital resource*. McGraw Hill.

López-Figueroa, J. C., Ochoa-Jiménez, S., Palafox-Soto, M. O., & Sujey Hernandez Munoz, D. (2025). Digital Leadership: A Systematic Literature Review. In Administrative Sciences. Vol. 15, Issue 4. Multidisciplinary Digital Publishing Institute (MDPI). https://doi.org/10.3390/admsci15040129

McCarthy, P., Sammon, D., & Alhassan, I. (2022). Digital transformation leadership characteristics: A literature analysis. *Journal of Decision Systems, 32*(1), 79–109. https://doi.org/10.1080/12460125.2021.1908934

Mihardjo, L. W. W. (2018). Does digital leadership impact directly or indirectly on dynamic capability: Case on Indonesia telecommunication industry in digital transformation? The journal of social sciences research. *SPI, 2*, 832–841. https://doi.org/10.32861/jssr.spi2.832.841

Ministry of Science, Education and Youth National Group for the Improvement of the Social Dimension of Higher Education (2024). Underrepresented and Vulnerable Groups in Higher Education in the Republic of Croatia - Updated List https://mzom.gov.hr/vijesti/objavljena-nova-verzija-popisa-podzastupljenih-i-ranjivih-skupina-u-visokom-obrazovanju/6477. Accessed:17 July 2025.

Misoc, I. (2017). The effects of cultural diversity on team performance: A comparison between the performance of Monocultural teams and the performance of multicultural teams in tourism organizations. *International Journal of Academic Research in Business and Social Sciences., 7*(7). https://doi.org/10.6007/IJARBSS/v7-i7/3090

Mitchell, A. (2023). Collaboration technology affords from virtual collaboration in the time of COVID-19 and post-pandemic strategies. *Information Technology and People, 36*(5), 1982–2008. https://doi.org/10.1108/ITP-01-2021-0003

Mitchell, R., Boyle, B., Parker, V., Giles, M., Chiang, V., & Joyce, P. (2015). Managing inclusiveness and diversity in teams: How leader inclusiveness affectsperformance through status and team identity. *Human Resource Management, 54*, 217–239. https://doi.org/10.6007/IJARBSS/v7-i7/3090

Molinsky, A. (2013). *Global dexterity: How to adapt your behavior across cultures without losing yourself in the process*. Harvard Business Review Press.

Mollah, M. A., Choi, J. H., Hwang, S. J., & Shin, J. K. (2023). Exploring a pathway to sustainable organizational performance of South Korea in the digital age: The effect of digital leadership on IT capabilities and organizational learning. *Sustainability (Switzerland)., 15*(10). https://doi.org/10.3390/su15107875

Moreno-Marcial, A., Moreno-Marcial, P., Mendoza-Romero, G., & Bueno-García, M. (2024). An analysis of leadership and proactivity to face the challenges of industry 5.0. *Web of Conferences., 532*. https://doi.org/10.1051/e3sconf/202453202007

Nishii, L. H., & Mayer, D. M. (2009). Do inclusive leaders help to reduce turnover in diverse groups? The moderating role of leader–member exchange in the diversity to turnover relationship. *Journal of Applied Psychology, 94*(6), 1412–1426.

Northouse, P. G. (2021). *Leadership: Theory and practice* (9th ed.). SAGE Publications.

Nurabadi, A., Triwiyanto, T., Gunawan, I., Valenda, O. Y., Pribadi, S. A., Darmawan, A., Nabila, A. W., Baharudin, A., Krisnafitriana, V., & Wardani, A. D. (2022). Assessing validity and reliability of an instrument for measuring student digital leadership: Example from Indonesia. Proceedings - international conference on education and technology, ICET, 2022-Octob. 288–292. https://doi.org/10.1109/ICET56879.2022.9990863

Oktaysoy, O., Topcuoglu, E., & Kaygin, E. (2022). A study on digital leadership scale adaptation. *In International Journal of Organizational Leadership., 11*. https://www.ijol.cikd.ca

Olsson, A. K., Eriksson, K. M., & Carlsson, L. (2024). Management toward industry 5.0: A co-workership approach on digital transformation for future innovative manufacturing. *European Journal of Innovation Management.* https://doi.org/10.1108/EJIM-09-2023-0833

Pacheco, D. A. J., & Iwaszczenko, B. (2004). Unravelling human-centric tensions towards industry 5.0: Literature review, resolution strategies and research agenda. *Digital Busines, 4, 2*, 1–13.

Padovano, A., Cardamone, M., Woschank, M., & Pacher, C. (2024). Exploring human-centricity in industry 5.0: Empirical Insights from a social media discourse. 5th international conference on industry 4.0 and smart manufacturing. *Procedia Computer Science, 232*, 1859–1868. https://doi.org/10.1016/j.procs.2024.02.008

Pandya, S. S., & Wang, J. (2024). Artificial intelligence in career development: A scoping review. *Human Resource Development International, 27*(3), 324–344. https://doi.org/10.1080/13678868.2024.2336881

Passalacqua, M., Pellerin, R., Magnani, F., Doyon-Poulin, P., Del-Aguila, L., Boasen, J., & Léger, P. M. (2025). Human-centred AI in industry 5.0: A systematic review. *International Journal of Production Research, 63*(7), 2638–2669). Taylor and Francis Ltd.. https://doi.org/10.1080/00207543.2024.2406021

Popescua, A. D., Borcaa, C., Fistisa, G., & Draghicia, A. (2014). Cultural diversity and differences in cross-cultural project teams. In *CENTERIS 2014 - conference on ENTERprise information systems/ProjMAN 2014 - international conference on project MANagement/HCIST 2014 - international conference on health and social care information systems and technologies* (Vol. 16, pp. 525–531). Science Direct, Procedia Technology. https://doi.org/10.1016/j.protcy.2014.10.120

Promsri, C. (2019). Developing model of digital leadership for successful digital transformation. *International Journal of Business Management., 02*(08).

Radonjić, A., Duarte, H., & Pereira, N. (2024). Artificial intelligence and HRM: HR managers' perspective on decisiveness and challenges. *European Management Journal, 42*(1), 57–66.

Rame, R., Purwanto, P., & Sudarno, S. (2024). Industry 5.0 and sustainability: An overview of emerging trends and challenges for a green future. *Innovation and Green Development, 3*, 4. https://doi.org/10.1016/j.igd.2024.10017

Randel, A. E., Dean, M. A., Ehrhart, K. H., Chung, B. G., & Shore, L. M. (2016). Leader inclusiveness, psychological diversity climate, and helping be haviors. *Journal of Managerial Psychology., 31*(1), 216–234.

Rannertshauser, P., Kessler, M., & Arlinghaus, J. C. (2022). Human-centricity in the design of production planning and control systems: A first approach towards industry 5.0. *IFAC-PapersOnLine, 55*(10), 2641–2646. https://doi.org/10.1016/j.ifacol.2022.10.108

Rejeb, A., Rejeb, K., & Zrelli, I. (2025). Industry 5.0 as seen through its academic literature: An investigation using co-word analysis. *Discover Sustainability, 6*, 307.

Renwick, D. W., Redman, T., & Maguire, S. (2012). Green human resource management: A review and research agenda. *International Journal of Management Reviews, 18*(1). https://doi.org/10.1111/j.1468-2370.2011.00328.x

Rice, M. (2010). *Diversity and public administration theory, issues, and perspectives*. Editura M.E Sharpe Inc.

Riedl, R. (2022). On the stress potential of videoconferencing: Definition and root causes of zoom fatigue. *Electronic Markets, 32*(1), 153–177. https://doi.org/10.1007/s12525-021-00501-3

Rijamampinina, R., & Carmichael, T. (2005). A pragmatic and holistic approach to managing diversity. *Problems and Perspectives in Management., 1*(2005), 109–117.

Roberson, Q. M. (2019). Diversity in the workplace: A review, synthesis, and future research agenda. *Annual Review of Organizational Psychology and Organizational Behavior., 6*(1), 69–88.

Roberson, Q. M. (2020). Access to justice as a human right, organizational entitlement and precursor to diversity and inclusion. *Equality, Diversity and Inclusion: An International Journal, 39*(7), 787–791.

Rockstuhl, T., & Van Dyne, L. (2018). A bi-factor theory of the four-factor model of cultural intelligence: Meta-analysis and theoretical extensions. *Organizational Behavior and Human Decision Processes. Volume, 148*, 124–144. https://doi.org/10.1016/j.obhdp.2018.07.005

Rüth, R., & Netzer, T. (2020). The key elements of cultural intelligence as a driver for digital leadership success. *Leadership, Education, Personality: An Interdisciplinary Journal., 2*(1), 3–8. https://doi.org/10.1365/s42681-019-00005-x

Sarkar, B. D., Shardeo, V., Dwivedi, A., & Pamucar, D. (2024). Digital transition from industry 4.0 to industry 5.0 in smart manufacturing: A framework for sustainable future. *Technology in Society, 78*. https://doi.org/10.1016/j.techsoc.2024.102649

Schwitter, N., Chatzitheochari, N., & Liebe, U. (2025). Disability discrimination in hiring: A systematic review. *Research in Social Stratification and Mobility, 98*, 1–14. https://doi.org/10.1016/j.rssm.2025.101069

Shore, L. M., Cleveland, J. N., & Sanchez, D. (2018). Inclusive workplaces: A review and model Lynn M. Shore, Jeanette N. Cleveland, Diana Sanchez. *Human Resource Management Review, 28*(2), 176–189.

Swart, K., Bond-Barnard, T., & Chugh, R. (2022). Challenges and critical success factors of digital communication, collaboration and knowledge sharing in project management virtual teams: A review. *International Journal of Information Systems and Project Management, 10*(4), 59–75. https://doi.org/10.12821/ijispm100404

Theng, B. P., Calen, C., & Toni, N. (2024). The role of digital leadership and employee empowerment on employees performance in the manufacturing industry in the era of society and industry revolution 5.0. *Journal Management Industri Dan Logistik., 8*(2), 167–184. https://doi.org/10.30988/jmil.v8i2.1501

Tigre, F. B., Curado, C., & Henriques, P. L. (2023). Digital leadership: A bibliometric analysis. *Journal of Leadership and Organizational Studies, 30*(1), 40–70. https://doi.org/10.1177/15480518221123132

Topcuoglu, E., Kobanoglu, M. S., Kaygın, E., Karafakıoglu, E., Erdogan, S. U., Torun, B. T., & Oktaysoy, O. (2023). The improving role of digital leadership in the impact of social loafing on job performance. *International Journal of Organizational Leadership., 12*(1), 22–40. https://doi.org/10.33844/ijol.2023.60347

Tót, A., Nagy, L., Kennedy, R., Bohuš, B., Abonyi, J., & b, Ruppert, T. (2023). The human-centric industry 5.0 collaboration architecture. *Methods, 11*. https://doi.org/10.1016/j.mex.2023.102260

Turan, A., & Doğan, O. (2025). Digital leadership in the context of society and state 5.0: Recommendations for policymakers based on public employees' perceptions. *The Academic Elegance., 12*(27) https://orcid.org/0000-0002-8109-4728

Uddin, A. S. M. A. (2023). The era of AI: Upholding ethical leadership. *Open Journal of Leadership, 12*(04), 400–417. https://doi.org/10.4236/ojl.2023.124019

Valette, E., El-Haouzi, H. B., & Demesure, G. (2023). Industry 5.0 and its technologies: A systematic literature review upon the human place into IoT- and CPS-b. *Computers & Industrial Engineering., 184*. https://doi.org/10.1016/j.cie.2023.109426

Vargas, R. V.; Barcaui, A. (2024). Beyond algorithms: The impact of emotional AI on corporate culture. London Business School Review.

Vidacek Hains, V, (2024). Study programs of informatics and adjustments for students with disabilities. MIPRO 2024 47th ICT and electronics convention proceedings, 727–730.

Wah, L. (1999). Diversity at Allstate. *Management Review, 88*(7), 24–30.

Wright, P. M., & McMahan, G. C. (2011). Exploring human capital: Putting 'human' back into strategic human resource management. *Human Resource Management Journal, 21*(2), 93–104.

Yadav, S., & Lenka, U. (2020). Diversity management: A systematic review. *Equality, Diversity and Inclusion: An International Journal., 39*(8), 901–929.

Yasmeen, K., Muhammad, D., Khan, L., & Imran, H. (2024). Exploring emotional intelligence, remote work dynamics, team collaboration, and adaptive leadership for enhanced success in the digital workplace. *Pakistan Social Sciences Review., 8*(2). https://doi.org/10.35484/pssr.2024(8-II)78

Yukl, G. (2013). *Leadership in organizations* (8th ed.). Pearson.

Zulu, S. L., & Khosrowshahi, F. (2021). A taxonomy of digital leadership in the construction industry. *Construction Management and Economics., 39*(7), 565–578. https://doi.org/10.1080/01446193.2021.1930080

Collective Leadership in Action: A Multi-Level Model for Adaptive and Collaborative Organizations

John R. Turner, Hoda Parvaneh Shirazi, Xiaoyan Xu, and Lisa Dinh-Trinh

Abstract Traditional leadership models rooted in hierarchical, command-and-control frameworks are increasingly inadequate in today's volatile, uncertain, complex, and interconnected environments. Organizations now operate as complex adaptive systems, requiring approaches that foster collaboration, adaptability, and shared accountability. This chapter introduces the Multi-dimensional Collective Leadership (MDCL) model, which integrates insights from collective, distributed, relational, shared, network, team, and complexity leadership theories into a cohesive framework for adaptive organizations. The MDCL model emphasizes three interdependent dimensions—co-creation (collaborative innovation and meaning-making), co-adaptation (collective agility and responsiveness to uncertainty), and co-governance (distributed authority and inclusive decision-making). Operating across micro (individual), meso (team/group), and macro (organizational/system) levels, these dimensions collectively enable organizations to navigate disruption while sustaining resilience and effectiveness. The chapter critiques the limitations of individual-centric, heroic models of leadership and advances the MDCL model as a comprehensive, multi-level framework that reflects the realities of modern organizations. Implications are offered for leadership practice and development, highlighting the need to design participatory structures, foster psychological safety, and cultivate adaptive mindsets that extend leadership capacities beyond formal roles. The MDCL model provides scholars and practitioners with a roadmap for reimagining leadership as a dynamic, distributed, and evolving process.

Keywords Collective leadership · Complexity · Co-creation · Co-adaptation · Co-governance · Adaptive leadership

---

J. R. Turner (✉) · H. P. Shirazi · X. Xu · L. Dinh-Trinh
Department of Educational Administration and Human Resource Development, Texas A&M University, College Station, TX, USA
e-mail: j.turner@tamu.edu

A. M. Waite et al. (eds.), *Leading in Industry 5.0*, Management and Industrial Engineering, https://doi.org/10.1007/978-3-032-19207-3_6

1 Introduction

Leadership practices were formalized during the Industrial Revolution, commonly referred to as Taylorism or scientific management, emphasizing command-and-control techniques. These methods were initially useful, as they provided efficiency during times of predictable and stable operations. Contemporary scholars argue that traditional hierarchical models, which assume that one leader can effectively direct and control all aspects of an organization, are inadequate in today's environment (Uhl-Bien et al., 2007; Pflaeging, 2014). However, today's leaders are trained in traditional leadership models, those designed for hierarchical and linear environments while operating in uncharted waters full of ambiguity, uncertainty, and unpredictability.

Organizations function as open systems, are entangled in global networks, while continuously responding and adapting to uncertainty and disruptive change. Leaders today face unprecedented levels of complexity, requiring them to rethink how leadership is conceptualized and practiced. This challenge demands moving beyond outdated methods and adopting leadership approaches capable of addressing systemic challenges, fostering collaboration, and enabling organizations to thrive within dynamic environments (Heifetz & Laurie, 1998; Berger & Johnston, 2015). Traditional leadership theories have significant limitations in managing complexity and uncertainty. Typically grounded in bureaucratic structures, these models focus on centralized decision-making and positional authority. Leadership within these frameworks is reduced to the control of processes, compartmentalization of tasks, and linear problem-solving (Dreier et al., 2019; Uhl-Bien & Marion, 2009). This perspective fails to acknowledge that today's organizations are complex adaptive systems where issues cannot be simplified or compartmentalized. The complexity of current challenges—such as climate change, economic instability, technological innovation, and global interconnectedness—necessitates leadership that is adaptive, distributed, and able to leverage collective knowledge across networks and systems. Moreover, the prevailing educational paradigm in leadership development still perpetuates a "heroic individual leader" myth, reinforcing unrealistic expectations that a single person can possess comprehensive knowledge and control (Senge et al., 2019). Such traditional frameworks neglect the critical roles of teamwork, collaborative decision-making, relational dynamics, and adaptive responsiveness that are fundamental in today's complex environments.

This chapter aims to address these limitations by introducing and elaborating on the "Multi-dimensional Collective Leadership" (MDCL) model. The MDCL Model integrates insights from collective leadership theories, offering a comprehensive, collaborative, and adaptive framework for leadership practice. This model reflects contemporary organizational realities and emphasizes shared responsibility, dynamic interactions, and multi-level adaptability. The primary objective of this chapter is to articulate clearly how collective, relational, shared, distributed, networked, and complexity leadership theories converge into the MDCL Model.

2 Overview of Collective Leadership Theories

This section provides a brief introduction to current leadership theories that belong to the "collective leadership" category. These theories include collective leadership (Ritchie et al., 2005), complexity fitness leadership theories (adaptive leadership, agile leadership, complex adaptive leadership, complexity capability, complexity leadership theory, sapient leadership, & system leadership), distributed leadership, network leadership, relational leadership, shared leadership, and team leadership.

3 Collective Leadership

Collective leadership theory adopts a dual agency-structure lens, emphasizing the mutual dependency between individual agency and structural conditions (Ritchie et al., 2005). Agency involves the actions of individuals and groups shaping and interpreting their roles, while structure refers to the systems and relationships that constrain or empower those actions (Ritchie et al., 2005). Collective leadership is often conflated with distributed leadership and is characterized by inclusive, emergent leadership networks (Döös & Wilhelmson, 2020; Kownacki et al., 2020).

Eckert (2019) redefined leadership in functional terms, focusing on task execution and responsibility rather than formal roles. Her model, applied to teacher leadership, included six factors: principal support, individual capacity, leadership task design, available resources, organizational context, and developmental opportunities (Eckert, 2019). This model also incorporated students and communities, aligning with Pino-Yancovic and Ahumada's (2020) concept of "leading upwards," which empowers lower-level actors through collaborative inquiry.

While not a unified theory, collective leadership serves as an umbrella for shared, distributed, team-based, and transformational models (Azeem & Mataruna, 2019). It centers on shared decision-making and mutual responsibility for enabling structures that support agency (Bernhardt, 2013; Ritchie et al., 2005). Power dynamics are inherent, as individuals use knowledge and resources to influence outcomes (Ni et al., 2018).

4 Complexity Fitness and Leadership Theories

Norris (2022) introduced "ninja training" in MBA programs to help students develop both traditional competencies and broader leadership capacities, such as mindsets and meta-skills. Unfortunately, most business education neglects complexity which runs the risk of leaving leaders unprepared (Norris, 2022). These theories highlight dual elements—technical skills and adaptive capabilities—that are essential for navigating today's complex environments. Berger and Johnston (2015) labeled this combination "complexity fitness" composed of the following theories: adaptive leadership, agile leadership, complex adaptive leadership, complexity capability, complexity leadership theory, sapient leadership, and system leadership.

4.1 Adaptive Leadership

Adaptive Leadership (Heifetz et al., 2009; Heifetz & Laurie, 1998) identifies six principles for leaders to follow: broad observation ("getting on the balcony"), identifying adaptive challenges, managing change-related distress, fostering disciplined attention, distributing responsibility, and protecting marginalized voices.

4.2 Agile Leadership

Agile Leadership (Hall & Rowland, 2016) addresses the gap between traditional training and real-world demands. Here, they use a "Nine Box" matrix to assess agility and performance that consists of a three-by-three matrix with agility along the horizontal and performance along the vertical. Agility ranged from low to high agility, and performance ranged from low to high performance. Leaders were rated regarding both measures and presented several skills for developing leaders based on their assessment which included rethinking current ways of doing business, identifying intent, exploring dilemmas and resolutions, and making use of media to name only a few (Hall & Rowland, 2016).

4.3 Complex Adaptive Leadership

Complex Adaptive Leadership (CAL; Obolensky, 2014) views leadership as being dynamic. A two-by-two matrix was generated to assess and develop leaders. This matrix was composed of people focus on the vertical axis and goal focus on the horizontal. Leaders are assessed as low-high people focused and low-high goal focused. The cells in this matrix feature four strategies: tell, sell, involve, and devolve. These features were adapted from models like situational leadership and Adair's (1988) leadership functions. While not empirically tested, CAL has been shown to have practical value for leadership training.

4.4 *Complexity Capability*

Complexity Capability (Omer, 2017) described five leader capabilities: negative (tolerating uncertainty), imaginal (creative synthesis), perspectival (navigating contradictions), autopoietic (self-organizing), and collaborative (shared world-making). A leader's capabilities evolve through transformative learning and as a leader experiences challenges that cause them to alter their current mental models.

4.5 *Complexity Leadership Theory*

Complexity leadership theory (CLT; Uhl-Bien & Marion, 2009; Uhl-Bien et al., 2007) conceptualizes leadership as arising from dynamic, multi-level social interactions. CLT sees leadership as emergent, contextual, and distributed across both people and ideas. It also distinguishes between "leaders" as individuals with influence, and "leadership" as a process embedded in relational systems. This theory outlines three leadership types, administrative, adaptive, and enabling, each with distinct roles and functions (Schophuizen et al., 2022) and are provided in Table 1. CLT seeks to reconcile the tension between innovation (exploration) and production (exploitation) within organizational systems (Tsai et al., 2019).

4.6 *Sustaining Leadership Types Through Organizational Interventions*

Schophuizen et al. (2022) proposed four organizational interventions (coordination, identification, reflection, and transformation) to support and sustain three forms of leadership (administrative, adaptive, and enabling) within an organization. Coordination involves implementing standardized procedures that support simultaneous leadership practices. Identification refers to replicating existing practices in new contexts. Reflection emphasizes practices that promote self-awareness and understanding diverse perspectives. Transformation involves integrating current practices with innovative or emerging approaches. These interventions guide how leaders make strategic decisions in complex environments and help maintain organizational adaptability by aligning top-down structures with bottom-up dynamics (Schophuizen et al., 2022).

Table 1 Three types of leadership in CLT

Leadership type	Description	Roles
Administrative leadership	The actions of individuals and groups in formal managerial roles who plan and coordinate organizational activities (the bureaucratic function)	• top-down oriented • formal with set patterns • planning and coordination toward outcomes • efficiency and effectiveness • focus on alignment and control • represented by hierarchical and bureaucratic functions
Adaptive leadership	An emergent, interactive dynamic that is the primary source by which adaptive outcomes are produced in a firm	• bottom-up oriented • informal and emergent • adaptive • creative learning actions that emerge from the actions of multiple agents in the organization
Enabling leadership	Serves to enable (catalyze) adaptive dynamics and help manage the entanglement between administrative and adaptive leadership	• top-down catalyst for optimal conditions for bottom-up adaptive capacity • leads entanglement between bureaucratic (i.e., administrative) and emergent (i.e., adaptive) forms of leadership

Notes Descriptions from (Uhl-Bien et al., 2007, p. 306); Roles from (Schophuizen et al., 2022, p. 473)

4.7 Sapient Leadership

Aneel and Gutman (2020) described leadership in the post-COVID world as perpetual (constant), pervasive (affecting multiple life domains), and exponential (rapidly accelerating). They defined sapient leadership as "leadership rooted in wisdom, discernment, and humanity amid rapid and often unsettling change" (p. 3). This leadership model is grounded in four key pillars:

1. Leader humility, authenticity, and openness to foster trust and psychological safety.
2. Psychological safety and trust to empower individuals and teams.
3. A culture of continuous learning to effectively navigate ongoing change.
4. Shared purpose and values to strengthen focus, unity, and resilience during turbulent times. (Aneel & Gutman, 2020, pp. 5–8).

Table 2 System leadership levels of analysis

Level of analysis	Description
The individual	The skills of **collaborative leadership** to enable learning, trust-building and empowered action among stakeholders who share a common goal
The community	The tactics of **coalition building and advocacy** to develop alignment and mobilize action among stakeholders in the system, both within and between organizations
The system	An understanding of the **complex systems** shaping the challenges to be addressed

Dreier et al., 2019, p. 13

4.8 System Leadership

Senge et al. (2019) identified three key capabilities for system leaders: the ability to understand broader systemic patterns, promote reflective and generative dialogue, and shift from reactive problem-solving toward co-creating the future (pp. 13–14). Though not formally defined, they emphasized these capabilities as being essential for navigating complexity. References to collective leadership were made but not explicitly defined. System leadership consists of four levels of analysis as shown in Table 2.

4.9 Distributed Leadership

Distributed leadership involves spreading leadership responsibilities across various organizational members, as opposed to centralizing it in one individual (Stuart, 2018; Lindahl, 2008). Rooted in a constructivist and agentic perspective, it emphasizes social interdependencies that are shaped through interactions (Printy & Marks, 2006; Lumby, 2009). Gronn (2003) described this as "concertive action," comprising spontaneous collaboration, intuitive understanding, and institutional arrangements. Distributed leadership is enacted by both formal and informal agents working toward shared goals (Leithwood & Mascall, 2008), expressed through shared responsibilities and collective agency (Turchi et al., 2023). Unlike shared leadership, which focuses on team roles, distributed leadership is more about the embedded processes across organizations (Lindahl, 2008; Khasawneh, 2011).

4.10 Network Leadership

Network leadership highlights interconnectedness, collaboration, and collective agency across sectors. It offers an alternative to traditional hierarchies by prioritizing relationship-building, mutual influence, and participatory governance (Tremblay, 2012; Poley, 1998). Inspired by social network theory and swarm intelligence, scholars argue that leadership arises through decentralized, dynamic interactions among agents (Bullington & Alford, 2019; McNulty et al., 2018). Network leadership supports system-wide alignment, distributed authority, and adaptive change in response to complexity (Chrispeels et al., 1999; Browne-Ferrigno & Bjork, 2018). Definitions of network leadership are provided in Table 3.

Table 3 Network leadership definitions

Definition	Source
Network leadership involves the production and transfer of information through dynamic communication networks. It emphasizes self-organizing structures where external interventions are minimal, fostering spontaneous organization and coordination within the network	(Baltaci & Balcı, 2017, p. 45)
Network leadership is defined as the ability to strategically use social networks to share expertise, learning, and resources across interacting units. This approach facilitates collaboration and improves organizational outcomes by leveraging connections effectively	(Bullington & Alford, 2019, p. 151)
Network leadership facilitates dynamic linkages across organizational networks, enabling the free flow of information and fostering collaboration. It involves creating systems that enhance communication and self-organization, which are essential for managing complexity within state-owned enterprises	(Donkor & Zhou, 2019, p. 156)
Network leadership involves fostering shared leadership models where collaboration among individuals, organizations, and systems is prioritized. This approach emphasizes participatorydecision-making and collective ownership of initiatives to align with shared visions and goals	(Poley, 1998, p. 20)
Network leadership emphasizes fostering collaborative inquiry and teamwork within and between schools. Leaders facilitate the exchange of knowledge and practices, aiming for systemic improvement rather than isolated success	(Pino-Yancovic & Ahumada, 2020, p. 3)
Network leadership is centered on creating and leveraging intentional interconnectedness among individuals, groups, and systems. This aspect distinguishes it from conventional leadership theories that may not prioritize interconnectedness as a core practice	(Tremblay, 2012, p. 37)

4.11 *Relational Leadership*

Relational leadership shifts the focus from individuals to relationships. Uhl-Bien (2006) distinguished between "entity" perspectives, which focus on leader traits and behaviors, and "relational" views, where leadership is socially constructed through interactions. McCauley and Palus (2021) described leadership as an emergent property of relationships, not individuals, placing leadership as an emergent construct that occurs through natural interactions in which meaning is constructed through shared meaning and dialogue (Webb, 2021). This approach favors constructionist inquiry and emphasizes collective processes over positional authority.

4.12 *Shared Leadership*

Shared leadership is a collaborative, emergent team process where leadership roles are distributed based on task needs and group dynamics (Carson et al., 2007; Slantcheva-Durst, 2014). While some define it as assigning leadership to capable members (Pearce & Conger, 2003; Avolio et al., 2009), others equate it with distributed leadership (Gibson et al., 2022). Theories grounding shared leadership include social learning theory and the resource-based view (Shoukat et al., 2023). Characteristics of shared leadership include decision-making, knowledge sharing, team empowerment, shared governance, accountability, communication, and collaboration (Pieczka & Casteltrione, 2020; Khasawneh, 2011; Torres et al., 2020). Table 4 provides some common definitions for shared leadership.

4.13 *Team Leadership*

Team leadership views leadership as a shared property among team members that emerges through the distribution of influence among members (Carson et al., 2007). While leadership in teams is often focused on a single leader managing the team, researchers have advocated for more studies on internal team leadership (Hirschy, 2011). Benefits include better decision-making, productivity, and innovation (Curry, 2014). Team leadership overlaps with collaborative and shared leadership but emphasizes equal participation and adaptability (Court, 2007). Key attributes include team and individual characteristics such as experience, roles, personality, and team structure (Lattner & Strehmel, 2023). Common definitions for team leadership are provided in Table 5.

Table 4 Shared leadership definitions

Source	Definition
(Afful & Hall, 2015; de Cruz, 2019, p. 215; Kocolowski, 2010; Wiens et al., 2024)	A dynamic, interactive influence process among individuals in groups for which the objective is to lead one another to the achievement of group or organizational goals or both. This influence process often involves peer, or lateral, influence and at other times involves upward or downward hierarchical influence. (Pearce & Conger, 2003, p. 1)
(Gibson et al., 2022, p. 3)	Reflects shared responsibilities
(Gu et al., 2018)	Shared leadership is a team property where leadership behaviors are displayed by multiple team members, rather than those at the top or in formally appointed roles
(Cobanoglu & Bozbayindir, 2019, p. 1266)	For educational organizations: The mutual efforts put forth by different groups in the organization… for improving the organization in their own fields of expertise or in a general sense
(Jakavonytė-Staškuvienė & Ignatavičiūtė, 2022, p. 3)	Cooperation, where every member of the community wants and can offer ideas, accepts and understands the ideas of others, and seeks to contribute to the best result created by the community
(Jakavonytė-Staškuvienė & and Ignatavičiūtė, 2022, p. 3)	A network of interactions between individuals and different situations
(Afful & Hall, 2015, p. 50)	A team process where leadership is carried out by the team as a whole, rather than solely by a single designated individual. (Ensley et al., 2006, p. 220)
(Stark, 2016, p. 23)	Begins with setting the conditions and good group processes for healthy, productive working relationships
(Turchi et al., 2023, p. 121)	A social process that influences and mobilizes the commitment of a group in view of the achievement of specific objectives, linked to a common vision (Harris, 2014)
(Shoukat et al., 2023)	A collaborative and cooperative mechanism between project team members and leadership to accomplish team goals and target achievement (Scott-Young et al., 2019; Wu et al., 2020)
(Torres et al., 2020, p. 2)	The combination of collaborative influence on instructional leadership and schoolwide decisions

(continued)

Table 4 (continued)

Source	Definition
(Vogel, 2022, p. 493)	Top-down, bottom-up and middle-out, relying less on positional power in senior executive roles and emphasizing trust in expertise
(Kalinovich & Marrone, 2017, p. 205)	Reflects the extent to which leadership influence is distributed across the members of a collective as opposed to leadership remaining centralized within one powerful figure

Table 5 Team leadership definitions

Source	Definition
Wolfinbarger (2022)	Team leadership is viewed as a dynamic and collaborative process rather than a positional role. It emphasizes mutually supportive relationships and collective engagement in decision-making
(Lattner & Strehmel, 2023, p. 3)	Team leadership involves creating structures and processes that foster collaboration, mutual support, and alignment of pedagogical goals among diverse team members. Leaders are responsible for maintaining team cohesion and ensuring effective communication, particularly in crisis situations
(Curry, 2014)	Team leadership is defined as a shared process where leaders and members work collaboratively to solve complex problems, leveraging diverse expertise and perspectives within the group

5 Multi-Dimensional Collective Leadership (MDCL) Model

The Multi-dimensional Collective Leadership (MDCL) model provides a contemporary perspective on leadership by integrating multiple theoretical insights into a cohesive, practical framework. Recognizing that traditional hierarchical and individualistic leadership approaches fail in today's complex and rapidly evolving environments, this model is relational and emphasizes adaptability, interactions, and shared governance. At its heart, the MDCL model is composed of three interrelated core dimensions: co-creation, co-adaptation, and co-governance. Each dimension captures essential capacities required for leadership effectiveness in dynamic, collaborative contexts. These dimensions are interdependent, meaning they overlap and influence each other continuously. Co-creation emphasizes collaborative innovation and collective meaning-making, co-adaptation highlights the importance of flexibility and responsiveness to change, and co-governance underscores shared accountability and distributed decision-making. A breakdown of each dimension related to the leadership theories presented in the previous section along with specific characteristics for each is provided in Table 6.

The interplay between these elements ensures organizational resilience and effectiveness, especially within complex and uncertain environments. The visual representation of the MDCL model (see Fig. 1) illustrates the overlapping nature of these

Table 6 Differences between collective theories

Theory	Micro/meso/macro	Co-adaptation/co-creation/co-governance	Characteristics
Collective leadership	Leader, organizational	Distributed, dynamic	Agency, structure, function
Complexity fitness leadership theories			
Adaptive leadership	Leader	Prescriptive	Entrepreneurialism, interpersonal dynamics
Agile leadership	Leader	Prescriptive	Leader's tasks
Complex adaptive leadership	Leader	Adaptive, dynamic. Prescriptive	Situational leadership
Complexity capacity	Leader, organizational	Prescriptive	Transformative learning
Complexity leadership	Organizational	Connective, distributed, dynamic, contextual	Administrative, adaptive, enabling leadership
Sapient leadership	Leader	Prescriptive	Leader pillars
System leadership	Leader, community, system	Prescriptive	Leader capabilities
Distributed leadership	Teams, organizations	Constructivism	Concertive action, shared leadership
Network leadership	Team-based environments, organizations	Collective action, collective intelligence, interconnectedness	Social network theory, swarm intelligence
Relational leadership	Leader-follower, leader-team, leader-dyad	Constructivism	Relational (attributes, behaviors, cognitions, perceptions)
Shared leadership	Team-based	Dynamic, interactive	Resource-based view (RBV), social learning theory
Team leadership	Individual (leader-team), team	Distributed, shared	Group dynamics, interpersonal relations, team climate/culture

three entangled dimensions within an organizations' structure (micro, meso, and macro). This depiction emphasizes how collective leadership dynamically operates at multiple scales, demonstrating leadership as a fluid, interactive, and adaptive process rather than a static, hierarchical function.

The Multi-dimensional Collective Leadership (MDCL) model offers a new perspective on leadership by merging critical aspects of existing leadership theories into a cohesive framework. Collectively, these theories emphasize that leadership

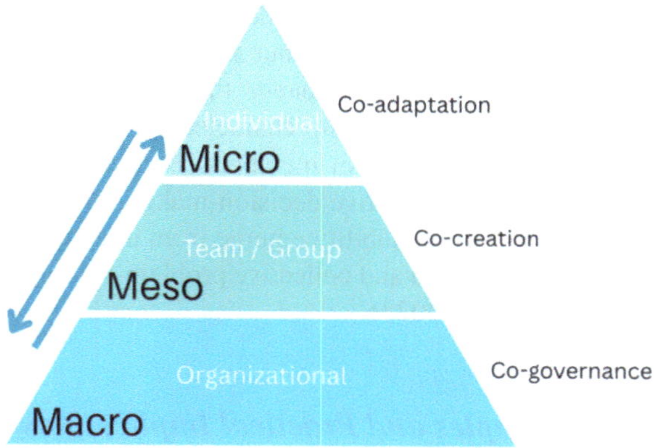

Fig. 1 Multi-dimensional collective leadership model

is most effective when it harnesses diverse knowledge, fosters dynamic interaction, and adapts to continuously evolving organizational contexts. The integration of these theoretical insights forms the foundational basis for the MDCL model.

The MDCL model acknowledges that traditional models, characterized by centralized authority and rigid hierarchical structures, fall short in addressing today's complex organizational environments. The MDCL model emphasizes adaptive, relational, and shared processes within a single model involving diverse organizational agents interacting across multiple levels of analysis. The core elements of the MDCL model are co-creation, co-adaptation, and co-governance. These dimensions are interconnected, overlapping, and continuously influencing each other. Collectively, they provide organizations with essential capacities to effectively navigate complexity, uncertainty, and disruptive change. This integration is visually represented in Fig. 1 with the two-directional arrows to illustrate the dynamic interplay among the dimensions and their operation at different organizational scales (micro, meso, macro).

6 Dimension 1: Co-Creation

Co-creation refers to the collaborative and interactive processes through which members generate new ideas, insights, and innovative solutions. It involves collective meaning-making and harnessing diverse perspectives to address organizational challenges creatively. Co-creation is deeply embedded within relational leadership theory and shared leadership theory.

- **Relational Leadership**: Emphasizes leadership as emerging from interactions and relationships among individuals (Uhl-Bien, 2006). Leadership is considered an ongoing relational process, constructed through mutual influence, trust-building, and open dialogue. The entity and relational approaches explained by

Uhl-Bien (2006) underline the importance of relationships, viewing leadership not as individual attributes but as a socially constructed phenomenon.

- **Shared Leadership**: Expands the concept of relational leadership to a team setting. Shared leadership characterizes leadership as dynamic, interactive, and distributed among team members (Pearce & Conger, 2003). This theory highlights how collaborative decision-making, mutual responsibility, empowerment, and shared accountability emerge from interactions among team members, facilitating innovation and collective problem-solving (Pieczka & Casteltrione, 2020; Shoukat et al., 2023).

6.1 Examples and Practical Implications

Organizations practicing co-creation often adopt agile methodologies, brainstorming workshops, and open innovation platforms to foster collective creativity. For instance, technology firms employing agile project management methodologies demonstrate co-creation through continuous dialogue and iterative idea generation among developers, users, and stakeholders.

7 Dimension 2: Co-Adaptation

Co-Adaptation captures the collective capacity of organizational members to respond, adapt, and thrive in uncertain and complex situations. This dimension prioritizes responsiveness, continuous learning, flexibility, and dynamic interactions to maintain organizational agility and resilience. Co-Adaptation aligns closely with complexity leadership theory (CLT) and adaptive leadership:

- **Complexity Leadership Theory (CLT)**: CLT presents leadership as an emergent, dynamic interaction within complex adaptive systems, advocating for informal networks and adaptive responses over centralized control (Uhl-Bien et al., 2007). It emphasizes enabling conditions that allow adaptive and innovative outcomes to arise naturally from interactions among agents within organizations. CLT distinguishes among administrative, enabling, and adaptive leadership roles, each necessary for managing organizational complexity effectively (Schophuizen et al., 2022).
- **Adaptive Leadership**: Adaptive leadership emphasizes diagnosing complex adaptive challenges, creating a safe and supportive environment, and maintaining disciplined attention on organizational issues. Leaders practicing adaptive leadership strategically distribute problem-solving responsibilities among organizational members, empowering them to address and navigate uncertainties and complexities collectively (Heifetz & Laurie, 1998).

Other complementary frameworks like agile leadership (Hall & Rowland, 2016), complex adaptive leadership (Obolensky, 2014), complexity capacity (Omer, 2017), and sapient leadership (Aneel & Gutman, 2020) further reinforce the significance of flexibility, responsiveness, and continuous learning as core capacities for organizations operating in uncertain and rapidly evolving environments.

7.1 Examples and Practical Implications

A practical example of co-adaptation is the rapid organizational response to the COVID-19 pandemic, wherein companies quickly shifted to remote operations, adapting communication strategies and being forced to learn how to navigate uncertainty to survive. Similarly, businesses in volatile markets continuously adjust their strategies through iterative learning processes, scenario planning, and feedback loops.

8 Dimension 3: Co-Governance

Co-governance describes the collective management of authority, decision-making responsibilities, and accountability. It emphasizes participatory governance structures, inclusive decision-making processes, distributed leadership roles, and collective ownership over organizational outcomes. Co-governance integrates principles from distributed leadership, network leadership, and functional leadership:

- **Distributed Leadership**: Describes leadership as a socially constructed, collective action distributed across multiple agents and organizational levels (Spillane, 2006; Lumby, 2009). It emphasizes relational interactions, interdependencies, shared responsibilities, and the collective enactment of leadership functions with a focus at the organizational level.
- **Network Leadership**: Accentuates the importance of interconnectedness, collaboration, and collective action across organizational boundaries. It stresses collective ownership, participatory governance, intentional connectivity, and alignment of efforts across diverse groups and stakeholders (Tremblay, 2012; Bullington & Alford, 2019) to resolve organizational issues.
- **Functional Leadership**: Eckert (2019) defined leadership based on tasks, responsibilities, and organizational functions rather than positions or roles. This theory promotes flexibility in role assignment, distributed accountability, and organizational structures designed to maximize collective efficacy and effectiveness for organizational processes.

8.1 Examples and Practical Implications

Practical manifestations of co-governance include shared governance practices in educational institutions, participatory budgeting in governmental organizations, and decentralized management structures in organizations (e.g., Spotify) where teams autonomously manage projects, make decisions collectively, and assume shared accountability for outcomes.

9 Multi-Level Integration: Micro, Meso, Macro

The MDCL model effectively operates across three interconnected organizational levels: micro-, meso-, and macro-levels. Each of these three levels is discussed next.

9.1 Micro-Level (Individual)

At this level, leadership is expressed primarily through interpersonal interactions, shared decision-making, and collective action among individual agents. These agents could be autonomous or part of a larger group or team. For agents to be successful, shared practices (collaborative decision-making, shared accountability, effective communication, mutual empowerment) that take place through interactions become essential (Khasawneh, 2011; Torres et al., 2020).

9.2 Meso-Level (Team/Group)

At the meso-level, leadership involves the coordination and alignment of teams and multiteam systems, focusing on integrating different functional areas and maintaining collaborative dynamics across organizational units. This requires shared governance practices, coordinated responses to changes, and distributed decision-making.

9.3 Macro-Level (Organization)

At the macro-level, leadership involves systemic integration, strategic alignment, and collaborative governance across the entire organization (intra- and inter-organizational networks). Network leadership and systems leadership become crucial for effectively managing large-scale complexity and uncertainty, emphasizing collective ownership, shared purpose, and participatory governance.

10 Summary

This chapter critiques the inadequacy of traditional, hierarchical, and individual-centric leadership models in today's volatile, uncertain, complex, and interconnected environments. It positions leadership as an emergent, relational, and adaptive process rather than a top-down function, drawing on a broad spectrum of collective leadership theories—such as complexity leadership, adaptive leadership, network leadership, relational leadership, shared leadership, and distributed leadership. Each theory contributes distinct principles: from adaptive problem-solving and agility to networked collaboration, shared accountability, and relational trust. This chapter outlines related models like sapient leadership that included wisdom and humanity; system leadership that highlighted systemic sense-making and generative dialogue; and functional approaches that view leadership by tasks rather than positions. Collectively, these perspectives converge on the need for leadership approaches that leverage diverse expertise as opposed to specializations, collaboration rather than command-and-control, and collective learning in contrast to following processes.

This chapter introduced the Multi-dimensional Collective Leadership (MDCL) model, a unifying framework integrating three interdependent dimensions, co-creation (collaborative innovation and shared meaning-making), co-adaptation (collective agility and responsiveness to complexity), and co-governance (distributed authority and inclusive decision-making). These dimensions operate dynamically across micro (individual), meso (team/group), and macro (organization/system) levels. The MDCL model entangles all three dimensions to take advantage of informal and formal networks across the organization, ensuring that there is alignment and coherence for organizational strategy at all levels within an organization.

The MDCL model also provided a preliminary map of the leadership theories that associated best with each of the three dimensions. The MDCL model offers a comprehensive structure for fostering creativity because agency and an organization's structure are entangled across all level of analysis providing coherence across organizational levels.

11 Implications for Practice—Leadership

The MDCL model offers practitioners a roadmap for transforming leadership practice from a positional and authority-based function to a collective, distributed, and dynamic structure that occurs across the whole organization. For leaders, this means shifting focus from directing tasks to creating the conditions where co-creation, co-adaptation, and co-governance can flourish at all levels equally (micro, meso, macro). From a pragmatic perspective, this requires designing participatory decision-making structures, embedding continuous learning processes, and cultivating psychological safety to encourage diverse voices in problem-solving and decision-making activities.

Leaders must act as boundary spanners rather than gatekeepers who bridge information across networks. Leaders must also provide intent and purpose to promote agency across the organizational levels, requiring cross-boundary collaborations and the removal of power structures. Leadership occurs through interactions among agents who develop trust among one another while distributing trust and responsibilities for tasks. This reconceptualization of leadership allows organizations to adapt and manage complexity and uncertainty better.

12 Implications for Leadership Development

The MDCL model reframes leadership development as a multi-level process that extends beyond the preparation of individual leaders. Instead of reinforcing the "heroic leader" paradigm, development initiatives need to develop the shared leadership capacities of teams, departments, and networks so that agents are capable of engaging in co-creation, co-adaptation, and co-governance. This involves cultivating adaptive mindsets, relational intelligence, and complexity thinking skills across the organization, not just in those with formal titles. Development programs should involve experiential, generative, and transformational learning.

Leadership development needs to address leadership in collective and shared contexts rather than viewing leadership as taking place in a few individuals. This supports the necessary competencies that support leadership as a collective construct such as distributed decision-making and collective problem-solving. Over time, these capabilities begin to support the embedded leadership structures across the organization's levels, further contributing to leadership that is distributed and shared throughout the organization.

13 Conclusion

This chapter presented the Multi-dimensional Collective Leadership (MDCL) model that views leadership as being a collective construct that is distributed across an organization involving micro (individual), meso (teams, groups), and macro (organization, network) levels. Leadership takes on different characteristics at each level which should be accounted for when looking at leadership within organizations. While it is becoming more and more apparent that the traditional command-and-control, or top-down, leadership structures that supported organizations through the industrial era will not successfully carry us through today's complex and entangled global landscape, it is unclear what will replace the leadership models and theories from yesterday.

The MDCL model provides one perspective that looks at including current leadership theories and research that view leadership as a collective construct rather than an individual construct. The MDCL model is a multi-dimensional and multi-level model because it provides multiple competencies for each level within an organization. Coherence is achieved across levels once agency becomes autonomous across boundaries, making leadership a distributed and shared process with minimal power structures. The MDCL model is an integrated leadership model that provides one perspective and option for organizations to implement. It is a journey that requires organizations to adapt and continuously evolve, all characteristics necessary for managing complexity.

References

Adair, J. (1988). *The action centered leader* (2nd ed.). Spiro Press.

Afful, D. & Hall, K. (2015). Enhancing effective administration at faculty level through shared leadership. *Journal of Education and Practice, 6*(11), 49–55.

Alanezi, A. (2016). The relationship between shared leadership and administrative creativity in Kuwaiti schools. *Management in Education, 30*(2), 50–56. https://doi.org/10.1177/0892020616643159

Aneel, C., & Gutman, R. (2020). *What it takes to lead through an era of exponential change*. Harvard Business Review.

Ashby, W. R. (1962). Principles of the self-organizing system. In Von Foerster, H. & Zopf, G. W. (Eds.), *Principles of self-organization: Transactions of the University of Illinois Symposium* (pp. 255–278). Pergamon Press.

Ashby, R. W. (2004). Principles of the self-organizing system. *Emergence: Complexity and organization, 6*(1–2), 102–126. https://doi.org/10.emerg/10.17357.6236a43bba1866fb2c8247c22a362dc9.

Avolio, B. J., Walumbwa, F. O. & Weber, T. J. (2009). Leadership: Current theories, research, and future directions. *Annual Review of Psychology, 60*, 421–449. https://doi.org/10.1146/annurev.psych.60.110707.163621

Azeem, M., & Mataruna, L. (2019). Identifying factor measuring collective leadership at academic workplaces. *International Journal of Educational Management*, (ahead-of-print). https://doi.org/10.1108/ijem-04-2018-0131

Baltaci, A., & Balcı, A. (2017). Complexity leadership: A theorical perspective. *International Journal of Educational Leadership and Management, 5*(1), 30. https://doi.org/10.17583/ijelm.2017.2435

Batistič, S., Černe, M., & Vogel, B. (2017). Just how multi-level is leadership research? A document co-citation analysis 1980–2013 on leadership constructs and outcomes. *The Leadership Quarterly, 28*(1), 86–103. https://doi.org/10.1016/j.leaqua.2016.10.007

Bennis, W. (2009). *On becoming a leader*. Basic Books.

Berger, J. G., & Johnston, K. (2015). *Simple habits for complex times: Powerful practices for leaders*. Stanford University Press.

Bernhardt, B. (2013). *Using data to improve student learning in elementary school*. Routledge.

Blanchard, K. (2008). Situational leadership. *Leadership Excellence, 25*(5), 19.

Blanchard, K. (2010). *Leading as a higher level: Blanchard on leadership and creating high performing organizations* (Expanded ed.). FT Press.

Bolman, L. G., & Deal, T. E. (2011). *Reframing organizations: Artistry, choice and leadership*. Jossey-Bass.

Bostanci, A. B. (2013). The prediction level of teachers' organizational citizenship Behaviors on the successful practice of shared leadership. *Eurasian Journal of Educational Research, 51*, 177–194.

Brown, D. F. (1994). Experiencing shared leadership: Teachers' reflections.

Browne-ferrigno, T., & Bjork, L. G. (2018). Reflections on education reform and team leadership. *Research in Educational Administration and Leadership, 3*(2), 339–347. https://doi.org/10.30828/real/2018.2.9

Bullington, T. S., & Alford, W. A. (2019). Teaching network leadership: Using collaborative structure and the remote associates test. *Journal of Leadership Education, 18*(4), 150–155. https://doi.org/10.12806/V18/I4/A1

Carson, J. B., Tesluk, P. E., & Marrone, J. A. (2007). Shared leadership in teams: An investigation of antecednent conditions and performance. *Academy of Management Journal, 50*(5), 1217–1234. https://doi.org/10.5465/amj.2007.20159921

Cawthorne, J. E. (2010). Leading from the middle of the organization: An examination of shared leadership in academic libraries. *The Journal of Academic Librarianship, 36*(2), 151–157. https://doi.org/10.1016/j.acalib.2010.01.006

Chrispeels, J. H., Martin, K. J., Harari, I., Strait, C. C., & Rodarte, M. A. (1999). Role conflict and role ambiguity: The challenges of team leadership at a middle School[1]. *Journal of School Leadership, 9*(5), 422–453. https://doi.org/10.1177/105268469900900505

Cobanoglu, N., & Bozbayindir, F. (2019). A study on shared leadership and positive psychological capitals of teachers at primary and secondary schools. *Universal Journal of Educational Research, 7*(5), 1265–1274. https://doi.org/10.13189/ujer.2019.070512

Çobanoglu, N., & Bozbayindir, F. (2019). A study on shared leadership and positive psychological capitals of teachers at primary and secondary schools. *Online Submission, 7*(5), 1265–1274.

Court, M. (2007). Changing and/or reinscribing gendered discourses of team leadership in education? *Gender and Education, 19*(5), 607–626. https://doi.org/10.1080/09540250701535642

de Cruz, N. P. (2019). A conceptual overview of attaining, maintaining, and regaining shared leadership in high performing teams. *Journal of Leadership Education, 18*(1), 213–226. https://doi.org/10.12806/V18/I1/T3

Curry, K. A. (2014). Team leadership. *Journal of Cases in Educational Leadership, 17*(2), 20–40. https://doi.org/10.1177/1555458914528909

Dachler, H. P., & Hosking, D. M. (1995). The primacy of relations in socially constructing organizational realities. In D. M. Hosking, H. P. Dachler, & N. K. J. GergenAnnunzio-Green (Eds.), *Management and organization: Relational alternatives to individualism* (pp. 1–29). Avebury.

Darban, M. (2021). Learning in virtual student teams: An examination of shared leadership. *Journal of Research on Technology in Education, 54*(5), 736–753. https://doi.org/10.1080/15391523.2021.1916800

DeChurch, L. A., Hiller, N. J., Murase, T., Doty, D., & Salas, E. (2010). Leadership across levels: Levels of leaders and their levels of impact. *The Leadership Quarterly, 21*(6), 1069–1085. https://doi.org/10.1016/j.leaqua.2010.10.009

Dionne, S. D., Gupta, A., Sotak, K. L., Shirreffs, K. A., Serban, A., Hao, C., Kim, D. H., & Yammarino, F. J. (2014). A 25-year perspective on levels of analysis in leadership research. *The Leadership Quarterly, 25*(1), 6–35. https://doi.org/10.1016/j.leaqua.2013.11.002

Donkor, F., & Zhour, D. (2019). Complexity leadership theory: A perspective for state-owned enterprises, *Ghana International Journal of Educational Leadership and Management 7*(2), 139. https://doi.org/10.17583/ijelm.2019.3647

Döös, M., & Wilhelmson, L. (2020). Changing organizational conditions: Experiences of introducing and putting function-shared leadership (FSL) into practice in schools and pre-schools. *Leadership and Policy in Schools, 20*(4), 672–689. https://doi.org/10.1080/15700763.2020.1734628

Dreier, L., Nabarro, D., & Nelson, J. (2019). *Systems leadership for sustainable development: Strategies for achieving systemic change*. Harvar Kennedy School.

Eckert, J. (2019). Collective leadership development: Emerging themes from urban, suburban, and rural high schools. *Educational Administration Quarterly, 55*(3), 477–509. https://doi.org/10.1177/0013161x18799435

Ensley, M. D., Hmieleski, K. M., & Pearce, C. L. (2006). The importance of vertical and shared leadership within new venture top management teams: Implications for the performance of startups. *The Leadership Quarterly, 17*(3), 217–231. https://doi.org/10.1016/j.leaqua.2006.02.002

Gibson, E. M., Thompson, J. M., Cook Sandifer, M. I., & Brant-Rajahn, S. N. (2022). A culturally affirming shared leadership framework. *Professional School Counseling, 26*(1). https://doi.org/10.1177/2156759X221134665

Goksoy, S. (2016). Analysis of the Relationship between Shared Leadership and Distributed Leadership [Paylaşılan Liderlik ile Dağıtımcı Liderlik Arasındaki İlişkinin Analizi]. *Eurasian Journal of Educational Research, 16*(65), 295–312. https://dergipark.org.tr/en/pub/ejer/issue/42412/510649

Gronn, P. (2003). *Education and social justice*. University of Wales Press.

Gu, J., Chen, Z, Huang, Q., Liu, H., & Huang, S. (2018). A multilevel analysis of the relationship between shared leadership and creativity in inter-organizational teams. *The Journal of Creative Behavior 52*(2), 109–126. https://doi.org/10.1002/jocb.2018.52.issue-210.1002/jocb.135

Hall, R. D., & Rowland, C. A. (2016). Leadership development for managers in turbulent times. *The Journal of Management Development, 35*(8), 942–955. https://doi.org/10.1108/JMD-09-2015-0121

Hambley, L. A., O'Neill, T. A., & Kline, T. J. B. (2007). Virtual team leadership: The effects of leadership style and communication medium on team interaction styles and outcomes. *Organizational Behavior and Human Decision Processes, 103*(1), 1–20. https://doi.org/10.1016/j.obhdp.2006.09.004

Harris, A. (2014). *Distributed leadership matters: Perspectives, practicalities, and potential*. Corwin Press.

Heifetz, R. A., Grashow, A., & Linsky, M. (2009). *The practice of adaptive leadership: Tools and tactics for changing your organization and the world*. Harvard Business Press.

Heifetz, R. A., & Laurie, D. L. (1998). The work of leadership. In *Harvard business review on leadership* (pp. 171–197). Harvard Business Review Press.

Hirschy, M. J. (2011). Virtual team leadership: A case study in Christian higher education. *Christian Higher Education, 10*(2), 97–111. https://doi.org/10.1080/15363751003676613

Hoffer, E. (1973). *Reflections on the human condition*. Hapre & Row.

Hogue, M., & Lord, R. G. (2007). A multilevel, complexity theory approach to understanding gender bias in leadership. *The Leadership Quarterly, 18*(4), 370–390. https://doi.org/10.1016/j.leaqua.2007.04.006

Hollander, E. P. (1958). Conformity, status, and idiosyncrasy credit. *Psychological Review, 65*(2), 117–127. https://doi.org/10.1037/h0042501

Jakavonytė-Staškuvienė, D., & Ignatavičiūtė, L. (2022). Experience of mentors and beginner primary school teachers in applying the principles of shared leadership during the school adaptation period: The case of Lithuania. *Cogent Education 9*(1). https://doi.org/10.1080/2331186X.2022.2070054

Kalinovich, A., & Marrone, J. A. (2017). Shared leadership: A primer and teaching recommendations for educators. *Journal of Leadership Education, 16*(1), 205–215. https://doi.org/10.12806/V16/I1/I1

Kezar, A., Carducci, R., & Contreras-McGavin, M. (2006). *Rethinking the 'L' word in higher education: The revolution of research on leadership: ASHE higher education report*. J. W. Sons.

Khasawneh, S. (2011). Shared leadership and organizational citizenship behaviour in Jordanian public universities: Developing a global workforce for the 21st century. *Educational Management Administration & Leadership, 39*(5), 621–634. https://doi.org/10.1177/1741143211408447

Kocolowski, M. D. (2010). Shared leadership: Is it time for a change? *Emerging Leadership Journeys, 3*(1). https://www.regent.edu/acad/global/publications/elj/vol3iss1/home.htm

Kosterelioglu, M. A. (2017). The effect of teachers' shared leadership perception on academic optimism and organizational citizenship behaviour: A Turkish case [article]. *International Journal of Leadership in Education, 20*(2), 246–258. https://doi.org/10.1080/13603124.2015.1066868
Kownacki, A., Barker, D., & Arghode, V. (2020). A grounded theory approach for exploring shared leadership: Evidence from urban primary schools in Pennsylvania. *International Journal of Leadership in Education, 26*(2), 201–222. https://doi.org/10.1080/13603124.2020.1804622
Kuhn, T. S. (2012). *The structure of scientific revolutions* (50th anniversary edition ed.). The University of Chicago Press.
Lattner, K., & Strehmel, P. (2023). Pandemic-related challenges for team leadership in ECEC centres in Germany. *International Journal of Child Care and Education Policy, 17*(1). https://doi.org/10.1186/s40723-023-00125-4
Leithwood, K., & Mascall, B. (2008). Collective leadership effects on student achievement. *Educational Administration Quarterly, 44*(4), 529–561. https://doi.org/10.1177/0013161x08321221
Lindahl, R. (2008). Shared leadership: Can it work in schools? *The Educational Forum, 72*(4), 298–307. https://doi.org/10.1080/00131720802361894
Lumby, J. (2009). Collective leadership of local school systems. *Educational Management Administration & Leadership, 37*(3), 310–328. https://doi.org/10.1177/1741143209102782
McCauley, C. D., & Palus, C. J. (2021). Developing the theory and practice of leadership development: A relational view. *The Leadership Quarterly, 32*(5), 1–15. https://doi.org/10.1016/j.leaqua.2020.101456
McNulty, E. J., Dorn, B. C., Goralnick, E., Serino, R., Grimes, J. O., Flynn, L. B., Cheers, M., & Marcus, L. J. (2018). Swarm intelligence: Establishing behavioral norms for the emergence of collective leadership. *Journal of Leadership Education, 17*(2), 19–41. www.journalofleadershiped.org/attachments/article/540/2017_0676Flood.pdf
Nazarpoori, A. (2017). A survey of the effect of leader's social intelligence on employee's perception of interpersonal justice: The mediating role of shared leadership. *International Journal of Leadership in Education, 20*(5), 618–631. https://doi.org/10.1080/13603124.2017.1283063
Ni, Y., Yan, R., & Pounder, D. (2018). Collective leadership: Principals' decision influence and the supportive or inhibiting decision influence of other stakeholders. *Educational Administration Quarterly, 54*(2), 216–248. https://doi.org/10.1177/0013161x17723427
Norris, J. (2022). Ninja training meets management education: Integrating Taijutsu into an MBA complexity leadership course. *Journal of Management Education, 46*(5), 951–974. https://doi.org/10.1177/10525629221090343
Obolensky, N. (2014). *Complex adaptive leadership: Embracing paradox and uncertainty* (2nd ed.). Routledge.
Omer, A. (2017). Imagination, emergence, and the role of transformative learning in complexity leadership. *Enlivening Edge*.https://enliveningedge.org/features/imagination-emergence-role-transformative-learning-complexity-leadership/
Pan, H.-L. W., Wiens, P. D., & Moyal, A. (2023). A bibliometric analysis of the teacher leadership scholarship. *Teaching and Teacher Education, 121*, 103936. https://doi.org/10.1016/j.tate.2022.103936
Parsons, K. (2014). *It started with Copernicus: Vital questions about science*. Promethus Books.
Pearce, C. L., & Conger, J. A. (2003). All those years ago: The historical underpinnings of shared leadership. In C. L. Pearce & J. A. Conger (Eds.), *Shared leadership: Reframing the hows and whys of leadership* (pp. 1–18). Sage.
Peter, D. W., Jori, S. B., KaaVonia, H., & Abigail, M. (2024). The complex web of teacher leadership: Examining the relationships between instructional support, shared leadership, and teacher satisfaction. *The Educational Forum, 88*(1), 94–106. https://doi.org/10.1080/00131725.2023.2207106
Pflaeging, N. (2014). Organize for complexity: How to get life back into work to build the high-performance organization..

Pieczka, M., & Casteltrione, I. (2020). AlcoLOLs, re-thinking drinking: Developing a shared leadership approach for alcohol education. *Health Education Journal, 79*(3), 346–361. https://doi.org/10.1177/0017896919883364

Pino-Yancovic, M., & Ahumada, L. (2020). Collaborative inquiry networks: The challenge to promote network leadership capacities in Chile. *School Leadership & Management, 40*(2–3), 221–241. https://doi.org/10.1080/13632434.2020.1716325

Poley, J. K. (1998). Creating shared leadership environments in institutional and international settings. *American Journal of Distance Education, 12*(2), 16–25. https://doi.org/10.1080/08923649809526991

Printy, S. M., & Marks, H. M. (2006). Shared leadership for teacher and student learning. *Theory Into Practice, 45*(2), 125–132. https://doi.org/10.1207/s15430421tip4502_4

Ritchie, S. M., Mackay, G., & Rigano, D. L. (2005). Individual and collective leadership in school science departments. *Research in Science Education, 36*(3), 141–161. https://doi.org/10.1007/s11165-005-9001-6

Sargut, G., & McGrath, R. G. (2011). Learning to live with complexity. *Harvard Business Review, 89*(9), 68–76.

Schlechty, P. C. (2005). *Shaking up the school house, reestablishing schools* (Y. Ozden, trans.). Nobel Yayincilik.

Schophuizen, M., Kelly, A., Utama, C., Specht, M., & Kalz, M. (2022). *Enabling educational innovation through complexity leadership?* Perspectives from four Dutch universities. https://doi.org/10.1007/s11233-022-09105-8

Scott-Young, C. M., Georgy, M., & Grisinger, A. (2019). Shared leadership in project teams: An integrative multi-level conceptual model and research agenda. *International Journal of Project Management, 37*(4), 565–581. https://doi.org/10.1016/j.ijproman.2019.02.002

Senge, P., Hamilton, H. A. L., & Kania, J. (2019). the dawn of system leadership. *Policy & Practice, 19426828), 77(1*, 12–19.

Shoukat, M. H., Shah, S. A., & Muneeb, D. (2023). Shared leadership and team performance in health care: How intellectual capital and team learning intervene in this relationship. *The Learning Organization, 30*(4), 426–445. https://doi.org/10.1108/tlo-12-2021-0146

Simon, H. A. (2019). *The sciences of the artificial [reissue of 3rd ed.]*. The MIT press.

Slantcheva-Durst, S. (2014). Shared leadership as an outcome of team processes: A case study. *Community College Journal of Research and Practice, 38*(11), 1017–1029. https://doi.org/10.1080/10668926.2012.727770

Smith, D. (2022). More pivots than a centipede on ice skates: Reflections on shared leadership in a post-secondary institution during COVID-19. *Canadian Journal of Educational Administration and Policy ,(200)*, 76–95.

Snowden, D. (2021). Cynefin: Weaving sense-making into the fabric of our world. *Cognitive Edge.*

Snowden, D. J., & Boone, M. E. (2007). A leader's framework for decision making. *Harvard Business Review,85*(11), 68–76. www.hrb.org

Spillane, J. (2006). *Distributed leadership.* Jossey-Bass.

Stacey, V. (2022). Shared leadership in higher education: An exploration of the composition of school leadership teams and school performance. *Journal of Higher Education Policy and Management 44*(5), 486–497.https://doi.org/10.1080/1360080X.2022.2109560

Stark, R. (2016). Cultivating shared leadership with docents and staff. *Journal of Museum Education, 41*(1), 22–28. https://doi.org/10.1080/10598650.2015.1134396

Stewart, G. L., & Manz, C. C. (1995). Leadership for self-managing work teams: A typology and integrative model. *Human Relations, 48*(7), 747–770. https://doi.org/10.1177/001872679504800702

Stuart, K. (2018). Value-based relational leadership practice in children's centres: An action research project. *Educational Action Research, 26*(5), 716–735. https://doi.org/10.1080/09650792.2017.1409129

Torres, A. C., Bulkley, K., & Kim, T. (2020). Shared leadership for learning in Denver's portfolio management model. *Educational Administration Quarterly, 56*(5), 819–855. https://doi.org/10.1177/0013161x20906546

Tremblay, C. W. (2012). Network leadership: An emerging practice. *College and University, 87*(4), 37–42.

Tsai, Y. S., Poquet, O., Gašević, D., Dawson, S., & Pardo, A. (2019). Complexity leadership in learning analytics: Drivers, challenges and opportunities. *British Journal of Educational Technology, 50*(6), 2839–2854. https://doi.org/10.1111/bjet.12846

Turchi, A., Greco, S., Morini, E., & Nencioni, P. (2023). Participatory communication and shared leadership: The European project learning leadership for change (L2C). *Research on Education and Media, 15*(1), 120–129. https://doi.org/10.2478/rem-2023-0016

Uhl-Bien, M. (2006). Relational leadership theory: Exploring the social processes of leadership and organizing. *The Leadership Quarterly, 17*(6), 654–676. https://doi.org/10.1016/j.leaqua.2006.10.007

Uhl-Bien, M., & Marion, R. (2009). Complexity leadership in bureaucratic forms of organizing: A meso model. *The Leadership Quarterly, 20*, 631–650. https://doi.org/10.1016/j.leaqua.2009.04.007

Uhl-Bien, M., Marion, R., & McKelvey, B. (2007). Complexity leadership theory: Shifting leadership from the industrial age to the knowledge era. *The Leadership Quarterly, 18*(4), 298–318. http://digitalcommons.unl.edu/cgi/viewcontent.cgi?article=1017&context=leadershipfacpub&sei

Webb, O. D. (2021). Unknown article. *Alberta Journal of Educational Research, 67*(2), 159–177. https://doi.org/10.11575/ajer.v67i2.68603

Webber, C. F., Jodi, N., Sarah, H., & and Braunberger, D. (2024). Contextualised shared leadership: A Canadian case study.*School Leadership & Management*,*44*(3), 318–337. https://doi.org/10.1080/13632434.2024.2303636.

Wiens, P. D., Beck, J. S., Hinton, K., & Moyal, A. (2024). The complex web of teacher leadership: Examining the relationships between instructional support shared leadership and teacher satisfaction. *The Educational Forum 88* (1), 94–106.

Wolfinbarger, K. G. (2022). Team leadership in engineering education. *New Directions for Student Leadership, 2022*(173), 53–61. https://doi.org/10.1002/yd.20479

Wu, Q., Cormican, K., & Chen, G. (2020). A meta-analysis of shared leadership: Antecedents, consequences, and moderators. *Journal of Leadership & Organizational Studies, 27* (1), 49–64. https://doi.org/10.1177/154805181882062

Yammarino, F. J., Dionne, S. D., Chun, J. U., & Dansereau, F. (2005). Leadership and levels of analysis: A state-of-the-science review. *The Leadership Quarterly, 16*, 879–919. https://doi.org/10.1016/j.leaqua.2005.09.002

Yammarino, F. J., Salas, E., Serban, A., Shirreffs, K. A., & Shuffler, M. L. (2012). Collectivistic leadership approaches: Putting the "we" in leadership science and practice. *Industrial and Organizationsl Psychology, 5*, 382–402. https://doi.org/10.1111/j.1754-9434.2012.01467.x

Zaccaro, S. J. & Klimoski, R. J. (2001). The nature of organizaitonal leadership: An introduction. In Zaccaro, S. J. & Imoski, R. J. K. (Eds.), *The nature of organizational leadership* (pp. 3–41). Jossey Bass.

Convertible Leadership in Industry 5.0

Chan Lee and Jae Eun Kim

Abstract This chapter presents Convertible Leadership, a framework for leaders navigating the dynamic demands of Industry 5.0. Rather than adhering to a single leadership style, convertible leaders intentionally shift between directive, collaborative, and coaching approaches based on situational needs and stakeholder expectations. The model is built on seven core capabilities: cognitive flexibility, emotional regulation, relational effectiveness, change readiness, performance consistency, digital proficiency, and systems thinking. Moving beyond traditional top-down control, convertible leaders engage teams in co-creating solutions and share leadership responsibilities fluidly. This approach enables organizations to remain adaptable and resilient while maintaining alignment with core values and ethical principles.

Keywords Convertible leadership · Leadership adaptability · Organizational agility · Leadership development · Strategic flexibility

1 Introduction: The End of Static Leadership

Over the past two decades, industry has witnessed a sweeping transformation driven by automation, big data, robotics, and artificial intelligence. This was the age of Industry 4.0, where efficiency, optimization, and machine-driven decision-making dominated strategic agendas. Yet, in the wake of global crises, climate urgency, demographic shifts, and ethical dilemmas posed by AI, a new paradigm has emerged—Industry 5.0.

Unlike its predecessor, Industry 5.0 is not merely a technological upgrade. It is a philosophical reorientation. At its core, Industry 5.0 is a human-centric, resilient, and sustainable response to the limitations of purely tech-driven systems. It places people, purpose, and planet at the center of organizational design. It values not only productivity, but meaning, empathy, trust, and ethical responsibility.

C. Lee (✉) · J. E. Kim
Department of Vocational Education and Workforce Development, Seoul National University, Seoul, South Korea
e-mail: chanlee@snu.ac.kr

A. M. Waite et al. (eds.), *Leading in Industry 5.0*, Management and Industrial Engineering, https://doi.org/10.1007/978-3-032-19207-3_7

This new era challenges leaders to evolve not only in skill but in identity. Organizations no longer require leaders with a single dominant style—whether authoritative, transactional, or transformational—but rather leaders who can shift fluidly between these modes based on context, culture, and change.

Convertible Leadership, a concept explored in depth in this chapter, proposes a leadership model built on strategic adaptability. Like a convertible vehicle that adjusts its configuration depending on road, weather, and destination, a convertible leader can modify their mindset, behavior, and influence strategies based on dynamic internal and external variables. They do not merely adapt; they redesign themselves in real time.

In the face of Industry 5.0, organizations need more than digital transformation—they require leadership transformation. The question is no longer, "What kind of leader are you?" but "Can you become the leader the moment requires?"

This chapter sets the stage for understanding how Convertible Leadership offers a framework to navigate this complexity. By examining how leaders must transform their perspective, leadership style, and organizational influence, we uncover a path to sustainable, inclusive, and human-centered impact in a rapidly changing world.

2 Why Convertible Leadership Matters Now

As the complexity of organizational life intensifies in the Industry 5.0 era, leaders are no longer called to fit a mold—they are expected to reshape themselves. In this environment, Convertible Leadership emerges as a model not of fixed identity, but of strategic self-redesign.

2.1 Definition

Convertible Leadership refers to the intentional capacity of leaders to adapt their leadership style, mindset, and behavior in response to evolving stakeholder needs, situational demands, and organizational purposes—while maintaining alignment with their core values and ethical principles.

The term "convertible" evokes the image of a car that can transform based on weather, road, or purpose. Likewise, a convertible leader can function in directive, participative, coaching, or visionary modes depending on what the moment demands.

This is not about being inconsistent or indecisive. Rather, it is a deliberate, principled agility—where adaptability is guided by emotional intelligence, strategic clarity, and ethical judgment.

> A convertible leader does not change who they are; they change how they serve.

Hyundai Motor Group's Chairman, Chung Eui-sun, offers a compelling example of Convertible Leadership in action. His long-standing support for Korea's national archery team goes beyond financial sponsorship—it reflects a deep belief in autonomy, trust, and process over outcome. Rather than imposing top-down directives or tying support to gold medals, Chung empowered the athletes and coaching staff to self-direct their routines, innovate their own feedback systems, and even reimagine the traditional reward structures. This leadership approach—firm in values (excellence, consistency), fluid in methods (autonomy, experimentation)—embodies the spirit of Convertible Leadership and shows how strategic adaptability can foster sustainable high performance.

2.2 *The Three Domains of Conversion*

Convertible Leadership unfolds through three sequential yet dynamic domains of transformation: Perspective Conversion, Leadership Behavior Conversion, and Organizational System Conversion. These layers reflect how leaders not only change how they act, but how they think, relate, and design sustainable systems around them.

1. Perspective Conversion

This domain focuses on how leaders reshape their internal frameworks—how they see themselves, others, and the world. It is the foundation for behavioral and systemic shifts.

- Paradigm Flexibility: The ability to shift between different worldviews depending on context.
- Weak Signal Detection: Developing sensitivity to early, often subtle indicators of change.
- Cognitive Agility: Navigating between analytical, intuitive, and systemic thinking modes.
- Self-awareness and Reflection: Recognizing one's assumptions and emotional drivers.

2. Leadership Behavior Conversion

At this level, leaders intentionally shift their leadership style and behavior to align with evolving contexts and stakeholder needs. In a recent internal survey on Convertible Leadership conducted across several Korean corporations, many leaders described themselves as flexible and adaptive. However, when asked whether they could easily shift into follower roles—depending on context and team needs—a noticeable portion expressed discomfort.

This tension suggests a gap between cognitive agility and behavioral convertibility, revealing how traditional leadership expectations may limit practical role fluidity.

- Cultural Code-Switching: Adapting leadership communication across generational, functional, and cultural differences.
- Relational Agility: Engaging with diverse stakeholders in ways that build trust and responsiveness.

- Collaborative Governance: Designing inclusive decision-making environments that empower others.
- Emotional Agility: Managing internal emotional responses to lead with empathy and presence.

3. Organizational System Conversion

The final domain addresses how leaders redesign structures, cultures, and ecosystems for long-term adaptability and resilience.

- Organizational Learning Design: Embedding reflection and feedback into daily operations.
- Commons-Based Innovation: Facilitating shared resources and collective value creation across boundaries.
- Regenerative Structures: Designing organizations that create positive impact for all stakeholders.
- Leverage Point Identification: Recognizing and activating small interventions that catalyze systemic change.

Each of these domains is interdependent. Perspective shifts open the door for behavioral adaptation, which in turn enables leaders to reimagine and co-design their organizational systems. Together, they form a dynamic architecture for navigating the complexity and human-centric demands of Industry 5.0.

2.3 *Why Traditional Models Fall Short*

Many traditional leadership models emphasize consistency of style—whether it be transformational, servant, authentic, or transactional. While each has merit, they often assume relatively stable contexts.

But Industry 5.0 is not stable—it is fluid, paradoxical, and diverse. The real-world leader must orchestrate multiple styles across stakeholders, generations, and geographies.

For example, a leader managing AI engineers, manufacturing teams, and social impact investors simultaneously must navigate vastly different expectations, languages, and values. There is no single style that fits all.

2.4 *Why Convertible Leadership Matters Now*

In Industry 5.0, leaders are not just responsible for performance—they are accountable for people, ethics, and societal value.

Convertible Leadership aligns with this reality by offering:

- Human-centric responsiveness.

- Strategic versatility in complex systems.
- Leadership maturity across emotional, digital, and intercultural dimensions.

It Is Not a Style—It Is a Philosophy of Becoming

But how do leaders develop this convertible capacity? The answer lies in cultivating a specific form of agility—one that serves as the inner engine of transformation.

3 The Agility Framework: Building the Inner Engine of Convertible Leadership

If Convertible Leadership is the vehicle, then agility is its engine. In an age defined by rapid shifts, complex systems, and human-machine integration, a leader's ability to adapt quickly, ethically, and strategically has become mission-critical.

Agility, in this context, goes beyond reactive flexibility. It is a structured capacity for learning, unlearning, and relearning—not only in knowledge but in identity, behavior, and values.

3.1 From Fixed Competence to Dynamic Learning

The accelerating pace of change has dramatically reduced the shelf life of professional competencies. Research from the World Economic Forum indicates that many technical skills become obsolete within five years. However, the meta-skill of adaptive learning—the capacity to continuously evolve one's capabilities—remains evergreen and increasingly valuable.

In Industry 5.0, leadership development is no longer about acquiring a fixed set of competencies. It is about cultivating the muscle of change—what we refer to as leadership agility.

3.2 Seven Dimensions of Leadership Agility

The following seven agility domains form the core capability set of a Convertible Leader, with each illustrated through Korean corporate leadership examples:

1. Cognitive Agility:
    - Ability to switch between different thinking modes: analytical, intuitive, strategic.
    - Helps leaders integrate multiple perspectives when facing complex decisions.

- Example: Naver's innovation approach, balancing data-driven search algorithms with intuitive user experience design, through strategic thinking that combines technical excellence with cultural understanding.

2. Emotional Agility:
    - Self-awareness and emotional regulation in complex social dynamics.
    - Helps leaders respond with empathy rather than reaction.
    - Example: CJ ENM's KCON strategy, creating emotional connections across diverse global audiences by understanding and adapting to local cultural sentiments while maintaining universal appeal.
3. Relational Agility:
    - Navigating diverse relationships across cultures, roles, and generations.
    - Critical for multigenerational and hybrid teams.
    - Example: LG Group's "Inhwa" (人和) philosophy, emphasizing harmonious relationships and human-centered management that bridges different generations, functions, and cultural backgrounds through trust-based collaboration and mutual respect.
4. Change Agility:
    - Psychological readiness to initiate and embrace change.
    - Leaders view change as a source of growth, not a threat.
    - Example: Olive Young's rapid digital transformation under female CEO leadership, making swift strategic pivots during market disruption while maintaining brand authenticity.
5. Performance Agility:
    - Maintaining high standards while adapting methods and approaches.
    - Balancing consistency in outcomes with flexibility in execution.
    - Example: Dunamu's operational efficiency, achieving exceptional productivity per employee while rapidly adapting platform features and user interfaces to meet evolving cryptocurrency market demands.
6. Digital Agility:
    - Competence in working with digital systems, AI, and chatbots.
    - Merging technical literacy with human-centered usage.
    - Example: Samsung Card's systematic integration of digital capabilities as one of five core competencies, embedding digital fluency across the entire organization through comprehensive training programs.
7. Systemic Agility:
    - Ability to identify leverage points within complex systems and design interventions that create transformational rather than incremental change.
    - Building cross-boundary coalitions and measuring systemic health indicators alongside traditional performance metrics.

- Example: Lotte Wellfood's Choco Pie adaptation for Indian markets, redesigning the product with vegetarian-friendly ingredients while understanding local religious and cultural preferences—demonstrating how small innovations can unlock new market ecosystems.

Agility is not about moving fast—it's about moving right.

3.3 *Application in Industry 5.0 Environments*

A manager overseeing AI ethics, remote teams, and circular economy models must be able to shift conversations, reframe problems, and re-engage trust—sometimes all within a single day. Without agility, this leader burns out or becomes rigid. With agility, they convert their approach in real time.

3.4 *Agility Is Trainable*

One of the most promising insights in leadership science today is that agility can be developed. Through reflection, coaching, cross-context experiences, and feedback loops, leaders can increase their conversion readiness.

Organizations that cultivate these seven domains create ecosystems of responsive leadership, rather than a hierarchy of control.

4 Leading for Industry 5.0: A Pillar-by-Pillar Connection

Industry 5.0 is not a mere continuation of Industry 4.0; it is a philosophical shift that repositions people, purpose, and sustainability at the center of industrial progress. At its foundation lie three pillars: human-centricity, resilience, and sustainability.

Convertible Leadership aligns with and amplifies these principles—not by adding new responsibilities to leaders, but by reshaping how leadership itself is defined and enacted.

4.1 *Human-Centricity: Leading with Empathy and Inclusion*

Industry 5.0 demands that leaders design systems not merely for humans but collaboratively with humans as co-creators. This requires building environments that acknowledge and leverage emotional intelligence, cognitive diversity, generational perspectives, and cultural differences as strategic assets rather than challenges to manage.

Convertible Leaders respond by:

- Practicing emotional literacy and deep listening.
- Recognizing when to lead or follow in cross-functional or hybrid teams.
- Co-designing solutions rather than imposing top-down mandates.
- Creating psychologically safe spaces where every voice matters.

The truly human-centered leader knows when to speak, when to step back, and when to amplify others.

4.2 *Resilience: Navigating Complexity with Conversion Readiness*

Resilience in Industry 5.0 is not just about bouncing back—it's about bending without breaking, and learning through uncertainty.

Convertible Leadership fosters resilience by:

- Enabling situational conversion in the face of disruption.
- Training teams to adapt roles fluidly—from leader to follower and back.
- Building relational trust that acts as an emotional shock absorber.
- Designing organizations with built-in structural flexibility.

This kind of leadership doesn't resist change; it invites change with strategic intention.

4.3 *Sustainability: Long-Term Thinking Through Leadership Fluidity*

Sustainability requires a shift from short-term performance to long-term system health. This includes environmental sustainability, but also relational, emotional, and cultural sustainability.

Convertible Leaders act as:

- Stewards of intergenerational vision, balancing today's decisions with tomorrow's impact.
- Cultural architects, shaping norms that support inclusion, equity, and meaning.
- Moral compasses, integrating purpose and profit, ethics and innovation.

Sustainable leadership is not about staying the same—it's about staying aligned.

4.4 Leadership as Role Fluidity: The Followership Imperative

In traditional models, leadership is positional. In Industry 5.0, it becomes relational and rotational.

Convertible Leaders embrace role fluidity by:

- Knowing when to defer, delegate, or disappear so others can lead.
- Creating horizontal networks where leadership circulates.
- Teaching teams that followership is not passive, but a powerful, conscious contribution.

This shift not only humanizes leadership, but democratizes innovation.

5 From Command to Co-design: Leading in Complexity

In a world of accelerating complexity, uncertainty, and interdependence, traditional command-and-control leadership is losing its relevance. Hierarchies are flattening, problems are becoming more adaptive than technical, and innovation emerges more often from the edges than the center.

Industry 5.0 demands not domination, but collaboration. Not rigid roles, but relational responsiveness. This calls for a shift from command to co-design—a move that lies at the heart of Convertible Leadership.

5.1 Beyond the Illusion of Control

Leaders have long been trained to decide, direct, and deliver. These functions were suitable in environments where predictability, repetition, and linearity were the norm.

But in Industry 5.0:

- The problems are often unstructured and systemic.
- The people are diverse in values, culture, and expectations.
- The pace of change outstrips any single person's ability to master it.

In such a world, trying to lead alone is not a sign of strength—it is a strategic weakness.

> Traditional command-and-control paradigms prove inadequate in complex adaptive systems. In such environments, influence through relationship-building, sense-making, and collaborative engagement becomes more effective than hierarchical authority.

5.2 The Leadership → Co-design Shift

Co-design is not merely consensus building or stakeholder engagement. It is the act of shaping the future together, by:

- Distributing leadership across boundaries and roles.
- Inviting multiple forms of intelligence—technical, emotional, experiential.
- Creating space for others to lead, rather than holding power tightly.
- Redefining leadership not as "the one who knows" but as "the one who convenes."

Convertible Leaders excel in this space because they are contextually aware and ego-flexible. They know that sometimes the best way to lead is to ask a better question, or to name the tension in the room.

They know when to speak, when to listen, and when to leave the room.

5.3 Leading with Psychological Safety and Cultural Humility

A precondition for co-design is psychological safety—the belief that one can speak up, disagree, or propose ideas without fear of retribution. Convertible Leaders cultivate this by:

- Being transparent about uncertainty.
- Modeling vulnerability and curiosity.
- Rewarding dissent and encouraging "failure-forward" learning.
- Acknowledging their positional privilege while inviting others into shared ownership.

They also demonstrate cultural humility—the ongoing process of self-reflection and learning across difference.

This makes them especially effective in diverse, global, intergenerational teams, which are becoming the norm in Industry 5.0.

5.4 Rotating Roles: Designing with, Not for

In co-designed systems, leadership is no longer positional—it is rotational. Authority flows to those with insight, not merely title. Leaders become facilitators of clarity, context, and collective ownership.

"In a co-designed system, the best ideas often come from those with the least formal power."

Convertible Leadership encourages:

- Dynamic role-switching between leadership and followership.
- Shared accountability, where everyone is a partial designer.

- Encouragement of localized autonomy, without losing systemic coherence.
- This is how organizations move from being machines to living ecosystems.

5.5 *A Leadership Practice for the Edge*

Complexity doesn't live at the center of the org chart—it lives at the edge: in customer feedback, in employee frustration, in ethical tensions with AI, in generational conflicts.

Convertible Leaders position themselves at the edges. They scan, sense, synthesize, and support. They become bridges between difference and convergence.

Rather than imposing vision from above, they activate vision from within—a distinctly Industry 5.0 leadership act.

5.6 *Convertible Leadership as System Innovation Orchestration*

In the context of Industry 5.0, the European Commission's Community of Practice emphasizes the role of organizations as hubs of system innovation. These hubs serve as orchestrators of transformation—not merely adapting to change but actively shaping it. Convertible Leadership offers a compelling model for this orchestration.

Convertible Leaders act as catalysts across boundaries, disciplines, and institutions. They do not manage change in isolation; they align diverse actors, build shared purpose, and create conditions for emergent solutions. Their agility allows them to operate not only as organizational stewards but also as ecosystem conveners.

Three key capacities of convertible system orchestrators:

1. Creating Learning Ecosystems:
   - Facilitate cross-sector knowledge exchange between industry, academia, civil society, and government.
   - Build governance capabilities that support anticipatory innovation and decentralized experimentation.
   - Promote commons-based resource models that democratize access to knowledge and tools.
2. Enabling Transformative Challenges:
   - Mobilize multi-stakeholder alliances around shared sustainability, well-being, and human-centric objectives.
   - Translate global challenges (climate, inequality, AI disruption) into local innovation opportunities.
   - Integrate open innovation and citizen participation into strategic problem-solving.

3. Demonstrating Alternative Paradigms:
   - Model new enterprise forms such as cooperatives, B-corporations, or steward-owned structures.
   - Embed values of inclusion, regeneration, and co-creation into organizational DNA.
   - Showcase how purpose-driven, human-centric approaches can deliver superior resilience and impact.

As system innovation orchestrators, Convertible Leaders do not merely react to the future. They convene, curate, and co-create it.

6 Strategic and Educational Implications

As Industry 5.0 redefines the very purpose of organizations—from maximizing output to generating shared value—leadership cannot be developed in the same way. Traditional competency models, static leadership pipelines, and one-size-fits-all programs fall short in cultivating the dynamic, self-converting leaders the future demands.

Convertible Leadership invites a fundamental redesign of how we identify, develop, and deploy leadership across the system.

6.1 Leadership Development Must Shift from Skill Acquisition to Identity Transformation

Leadership is not merely about knowing more—it is about becoming different. The development of a convertible leader involves:

- Encouraging deep self-awareness, not just 360-degree feedback.
- Creating experiential learning zones across functions, cultures, and generations.
- Supporting leaders through identity dissonance, not avoiding it.
- Emphasizing reflective practice over technical certification.

"Effective leadership development transcends traditional curriculum-based approaches. It requires experiential learning environments that serve as crucibles for transformation—spaces where leaders can safely experiment with new identities, receive real-time feedback, and reflect deeply on their evolving practice."

6.2 *Organizations as Learning Ecosystems*

To enable Convertible Leadership to flourish, organizations must evolve into learning ecosystems—environments that support continuous transformation across individuals, teams, and systems. This transition requires strategic shifts in structure, culture, and evaluation practices.

Structural Redesign:

- Move from rigid hierarchies to fluid, networked collaboration models.
- Establish regenerative governance structures that allow decision-making to adapt at multiple levels.
- Create shared knowledge platforms that support collective intelligence across functions and boundaries.

Cultural Evolution:

- Encourage intelligent risk-taking and experimentation as part of daily work.
- Build a culture of reflection and mutual learning, where feedback loops are normalized.
- Cultivate a proactive mindset for sensing, interpreting, and responding to future-oriented signals.

Measurement Reimagined:

- Integrate human-centric, sustainability, and resilience metrics alongside performance indicators.
- Track learning capacity through indicators such as team adaptability, participation in co-creation, and psychological safety.
- Use data not only to measure output, but to shape culture and decision-making.

Some organizations are even exploring crypto-economic models and blockchain-based reputation systems to create more transparent and participatory feedback mechanisms, though these approaches are still in experimental stages within traditional corporate structures.

HR systems must also transform:

- Shift from static performance management to agility and growth tracking.
- Replace fixed leadership pipelines with adaptive capability networks.
- Embed Convertible Leadership principles into leadership development, coaching, and talent mobility strategies.

By designing organizations as living, learning systems, Convertible Leadership becomes scalable, sustainable, and regenerative.

6.3 System Model for Convertible Leadership

Drawing from Industry 5.0's core principles—Human-centricity, sustainability, and resilience—the System Model for Convertible Leadership (Fig. 1) offers a dynamic framework for assessing and developing leadership capabilities. Unlike static competency models, this framework structures leadership development through a systems-thinking lens, presenting leadership as a set of evolving capabilities shaped by context.

This model adapts Industry 5.0 principles into a System-based Model composed of three categories: input, process, and output across three leadership dimensions. It guides organizations in designing interventions and tracking the growth of leadership agility and adaptability over time.

Human-centricity emphasizes internal emotional awareness and interpersonal sensitivity as foundational inputs, supported by processes that promote psychological safety. These conditions enable higher outputs such as engagement and shared ownership of leadership.

Sustainability requires future-facing vision as an input, integrated through operational processes that embed purpose into day-to-day execution. The result is not just alignment but mission execution—leaders who act decisively in service of long-term goals.

Resilience stems from signal detection—a leader's ability to sense weak signals in turbulent environments. Through processes that enhance adaptive capacity, this domain culminates in decisive activation—the readiness and will to act swiftly and strategically in uncertainty.

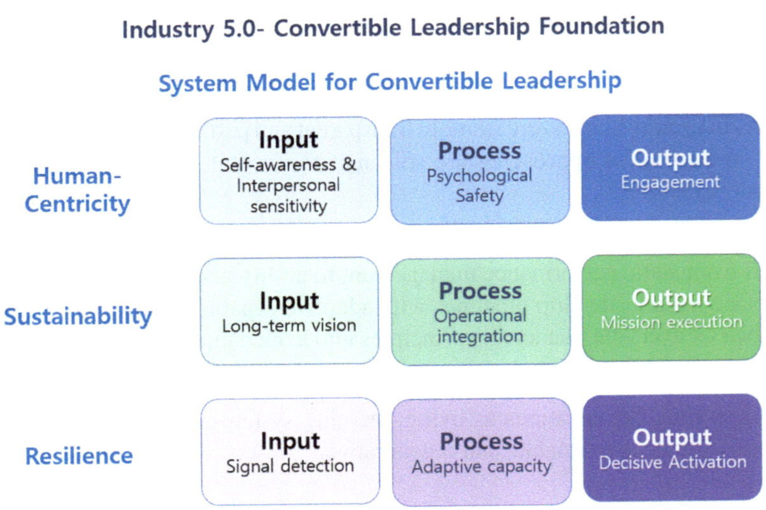

Fig. 1 System model for convertible leadership

This model allows for phased assessment of a leader's development path—from reactive rigidity to full convertible agility. It supports evaluation, design, and feedback loops within organizational learning systems.

Following this system framework, leadership development must evolve beyond static curricula toward dynamic, real-time capability building:

Blend interdisciplinary insight: Integrate technology, psychology, systems thinking, and ethics in leadership development.

Simulate high-stakes ambiguity: Use scenario-based learning to prepare leaders for complex, uncertain environments.

Develop sense-making, not just sense-giving: Encourage leaders to interpret and navigate complexity, not merely deliver instructions.

Leverage advanced tools: Incorporate generative AI and other digital technologies as reflective collaborators, not just efficiency boosters.

As such, the System Model for Convertible Leadership offers a structured approach to operationalizing Industry 5.0 principles within contemporary leadership paradigms, enabling organizations to cultivate adaptability, purpose, and systemic impact.

6.4 The Role of Middle Managers as Conversion Catalysts

Middle managers often feel squeezed—under pressure from above and pulled by teams below. But in a Convertible Leadership model, they are reframed as conversion catalysts:

- Facilitating sense-making between strategic vision and on-the-ground reality.
- Acting as adaptive shock absorbers during organizational change.
- Leading with horizontal influence, not vertical authority.

Investing in this group has the highest return for agility, retention, and engagement.

6.5 Ethics and Purpose as Anchors in a Convertible World

Amid all this fluidity, Convertible Leadership must not devolve into opportunism or shapelessness. The center must hold. That center is ethical clarity and purpose alignment.

- Education and organizational strategy must ask:
- What values remain unchanged, even as methods evolve?
- What long-term commitments are we willing to prioritize over short-term wins?
- How do we ensure human dignity is not sacrificed in the name of transformation?

In this way, convertible leadership becomes a tool for strategic morality, not just strategic success.

7 Conclusion: Leadership as a Moving Form

Industry 5.0 is not a destination—it is a direction. It asks us not only to integrate new technologies, but to reimagine what it means to be human in systems of production, power, and purpose. And in this evolving landscape, leadership is no longer a title, a style, or a set of techniques. It is a moving form.

Convertible leadership offers a compass, not a map. It provides a philosophy of becoming, rooted in agility, guided by ethical purpose, and responsive to complexity. It recognizes that leadership today is not about having all the answers, but about becoming the kind of person who can live the questions with courage and clarity.

7.1 The Paradox of the Modern Leader

Leaders must now:

- Be firm in values, but fluid in action.
- Hold power, while distributing agency.
- Move fast, but listen slowly.
- Think systemically, but act locally.
- Lead with authority, but follow with humility.

This is not a contradiction—it is a competency. And it is trainable.

7.2 From Transformation to Transcendence

In an era of exponential change, what if leadership is not merely about transformation, but transcendence?

- Transcending roles to inhabit purpose.
- Transcending fear to enact responsibility.
- Transcending ego to serve something larger than self.

Such leaders do not just navigate Industry 5.0—they shape it.

7.3 Final Reflection: A Question to Carry

As you leave this chapter, take this question with you—not to answer it immediately, but to live with it:

> In the moments that matter most, can I become the leader the moment needs?

The world does not need more perfect leaders. It needs more present, responsive, and ethical ones.

To lead in Industry 5.0 is to lead in motion.

To be convertible is to be deeply, unshakably human.

Effectively Leading in Industry 5.0: A Holistic Approach

Leadership Strategy for Building Industry 5.0: Human-centric Societal Organizations

Jeffrey Moore and William Hanson

Abstract Leadership strategies play an essential role in operationalizing Industry 5.0 by prioritizing human-centricity, sustainability, and resiliency. We use the European Commission's Industry 5.0 framework, the Burke–Litwin Model, and Complexity Leadership Theory to show what is required to drive transformation. These models reveal how mission, culture, and leadership must align to drive the transformation. Traditional organizational top-down approaches are no longer effective in today's volatile and complex environments. Emerging is a new employee-centric leadership that transforms cultures; it empower employees by encouraging flexibility, strengthening engagement, and fostering collaboration. These organizations become adaptive learning systems. They are dependent on employees' engagement to solve business problems and navigate constant change. A case study of a North American manufacturing plant demonstrates how employee-centric leadership enabled alignment with Industry 5.0. It demonstrates how employee-centric leadership creates cultural change, employee empowerment, and community integration. Findings reveal practicing shared leadership, cultivating open communication, and living out core values created a more resilient, connected, sustainable, and thriving workplace. Industry 5.0 ultimately requires organizations to become adaptive systems where human-centricity and business operations integrate. The result is a human-centric societal organization.

Keywords Industry 5.0 · Human-centric leadership · Employee engagement · Cultural transformation · Thriving workplace

J. Moore (✉) · W. Hanson
College of Business and Economics, Anderson University, Anderson, SC, USA
e-mail: jmoore@andersonuniversity.edu

A. M. Waite et al. (eds.), *Leading in Industry 5.0*, Management and Industrial Engineering, https://doi.org/10.1007/978-3-032-19207-3_8

1 Introduction: Contextualizing the Shift to Adaptive Organizational Leadership Models

Back in the industrial days, bosses ran the show like master planners, splitting up work and directing people like pieces on a chessboard to churn out products or services as cheaply and efficiently as possible. This old-school, top-down method is all about control of tasks and the management of workers, with strict hierarchies calling the shots to keep costs low and profits high (Giannantonio & Hurley-Hanson, 2011; Taylor, 1967). But by the late 1900s, this approach became less effective and did not keep companies competitive in a fast-changing world. Changing methods or innovating new solutions were slow; adapting to new environments was untested. Workers, especially younger ones, want more than a paycheck; they crave a workplace that values their ideas and builds real connections. This makes employee-focused leadership key to keeping people happy, sparking new ideas, and helping companies adapt to change.

As the industrial models began to lose steam, there arose the information age, shaking up old standards and expectations with intense competition and new technology: this changed the nature and priorities of jobs overnight (Gleick, 2011; Nicholson, 1996; Wu, 2011). New technology increased access to data and it's processing which enabled organizations to obtain and analyze large datasets, thus improving decision-making. Technological advances also improved operational efficiencies and increased responsiveness to market changes and digital communications. In turn, these changes reshaped organizational interactions and workflows (Brynjolfsson & Hitt, 2000; Leonardi, 2011; Sambamurthy et al., 2003). Today's workplaces are turbulent and full of surprises, uncertainty, and complexity, often referred to as Volatile, Uncertain, Complex, and Ambiguous (VUCA); it includes conditions created from competitors, governmental regulation, natural disasters, and unpredictability of market forces. Even more recent methods seemed to have lost their effectiveness in today's environment. To remain viable, companies need to be adaptive, embracing new leadership models that deal with global changes (Arena & Uhl-Bien, 2016; Bennett & Lemoine, 2014;Collins, 2002; Uhl-Bien et al., 2007). To that end, later we will introduce a newer model termed Complexity Leadership, addressing how organizations can nurture factors to become and remain adaptive to changing conditions.

Normally, changes emerge from various sectors affecting business and business responds in kind (Aguilar, 1994; Porter, 2008). But sometimes, the force for change emerges from governmental actions. One example is the recent move by the European Commission to nurture industry that is adaptable, inclusive, and aligned with the needs of society (Breque et al., 2021; Nahavandi, 2022). The European Commission's Industry 5.0 Report provides our starting point for understanding the leadership strategy necessary to creating human centric, sustainable and resilient organizations. Industry 5.0 begins with a human-centric concept of the integration of business and society. It recognizes the importance of organizational innovation, sustainability and adaptation, not only for business to compete, but for industry living in harmony with society. One such alignment between industry and society is Walgreens' disability employment initiative and their collaboration with societal support systems (Moore & Hanson, 2023). Their pro-social partnerships saw local non-profits and government agencies work together to provide public transportation to the workplace, recruitment of job candidates, onboarding and on the job training (Maxey & Moore, 2017; Moore et al., 2020; Walgreens, n.d.).

To elaborate some key aspects of the Commission's report, we use the Burke–Litwin Model (BLM) to identify related strategic organizational elements that are essential to create long-term change (Burke, 2018; Burke & Litwin, 1992). In particular, we will discuss how culture needs to be aligned with the strategic and mission-oriented elements, critical for successful transition to Industry 5.0. We will analyze these elements using a Complexity Leadership Theory (CLT) theoretic lens, discussing how culture transformation is operationalized. Finally, we will conclude with a vignette from a manufacturing organization that successfully changed its outdated management culture to be human centric, modeling Industry 5.0.

2 Leading the Future: Change Models and Leadership for Industry 5.0

Industry 5.0 seeks to shift European focus from exclusive efficiency and productivity key performance indicators (KPI) to comprehensively integrating human-centricity, sustainability, and resilience across all industrial operations (Mourtzis et al., 2022). This new paradigm emphasizes a value-driven approach, integrating collaboration between government, business and society, for human betterment. It allows integration of advanced technologies such as Artificial Intelligence and collaborative robots (cobots) to enhance creativity and well-being. This includes promoting eco-friendly practices like circular economy and energy efficiency and ensuring adaptable operations to withstand disruptions (European Commission, 2020). Key elements of this vision include leading digital and environmental transitions, investing strategically in skills development, and actively promoting talent, diversity, and empowerment within the workforce. This leaves businesses with a tall order.

Figure 1 provides a visual representation of how Industry 5.0, BLM and CLT interact for human-centric organizational transformation. Beginning at the base of Fig. 1, we start with the Industry 5.0 report which sets the vision for integration between industry and society (European Commission, 2020). Next the BLM is the change model necessary to identify key organizational elements needing focus to bring about the necessary change. Finally, CLT is the leadership model that operationalizes how managers can practically lead in a different way to bring about an employee-centric organization.

ALText. Conceptual diagram showing how Industry 5.0, the Burke–Litwin Model, and Complexity Leadership Theory align for organizational transformation. Industry 5.0 provides the vision for integrating business and society. The Burke–Litwin

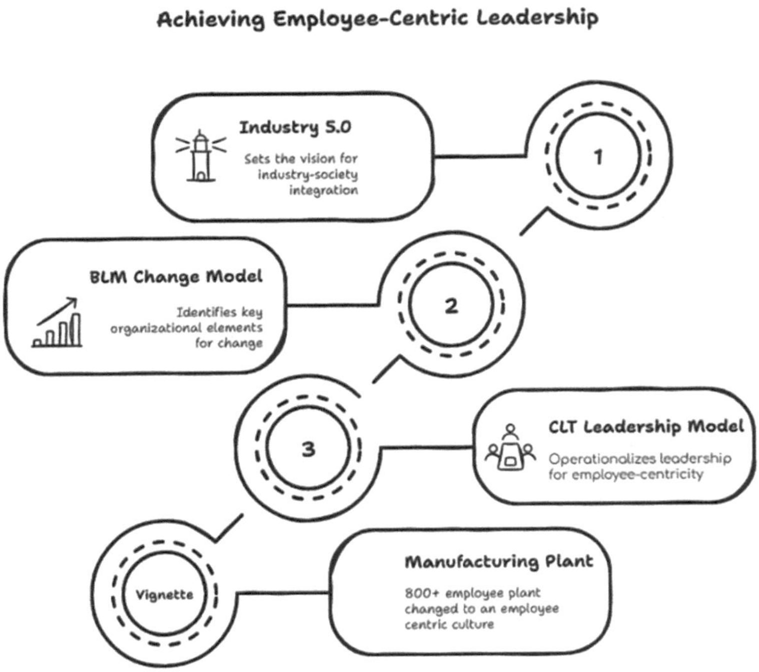

Fig. 1 Organizational transformation for Industry 5.0

Model identifies organizational elements that need change. Complexity Leadership Theory demonstrates how leaders implement these changes to build human-centric organizations.

Changing company culture to embrace an employee-centric approach is a serious issue because it transforms the workplace from a rigid, top-down setup into a collaborative, flexible environment. This is like moving from a micromanaging boss to a team leader who encourages everyone to contribute ideas. It is participative based and allows greater adaptability. Companies must have closely tied strategies to change their culture and develop an adaptive organization. These strategies would be the end-state for each major effort. For example, important goal mentioned in the Industry 5.0 report would include sustainability; the strategy might include the activities and initiatives required to achieve it. It must be co-developed between stakeholders such as managers, employees and requisite community partners in order to last. Strategies must directly align with the company's primary purpose or mission (Burke & Litwin, 1992).

An employee-centric culture offers key benefits. It promotes adaptability, empowering teams to respond quickly and creatively to challenges, like solving tech issues during a major project. This flexibility allows businesses to thrive in unpredictable environments by granting employees more leeway and authority in solving problems, as well as co-creating policies that support human development. It also boosts employee satisfaction by making individuals feel valued and supported through active listening, flexible work policies, and mental health resources. It reduces burnout and encourages innovation as second nature aspect of work. Such a culture fosters resilience, encouraging collaboration, consensus, idea creation, organizational learning, interdependence and bottom-up managerial support (Uhl-Bien et al., 2007; Uhl-Bien & Marion, 2009; Senge, 1990).

This cultural shift empowers workers by fostering adaptability, boosting engagement, and promoting teamwork, aligning with Industry 5.0's focus on people-first workplaces that thrive in a chaotic, ever-changing world. To thrive in this landscape, organizations must evolve into adaptive learning systems, demonstrating the agility required to navigate these dynamic challenges effectively (MacGregor, 1978). These cultural shifts require a particular leadership strategy for building human-centric societal organizations.

3 Leadership Strategy for Long-Term Change to Human-Centricity

Central to this leadership strategy needed to create long-term human-centric change we utilize a change model. As we mentioned earlier, the Burke–Litwin Model (1992) explains how different parts of a company work together when big changes are needed. Created by Warner Burke and George Litwin, this model points out important elements of a company, splitting them into two groups: transformational factors, which are shaped by what's happening outside the company, and transactional factors, which are about the day-to-day operational activities inside the company. Consideration of these factors contributes to the success rate of organizational change.

Transformational factors are key drivers that challenge a company to change. The organization must align with the external environment. It must develop an updated mission and strategy, leadership model, and culture, one inseparable from the other. The outside world (external environment), like new tech or customer demands, pushes the company to rethink its game plan (mission and strategy). Burke and Litwin (1992) describe the business environment as chaotic, involving a "confluence of processes that are extremely difficult to predict and almost impossible to control" (p. 523). This is an environmental norm rather than an exception. Leaders, responding to the external environment, step up with a new direction and strategy, bringing everyone together in a new direction. However, when leaders choose a new direction, they need employee buy-in for success. The company's culture (its shared values and sentiments) either supports the change or fights it. For example, when Netflix faced streaming competition (external environment), it shifted its mission from DVDs to online content (Hastings & Meyer, 2020). The company succeeded because leaders rallied employees and created a culture of innovation to make it stick.

In order to create long-term change within an organization its transformational factors must be aligned with its transactional factors. Thus, transformational factors of organizational culture must be aligned with the mission and strategy to bring about a change in human-centricity. In the context of Industry 5.0, this requires redefining mission statements to explicitly prioritize human well-being, resilience, and sustainability. Strategies must integrate these values into concrete initiatives. Burke and Lewin (1992) stress clarity of purpose and direction where human-centricity is not just a good thing but an essential part of the new organization. Through collective planning, mission and strategy need to align with external forces and adapt to societal needs. However, strategy is not a distant concept disconnected from real life but an integral part of guiding day-to-day operations in order to achieve long-term objectives.

While the BLM provides a change model for the organizational system, Complexity Leadership Theory (CLT) offers a practical leadership model for making it happen. CLT comes from research on Complex Adaptive Systems (CAS) where these systems are self-adaptive. These systems vary, composing a group of things—people, technology, or even machines—that work together, learn from each other, and change their behavior based on what's happening around them (Holland, 1995). Think of a CAS like a big dance party. Each dancer (agent) follows simple steps, but together they create an awesome dance routine (emergence) without a choreographer bossing them around. If someone trips, others keep dancing, and the party goes on (resilience)! From this work, Complexity Leadership Theory (CLT) emerged showing how leadership can create an environment where people, ideas, technology and machines can collectively adapt to achieve goals, learn, innovate, and handle challenges (Marion, 2008; Uhl-Bien et al., 2007; Uhl-Bien & Marion, 2009).

CLT breaks leadership into three functions that blend together to help companies handle a constantly moving, fast-changing world. An employee-centric culture thrives through adaptive, enabling and administrative leadership styles. **Adaptive leadership** encourages front-line employees to generate creative solutions, promoting innovation and teamwork, like baristas helping design a new drink menu. **Enabling leadership** champions collaboration by breaking down silos and fostering open communication, allowing departments to co-create and learn from mistakes, like IT and marketing teams sharing app launch tips in a group chat. The *enabling* leaders remove barriers to employeeEmployee goals and champion innovative ideas to company executives (Uhl-Bien et al., 2007). These leaders also support the adaptive function—ensuring employees can innovate and create solutions to front-line issues formally and informally. Together, these elements create a dynamic workplace where people feel energized, engaged, and ready to contribute. The result is a more resilient, innovative, and people-first organization that's prepared to adapt and thrive in the rapidly evolving world of work. **Administrative leadership** supports these efforts by simplifying rules, aligning resources with team goals, and reinforcing the organization's purpose—ensuring good ideas, like a recycling initiative, are supported and sustained. Through a complexity leadership lens, the *administrative* function of business focuses on a support role. Staff provides support to the front line, to include coordination of resources and activities, as well as creating human-centric policies. These leaders nurture employee success via developmental training and co-creation of company values and rules. Together, these approaches build a flexible, empowered, and united workforce. By juggling these three roles, leaders help their company roll with the challenges, navigating stormy markets, by encouraging teamwork and smart ideas. They create a powerful, thriving organization, quickly adaptive to environmental challenges. They instill employees with a sense of flourishing and well-being, ensuring both the company and its people grow stronger together.

Achieving long-term human-centric change requires the integration of a clear change model and a practical leadership strategy. The Burke–Litwin Model provides the structural lens to understand how transformational and transactional factors must align, ensuring mission, strategy, culture, and leadership move in concert toward human-centricity. Complexity Leadership Theory then translates this alignment into action, equipping leaders to foster adaptability, collaboration, and innovation in rapidly shifting environments. We now turn to a practical example of an 800 employee manufacturing plant that accomplished this change.

4 Practical Example Industry 5.0—Transformation of an Automotive Manufacturing Plant

The case study of a manufacturing plant in the Southeastern part of North America embodies Industry 5.0's human-centric pillar by prioritizing the well-being and empowerment of the workforce (Moore & Hanson, 2022). Prompted by issues like high turnover and low satisfaction, the intervention directly addressed a failing sense of employee value and respect. By promoting a complexity leadership approach, the initiative fostered an environment where workers were empowered to participate in decision-making and self-organize, moving away from autocratic styles. With assistance in the revision of organizational strategy and mission, supervisors shifted from controlling figures to supportive facilitators, leading to increased sense of union with the organization. By revising company values with employee participation, leaders contributed to employee involvement and the creation of an "employment bridge" for professional development. These outcomes of shared decision-making increased employee engagement, improved job satisfaction and reduced turnover. Shared buy-in also served as a competitive advantage in attracting talent, directly reflecting Industry 5.0's goal of placing the worker at the heart of production.

Furthermore, the case study illustrates key elements of Industry 5.0's resilience focus through its promotion of human-centric adaptability and agile organizational structures. The application of Complexity Leadership Theory facilitated the emergence of an adaptive organization that relied on employee participation, empowering employees to innovate internally, and participate in decision-making crucial for responding to environmental changes.

This cultural shift, moving from a top-down to a participative approach, enhanced the plant's ability to achieve worker buy-in and develop more effective, team oriented solutions, thereby increasing its innovation and resilience in solving new challenges. The strengthening of values and behaviors fostered teamwork, shared problem-solving, and increased trust, all contributing to the organization's capacity to remain united and prosperous during challenging times. Furthermore, the company integrated external factors through increased community outreach and sponsored fundraisers for local charities, as well as planting trees in the local town. University internships were developed as a result of the culture transformation, allowing young talent to participate in the growing economic benefits of the company, further increasing the sustainability and investment in society.

To drive meaningful change across the organization, as seen in Fig. 2, the plant invested in leadership development and revised organizational values that focused on empathy, collaboration, and employee empowerment. Through hands-on training and culture shifts, leaders began to prioritize trust-building and inclusivity, paving the way for a more engaged and resilient workforce. Employees became co-creators in problem-solving committees and a formal part of company all employee quarterly meetings. Furthermore, they joined a culture change committee that worked on plant crises and reported directly to the General Manager of the plant.

ALText. Framework illustrating culture transformation in a manufacturing plant toward human-centricity. The model highlights leadership training that develops empathy and trust, adaptive leadership empowering front-line employees, enabling leadership fostering co-creation, and administrative leadership supporting staff with clear systems. It also shows sustainability projects and community outreach as reinforcing practices for a human-centric culture.

Each aspect of the model was critical for organizational human-centricity. We elucidate each one of these here (Moore & Hanson, 2022):

- Through **leadership training** managers were trained to understand how their behavior impacted others. They practiced empathy, listening, and respectful communication. Twenty hour scenario-based learning workshops were developed where real-life workplace situations were used to help leaders rehearse better responses to conflict, giving good feedback, creating positive team dynamics and how to coach instead of dictate. Leaders began seeing their role not just as enforcing rules, but as building trust and supporting their team's growth. This was needed to prepare leaders for human-centric change.
- The plant embraced **Participative** (adaptive leadership) leadership by giving front-line workers more voice in solving problems. Instead of just taking orders, employees were encouraged to share ideas and help shape daily operations, like choosing better meeting spaces or suggesting ways to recognize hard work. This empowered them to try new things and feel part of the solution. Front-line employees became part of a culture committee tasked with understanding the different workplace issues across departments and shifts within the plant. Employees felt heard and valued which was essential in moving the culture toward Human-centric change.
- **Co-creator and Champion of Solutions** (enabling leadership) by breaking down barriers between managers and employees, creating safer spaces for open conversations, and encouraging mutual respect. Front-line workers were invited to join problem-solving teams selected from the culture committee. The General Manager (GM) assigned one of his directors to be the resource for each of the

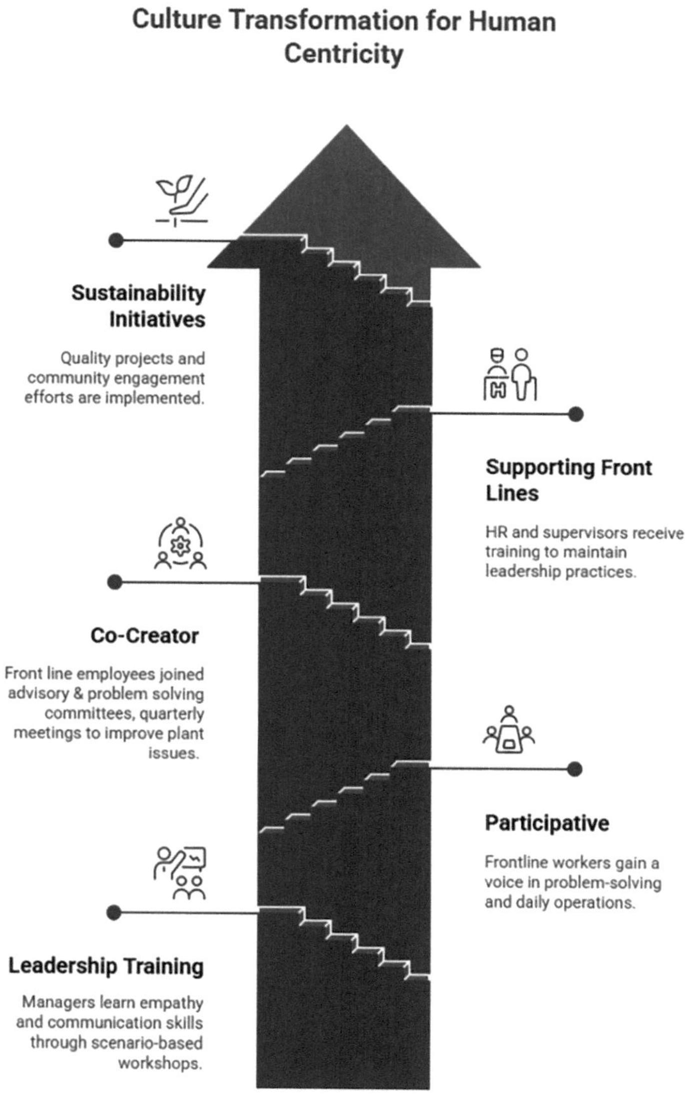

Fig. 2 Culture transformation for human-centricity

six teams tackling one of the six plant issues. The culture in the plant dramatically changed based on the intentional support and investment of the GM, further moving to a human-centric culture.

- **Supporting the Front Lines** (administrative leadership) became a manager focus as they devoted time for meetings with front-line employees and their training. They clarified objectives by setting clear expectations and integrating them into workplace systems. The human resource team underwent training so they could continue after the training consultants left. They didn't just talk about respect; they built it into the job by training front-line supervisors and managers throughout the plant. Leadership training became a tool for building trust, allowing people to learn from mistakes, be flexible, and work together in new ways. This approach focused on encouraging respect and caring by training managers to treat employees with dignity, listen to their concerns, and offer constructive feedback which anchored the human-centric culture in the organization.

- **Sustainability** New cross departmental quality projects by engineering with front-line workers reduced waste by reusing scrap, studied cycle time of machines to reduce energy consumption and raw materials. Furthermore, fundraisers were hosted by Human Resources on the premises for local charities quarterly, blood banks come at the facility for blood donations, and reforestation efforts were made to beautify nature. University internships were revitalized by the Engineering department and where managers were proud to introduce their department to aspiring young interns, increasing their pro-social behaviors. Investing in the well-being of others in the organization and outside the organization is essential to becoming a human-centric societal organization.

The results of applying complexity leadership in the manufacturing plant were both positive and transformative. After the training and cultural interventions, supervisors began showing more respectful and supportive behaviors, such as holding team meetings in quieter spaces to encourage participation and recognizing employee achievements more frequently. These changes helped rebuild trust and made employees feel more valued and heard. This shift in leadership focused on building trust, emotional intelligence, and shared problem-solving. As a result, the organizational culture began to transform—from one that was toxic and production-focused to one that was more respectful, inclusive, and aligned with the company's core values.

5 Conclusion: Organizations Prioritizing Employee Well-Being Are Transformed

In conclusion, the transformation seen in this case study highlights how the Burke–Litwin Model and Complexity Leadership Theory can work together to bring real cultural change to the workplace. As this paper shows, Complexity Leadership Theory aligns with Industry 5.0 paradigm. By shifting from a top-down, production-focused approach to one that values people, the organization began to reflect the goals of Industry 5.0. A future where human well-being, creativity, and adaptability are just as important as technology. This cultural shift empowers workers by fostering adaptability, boosting engagement, and promoting teamwork, aligning with Industry 5.0's focus on people-first workplaces that thrive in a chaotic, ever-changing world. To thrive in this landscape, organizations must evolve into adaptive learning systems that are employee centric, demonstrating the agility required to navigate these dynamic challenges effectively. Through shared leadership, open communication, and a focus on values in action, this company became an example of how modern leadership can lead to a stronger, more connected workplace. It is important to note that these transformations, in the case study (Moore & Hanson, 2022), and another involving a Walgreens distribution center (Moore et al., 2020), produced greater productivity than before their adoption of these attributes.

References

Aguilar, F. J. (1994). *Managing corporate ethics: Learning from America's ethical companies how to supercharge business performance*. Oxford University Press.

Arena, M. J., & Uhl-Bien, M. (2016). Complexity leadership theory: Shifting from human capital to social capital. *People & Strategy, 39*(2), 22–27.

Bennett, N., & Lemoine, J. (2014). What VUCA really means for you. *Harvard Business Review, 92*.

Breque, M., De Nul, L., & Petridis, A. (2021). Industry 5.0: Towards a sustainable, human-centric and resilient European industry. *European Commission, Directorate-General for Research and Innovation*. https://doi.org/10.2777/308407

Brousseau, K. R., Driver, M. J., Eneroth, K., & Larson, R. (1996). Career pandemonium: Realigning organizations and individuals. *Academy of Management Executive, 10*(4), 52–66. https://doi.org/10.5465/AME.1996.3145319

Brynjolfsson, E., & Hitt, L. M. (2000). Beyond computation: Information technology, organizational transformation and business performance. *Journal of Economic Perspectives, 14*(4), 23–48. https://doi.org/10.1257/jep.14.4.23

Burke, W. W. (2018). The rise and fall of the growth of organization development: What now? *Consulting Psychology Journal: Practice and Research, 70*(3).

Burke, W. W., & Litwin, G. H. (1992). A causal model of organizational performance and change. *Journal of Management, 18*(3), 523–545.

Collins, J. C. (2002). *Built to last* (3rd ed.) HarperBusiness.

European Commission: Directorate-General for Research and Innovation. (2020). *Industry 5.0—Human-centric, Sustainable and Resilient*. Publications Office. https://data.europa.eu/doi/10.2777/073781

Giannantonio, C. M., & Hurley-Hanson, A. E. (2011). Frederick Winslow Taylor: Reflections on the relevance of the principles of scientific management 100 years later. *Journal of Business and Management, 17*(1), 7–10. https://doi.org/10.1504/JBM.2011.141187

Gleick, J. (2011). *The information: A history, a theory, a flood*. Vintage.

Hastings, R., & Meyer, E. (2020). *No rules rules: Netflix and the culture of reinvention*. Random house.

Hurley-Hanson, A. E., Giannantonio, C. M., & Dehning, B. (2011). Emerging issues of management education in the 21st century.

Holland, J. H. (1995). *Hidden order: How adaptation builds complexity*. Basic Books.

Leonardi, P. M. (2011). When flexible routines meet flexible technologies: Affordance, constraint, and the imbrication of human and material agencies. *MIS Quarterly, 35*(1), 147–167. https://doi.org/10.2307/23043493

MacGregor, B. J. (1978). *Leadership*. Torchbooks.

Marion, R. (2008). Complexity theory for organizations and organizational leadership. In M. Uhl-Bien & R. Marion (Eds.), *Complexity leadership, part 1: Conceptual foundations* (pp. 1–15). Information Age Publishing.

Maxey, E., & Moore, J. (2017). Impetus for culture transformation: Pre-hire training for employees with disabilities. In S. Frasard & F. C. Prasuhn (Eds.), *Training initiatives and strategies for the modern workforce* (pp. 119–130). IGI Global. https://doi.org/10.4018/978-1-5225-1808-2.ch006

Moore, J., & Hanson, W. (2022). Improving leader effectiveness: Impact on employee engagement and retention. *Journal of Management Development, 41*(7/8), 450–468.

Moore, J., & Hanson, W. (2023). Sephora—Journey to inclusive workplace—Let us belong philosophy (chapter 13). In J. Ingold & P. McGurk (Eds.), *Employer engagement: Making active labour market policies work*. Bristol University Press.

Moore, J., Maxey, E. C., Waite, A. M., & Wendover, J. D. (2020). Inclusive organizations: Developmental reciprocity through authentic leader-employee relationships. *Journal of Management Development, 39*(9/10), 1029–1039. https://doi.org/10.1108/JMD-05-2019-0211

Mourtzis, D., Angelopoulos, J., & Panopoulos, N. (2022). A literature review of the challenges and opportunities of the transition from industry 4.0 to society 5.0. *Energies, 15*(17), 6276. https://doi.org/10.3390/en15176276

Nahavandi, S. (2022). Industry 5.0—A human-centric solution. *Sustainability, 14*(3), 1287. https://doi.org/10.3390/su14031287

Nicholson, N. (1996). Career systems in crisis: Change and opportunity in the information age. *Academy of Management Executive, 10*(4), 40–51. https://doi.org/10.5465/ame.1996.3145318

Porter, M. E. (2008). The five competitive forces that shape strategy. *Harvard Business Review, 86*(1), 79–93.

Sambamurthy, V., Bharadwaj, A., & Grover, V. (2003). Shaping agility through digital options: Reconceptualizing the role of information technology in contemporary firms. *MIS Quarterly, 27*(2), 237–263. https://www.jstor.org/stable/30036530

Senge, P. (1990). Peter Senge and the learning organization. *Dimension, 14*.

Taylor, F. W. (1967). *The principles of scientific management*. Norton.

Uhl-Bien, M., & Marion, R. (2009). Complexity leadership in bureaucratic forms of organizing: A meso model. *Leadership Quarterly, 20*(4), 631–650.

Uhl-Bien, M., Marion, R., & McKelvey, B. (2007). Complexity leadership theory: Shifting leadership from the industrial age to the knowledge era. *Leadership Quarterly, 18*(4), 298–318.

Walgreens. (n.d.). *Disability inclusion*. Walgreens. https://www.walgreens.com/pdf/diversityreport2013.pdf

Wu, T. (2011). *The master switch: The rise and fall of information empires*. Vintage.

Shadow Leadership: An Emerging Organizational Influence, Hiding in Plain Sight

E. Scott England and Ryan Donlan

Abstract Leadership is a complex interplay between the prominent leader (i.e., tasked with guiding an organization) and leadership peers—those who influence outcomes both internally and externally. This relationship requires harmony, much like music, where differences blend to create strength and balance. However, when harmony falters, shadow leadership can emerge. Shadow leadership represents the subtle-to-overt divergence of leadership peers from the prominent leader's vision, evolving through stages metaphorically aligned with musical elements: descant, departure, dissonance, discord, and destruction. These stages illustrate how misaligned agendas gradually erode trust, destabilize culture, and diminish the authority of the leader. While early stages allow opportunities for corrective action and restoration of harmony, later stages often culminate in the leader's loss of prominence or removal. By framing leadership through the lens of harmony and shadow leadership, this chapter underscores the necessity of reciprocity, respect, and awareness in leadership dynamics. Effective leaders must recognize early signs of shadow leadership and respond with openness, collaboration, and authenticity to sustain organizational stability and success.

Keywords Leadership · Shadow leadership · Organizational influence

1 Introduction

Leadership is more than a concept. Leadership is complex, comprised of many different dynamics, factors, and actors. Ultimately though, the chief leadership position is comprised by an individual selected by constituents to become the prominent

E. S. England (✉)
University of Maryland Eastern Shore, Princess Anne, MD, USA
e-mail: esengland@umes.edu

R. Donlan
Indiana State University, Terre Haute, IN, USA

A. M. Waite et al. (eds.), *Leading in Industry 5.0*, Management and Industrial Engineering, https://doi.org/10.1007/978-3-032-19207-3_9

leader. Whether this is a president of a bank, a CEO of a company, or a superintendent in education, a prominent leader is chosen to be the face of the organization. Each leader enters their role with a personal vision, a certain leadership style that impacts decision-making, and other various approaches that both directly and indirectly navigate the organization and equip those who serve under the prominent leader.

At the same time, those who govern the prominent leader—in this case, we use board of directors as a universal term—are also impacted by the direction the prominent leader is leading the organization and those within it. This further amplifies the complexity of leadership, in that it adds another layer of constituent feedback in terms of expectations, shared authority, and the willingness to support or challenge the prominent leader.

To help clarify our position moving forward, we categorize leadership into two areas. We have the prominent leader who is overall tasked with being in charge of the organization both in day-to-day and overall mission and vision maintenance. We will refer to everyone else as leadership peers. Leadership peers are those within the organization (i.e., non-prominent leaders, subordinates, directors, etc.) as well as constituents outside of the organization who still share influence in some nature (i.e., community members, shareholders, consumers, etc.). Further defining these two leadership areas will enable us succinctly to highlight potential disruptions to an organization, and ultimately, the prominent leader.

We establish the variety of facets within leadership to highlight the reciprocity required to lead an organization to success. Leadership cannot be one sided; it is rather a balance between the prominent leader and leadership peers—a direct exchange in communication, decision-making, and overall direction that serves as one of the pillars for success.

Effective organizations also refer to the prominent leader and leadership peers as working in harmony. As in music, harmony in the workplace offers balance and unification—a means to complement rather than create conflict. Harmony is never about being on the same page; rather, it is about combining differences between the prominent leader and leadership peers to create like-minded outcomes that strengthen the organization.

Harmony elicits thoughts of music, which is where we draw the foundation of our topic. Harmony in music involves two or more individuals operating on their own path but with the similar goal of creating pleasing sounds that bolster the arrangement. Like music, working in harmony revolves around the different skillsets and leadership perspectives to work toward shared success for the organization.

Harmony, in both music and the workplace, must also feature collaboration and cadence as critical characteristics. Without them, different *tunes* begin emerging. This leads to a fragmentation of the arrangement in music, or the environment in the workplace—which leads to another essential element of harmony, respect.

When the prominent leader and leadership peers understand each other's roles, ideas, and intentions, a mutual respect begins to develop. This serves as the catalyst in binding various viewpoints when conflicts emerge. It goes without saying that disagreements will occur. We cannot advocate for a work environment free from such; rather, we strive for an understanding of how resolutions to conflicts can be openly communicated with each party feeling safe, heard, and valued.

When there is a divergence from harmony in the workplace, the prominent leader and leadership peers set off on paths that, while similar in the beginning, can often result in a collision course. The most adversely impacted is often the prominent leader. The phenomena we speak of in this case often operates from the shadows, as in, difficult to notice if close attention is not paid. We refer to this as shadow leadership.

2 Shadow Leadership

When prominent leaders and leadership peers operate within the same domain, the dynamics between the groups become even more intricate. When these dynamics are efficiently navigated, the strengths of both groups will complement each other. When the dynamics are mismanaged, the organization can become fragmented, thus creating separate agendas that clash—with trust and confidence in the prominent leader slowly eroding.

The stages of shadow leadership draw upon musical elements to serve as metaphors for the stage names. Music often serves as a means to convey a message, and, by design, this auditory message can elicit a visual image. Many times, the untrained ear may struggle to discern subtle changes within a musical composition. It is only when the piece changes so dramatically that a listener becomes aware.

In organizations, shadow leadership emerges from a diverging agenda of leadership peers from the prominent leader. Shadow leadership can manifest as behaviors, attitudes, or even actions that serve to undermine the prominent leader. It threatens to disrupt the organization from being effectively managed.

When leadership peers shift into shadow leadership, they begin to impinge upon the goals and progress of the organization. They often do so with a hidden agenda that diverges from the prominent leader. This slowly erodes the trust in the prominent leader as others begin to question their intentions. This lack of trust begins to impact other areas such as productivity, unification, and the ability to effectively manage conflict. Once shadow leadership begins to take hold, the prominent leader is forced to juggle between addressing a continuous cadence of deleterious divergence, while also still managing the organization as part of their daily duties.

The beginning stage of shadow leadership can be subtle and seem aligned with the prominent leadership, which can be difficult to discern. To keep with the theme of music, we constructed the stages of shadow leadership with musical metaphors to highlight how the lack of harmony would look. These stages, which will be defined and expanded upon, are descant, departure, dissonance, discord, and finally, destruction.

But as in music, even the untrained ear, with slight guidance, can begin to pick up on the subtleties. With proper knowledge of shadow leadership, a prominent leader can pick-up-on when shadow leadership is beginning to expose itself.

3 Descant

In music, a descant is a melody that is played or sung above the main melody (MacDonald, 2024). The etymology of the word descant has taken on several variations over many time periods but has always meant *two parts* (Apel, 1969). It is often used as a musical element to elevate the melody in terms of complementing or adding a pleasing layer. It can, however, be used to contrast the main melody, which is where we align the first stage of shadow leadership.

The shadow leadership descant serves as a launch point in disrupting the organization, with a main goal of deviating from the path of the prominent leader. While we give respect to the fact that a descant in the organizational setting could be one that adds richness to the overall purpose, we choose to focus instead on it being a clash with the prominent leader, creating a tension not previously felt.

While the descant serves as the first stage of shadow leadership, it can also be one of the most difficult to identify. The shadow leadership descant should not be confused with harmony—it is essentially two narratives being carried out, and though different, these narratives often give the appearance of moving toward the same objective. So, the shadow leadership descant can give the impression of everyone still being like-minded while in reality, the shadow leaders are establishing their framework for a different agenda.

It is worth noting that at this point, the prominent leader still retains prominence in the organization. Many leadership peers not involved in the launching of shadow leadership may not even be aware a descant is appearing. Those involved in a shadow leadership launch could be as few as just one dissenting leadership peer or a group. At this stage, it is typically a relatively low number as their descant is done discreetly so as not to alarm the prominent leader.

3.1 Identifying Descant

If the goal of the descant is to begin a new melody (i.e., narrative) without the knowledge of such, then identifying this stage can be particularly difficult. Especially if all gives the appearance of maintaining harmony or even status quo. The intuitive prominent leader will keep a keen awareness of what is going on around them to determine if a shadow leadership descant is beginning.

To do this, one must be aware of what a shadow leadership descant could look like. While these are not an exhaustive list of examples, we present scenarios in which illustrate how to discern when descanters are present.

3.1.1 Scenario 1

A prominent leader at a marketing firm announced a new initiative centered on digital innovation. It was made clear that the organization would invest in further technology, create concise campaigns, and strengthen interaction between the organization and potential clients. It was the understanding that leadership peers would be supportive of this rollout.

As the rollout began though, a descant began in the organization. Some leadership peers began voicing quiet concerns about the rapidness of such a drastic change. This was done with the appearance of still being supportive and on the side of the rollout. There were also concerns that such a sudden change and investment into technology would lead to further automation, giving the appearance that potential layoffs could be in the future. The leadership peers launching into a descant created subtle resistance to the rollout, slowing progress just enough to be scantly perceptible.

3.1.2 Scenario 2

A prominent leader within a large hospital implemented a plan to improve patient care by standardizing procedures across departments. The prominent leader spoke of the plan and how it was designed to emphasize safety while also providing more consistent and efficient care to patients.

The descant emerged when a few members of the nursing staff (i.e., leadership peers) began to quietly express concerns of how the plan sounded good in theory but lacked practicality to align with patient needs along with the fast pace of the hospital environment. This descant was led through informal conversations, often in the break room, on a rare occasion of downtime at the nurse's station, or during shift changes when nurses scrapped their scrubs for civilian clothes.

These descants from the scenarios did not necessarily arise due to outright resistance. Both were formed from previous experiences of the leadership peers, some may even say institutional "Guardians," who believe they are compelled to protect the organization's past and present from the future (Gruenert et al., 2018, in Frederick et al. 2018) felt as though the prominent leader was making changes either too quickly or too informally without the necessary information, potentially placing some at risk.

An astute prominent leader remains aware when there are subtle shifts in the leadership peers. The descant, while difficult at times, is not impossible to identify. Slight shifts in leadership peers' body language, tone, and overall demeanor may indicate that a descant is in process. The descant stage is also one of the easiest stages to restore back to harmony.

3.2 *Corrective Action*

When a descant is observed, the prominent leader should work toward a corrective action that will allow harmony to resume. This is done by calling attention to the descant in a respectful manner that allows the dissenting leadership peers to feel heard and understood. It acknowledges that foundational needs of safety and security in the organization are being considered. The prominent leader must avoid discrediting the descant, but instead one should remain open, transparent, and communicative.

Avoiding a descant in an organization is nearly impossible. When new ideas are presented, it can be expected that a descant will appear in some form. Every organization will ebb and flow in and out of descant and harmony. It is when a descant goes unaddressed that the next stage of shadow leadership becomes present.

4 Departure

While departure is not a foundational musical element like melody or harmony, it is still recognized as a musical structure that alters the main piece (Nobile, 2011). Departure creates a contrast to the original melody. It could be a new melody that is played or a modulation to a new key.

Departure in organizations is when the shadow leadership begins to interject narratives that are distinctly different from the prominent leadership. Characteristics of departure that manifest in individuals may include differences in body language or tone of voice, while more structural evidence could include attempting to usurp a meeting's goal or direction of conversation. These characteristics would be present during meetings between the shadow leadership and the prominent leadership, but they may also be present during other social situations such as encounters within the community.

At the departure stage, shadow leaders often begin recruiting more leadership peers to join them in their divergent mission. The departure signals a clear moving away from the prominent leader's goals, objectives, and mission. It is when the climate of an organization has a clear shift while the culture of the organization becomes threatened.

4.1 *Identifying Departure*

Unlike the descant, the departure stage of shadow leadership is much easier to discern. It could be as simple as the prominent leader noticing informal meetings taking place. It could also be nuanced insubordination masking itself as unawareness or others professing a lack of resources stymying forward progress on a directive or decision made. Departure rarely begins with outright rebellion. Identifying departure can be as simple as noticing a change in motivation and engagement within the organization. Other behavioral shifts may indicate departure in shadow leadership is taking place.

4.1.1 Scenario 1

A prominent leader at a retail corporation introduced aggressive cost cutting as the organization's top priority. To do this, the prominent leader emphasized the reduction of employee hours, tougher sales quotas, and the elimination of perks that were previously enjoyed by leadership peers. While the idea was more efficiency and creating larger profit margins, the leadership peers began to feel increasing pressure.

As the descant was ignored, the departure began at the store levels. Supervisors began voicing their frustrations to other leadership peers in team meetings and in private conversations. To help finesse, or even through professional peer pressure, coerce the other leadership peers to their side, the shadow leaders quietly bend the rules in clear opposition of the prominent leader. This was done through offering to schedule more staff during busy times and creating flexibility in the quotas.

4.1.2 Scenario 2

A prominent leader of a national warehouse chain wanted to create efficiency across all regional warehouses. A plan was implemented that included mandatory overtime and a more invasive form of monitoring employee output. While the plan seemed fool proof in the formulation stages, the perception of execution of it was done with visible disregard to leadership peers, while in reality, unintentional, or at best, through ill-advisement or incomplete analysis.

Shadow leaders emerged, creating private forums through social media outlets to let other leadership peers across other regional warehouses vent their frustrations. Talks of burnout and other unfair practices became a common theme. Shadow leaders also enabled leadership peers to follow alternative paths to the prominent leader's proposed plan. It became their way of recruiting more to their side in the shadows.

To be clear, departure is not an outright rebellion. While some actions of the shadow leaders may seem rebellious, the decisions they are making are done so more to engage and recruit other leadership peers to their purpose. This uprising is definitely noticeable among the prominent leader and the leadership peers as a sense of unsettledness between the two groups begins to create minor waves in the organization.

At this stage, the prominent leader is more aware of the shift against their leadership. A natural reaction to this is to dig one's heels into their position or decision-making to assert themselves as the prominent leader. While remaining calm at any stage is important, it is extremely important to leverage calmness at departure. Still in the early stages of shadow leadership, recovering from departure can be relatively easy with calm patience and effort by the prominent leader.

4.2 Corrective Action

Unmanaged departure in shadow leadership can begin to feel like chaos. But the goal of departure, as like in music, is to offer a return. Return in this instance may be a return to descant rather than immediately back to harmony.

Being present in an unassuming manner is an important first step in corrective action. This can include even simple conversations with those presenting the shadow leadership. Of equal importance to the prominent leader during this time is to be considerate of one's own body language and tone. Much like the negative changes in body language and tone of the shadow leaders, a prominent leader can demonstrate compassion and empathy by remaining unassuming.

Acknowledgment is another key component of corrective action of departure. Shadow leadership often emerges when leadership peers do not feel heard, seen, or valued. Natural instinct is often to dismiss those who dissent, but an effective prominent leader will acknowledge when decisions were made without full input from leadership peers or when decisions were seemingly made too quickly or without transparency.

The goal of the corrective action is to move the departure back to the descant stage where conversations can continue in a less hostile manner. But without corrective action on this stage, the prominent leader not only is at risk of shadow leadership moving to the next phase, they are at risk of losing their prominence in leadership.

5 Dissonance

Dissonance in music is easy to identify even to the novice listener. The noticeable clash of notes in the melody creates instability to the piece. With this, a sense of tension or unease appears as the musical arrangement descends into more chaotic sounds.

The dissonance stage of shadow leadership follows the same path as the musical element. This noticeable clash between the shadow leaders and prominent leader creates instability in the organization as the shadow leadership has been allowed to move past the departure uprising into the beginning of a recognized rebellion.

It is at this stage when the prominent leader begins to lose control if uncorrected. To exemplify this, we propose that the prominent leader no longer remains prominent but instead becomes the designated leader, such as one with titular authority yet with actual authority waning. This is to say that they are still the top leader of the organization, but their role, influence, and authority are under attack.

5.1 *Identifying Dissonance*

Each stage of shadow leadership becomes easier to identify. The now designated leader is tasked with maintaining their position while also preventing an insurrection within the organization. Leadership peers no longer holding quiet meetings but speaking out loud their concerns and frustrations is one way to identify dissonance of shadow leadership.

Another telling sign of dissonance is the actions of the leadership peers not under direct influence of the shadow leaders. While they may still give the impression of neutral in their position, the designated leader may notice eye contact being avoided or conversations that were previously rich, enjoyable, and lighthearted becoming more curt or nonexistent.

5.1.1 Scenario 1

A designated leader at a non-profit organization implemented aggressive fundraising goals that gave the appearance to leadership peers as more of a moonshot goal rather than attainable. It was also a direct shift from previous goals which highlighted the organization's commitment to community outreach and needs.

The sudden shift on financial prosperity was unsettling to leadership peers who formed a shadow leadership initiative in response. Dissonance was finally achieved after the shadow leadership progressed through the initial phases unaddressed. Shadow leaders began to call attention to the fact that the organization emphasized fundraising campaigns and meeting the needs of donors rather than constituents in need of support. The organization's mission and vision were called into question as the delegated leader was accused of creating an agenda opposite of the organization's original focus.

Along with declining staff morale, turnover began to rise. The reputation of the organization began to decline as the focus on financial security became more noticeable. The culture of trust began eroding at a pace that sent clear signals that a change in leadership was warranted.

5.1.2 Scenario 2

A tech company has always thrived on risk taking. Leadership peers were encouraged to present ideas and experiment without the fear of reprimand if the idea failed. The now designated leader—a former prominent leader brought in to stabilize the company—had implemented structures that restricted the previously enjoyed freedom. Rather than experimenting, leadership peers had to seek approval from the designated leader who routinely dismissed ideas that seemed too risky. Along with this came performance reviews that further hampered the previously enjoyed experimentation and risk taking.

The actions of the designated leader did not take long for shadow leaders to emerge and swiftly move into the dissonance stage. Shadow leaders felt that the foundations of the company had been fractured, moving the organization in more alignment with other tech organizations that had the appearance of being stagnant in the industry.

Shadow leaders openly vented frustrations to anyone who would listen. At meetings led by the designated leader, shadow leaders called into question the competence of the designated leader to be in charge of an organization previously known for their innovation and rapidly changing services and products. The morale of the organization noticeably declined as talks of walkouts surfaced.

The key characteristic of any dissonance is a shift downward in the climate, and the increasingly binding nature of social capital in the organization's culture. Morale begins to wane as the culture choreographs attention to a place of survival on the part of the leadership peers. The designated leader will begin to feel isolated in the dissonance stage. They begin to enter a different type of survival, themselves, as they are more than likely experiencing for the first time a threat to their job security.

Corrective action is necessary, but it should be made known that it will be difficult. One cannot expect to make a couple of changes or deliver a few positive messages and reverse the damage that has been done to this point. The idea of corrective action is to stabilize first, reverse the stages second.

5.2 Corrective Action

In the dissonance phase, it is possible to recover yet increasingly difficult for designated leaders. One corrective action measure could be staging meetings with the shadow leaders in a safe environment where open dialogue is encouraged. This must be done with the understanding that safe dialogue does not directly translate to noncontentious dialogue. A designated leader must recognize that significant damage has been done and that while healing can take place, it must also take time. The purpose of these meetings could be revisiting the organization's mission and vision, without interjecting personal goals. The focus should not be on individual agendas; rather, the discussions should have a central focus on the organization in which everyone serves.

These discussions can parlay into personal objectives overtime so that everyone may gain insight into the direction each is wanting to go. While the dissonance stage is beyond a simple misunderstanding, thoughtful dialogue among parties may help each other better understand why decisions are made. The designated leader must understand that blame will be placed on them—that is often the nature of the top leadership position—but to try to adamantly defend one's position will only lead to further dissent among the shadow leaders.

If a designated leader wants full recovery and transition back to being the prominent leader, they must acknowledge that they will have to accept their missteps and even backtrack on previous decisions made. This does not mean they must give in every time leadership peers rebel. It instead should be used as a learning opportunity on how to leverage shared decision-making and executing with clear intentions.

Without patient and authentic corrective action, the designated leader will watch the shadow leadership continue to gain momentum and leverage their power as the now prominent leadership in the organization.

6 Discord

Discord in music is often closely tied to dissonance with the distinct difference being that the tone is more significantly harsh. If dissonance was unpleasing to the listener's ear, discord would potentially drive the listener away completely. While discord can happen within an arrangement, an easy illustration of discord could be when individual musicians tune their instruments at once. The sound is often unpleasing and not enjoyable.

The discord stage of shadow leadership is a further moving away from harmony in the organization. Discord is disruptive to the point that even an organization's reputation or financial well-being (e.g., stocks) are impacted. The once-prominent, then designated leader at this point can be considered a damaged leader. No longer is the shadow leadership an internal battle, discord spreads externally to outsiders to give example of the struggles the organization is facing. In metaphorical terms, discord can be equivalent to saying that the vultures are circling.

6.1 Identifying Discord

Discord is identifiable by everyone both within and without the organization. If not done so already, discord often activates the fight, flight, or freeze component of the damaged leader's brain. Other influences may take place such as routine decision-making taking place outside of the damaged leader's control or authority.

Another characteristic of discord is the feeling of shame by the damaged leader. Even when in public or private circles, a sense of uncomfortableness is often present. Others seek avoidance of them, as they are often viewed as a lame duck in a position of leadership without true authority.

6.1.1 Scenario 1

A superintendent (i.e., the damaged leader) of a school district began implementing initiatives based on successes at a previous school district rather than identifying true needs and abilities within the current district. This created a shadow leadership as people felt undervalued and unheard. The damaged leader was also accused of being out of touch with the needs and priorities of the district.

At monthly board meetings, leadership peers and other constituents publicly spoke out against the damaged leader. Name calling and character denigration—whether true or not—began to take center stage. At separate times during these disputes, the damaged leader would attempt to prove authority (fight), withdraw from the conversations (flight), or remain silent not only during meetings but throughout day-to-day operations (freeze).

Eventually, shadow leaders gained enough support and power to issue a public vote of no confidence. Local media outlets then began to investigate the reasons of the dissent, publishing often one-sided narratives driven by the shadow leadership at play.

6.1.2 Scenario 2

A damaged leader of a large equipment manufacturer took the position of a tough representative in contract negotiations. They felt that previous leaders of the organization were too lenient, agreeing to contracts that diminished the authority of the top leadership positions. The damaged leader felt in order to regain the confidence of shareholders, they had to remain firm and unyielding in the contract discussions taking place.

The shadow leadership was initiated when informal talks between the damaged leader and the leadership peers from the negotiating team started. The overall tone was that the leadership peers' positions and ideas were not going to be considered. The unyielding stance of the damaged leader rapidly progressed the shadow leadership phases to the point of discord.

Open letters were distributed by the shadow leaders both inside and outside of the organization. A call for the removal of the damaged leader was issued. As further breakdowns occurred, the previously discussed strikes were launched into action. Picket lines shut down operations of the organization, creating further instability and allowing competitors to capitalize from unsureness in consumers.

Discord cannot be sheltered to an internal issue only. Discord often goes public with the shadow leaders enhancing their position through the use of social media and other traditional media outlets. Even without walkouts and strikes, productivity often wanes with shadow leaders going through the motions and operating with complete disregard to previously issued directives.

Another characteristic of discord is a transition from passive aggressive to outright aggressiveness that seeks to ruin the reputation of the damaged leader. Sarcasm and the absence of any communication begin to run rampant. The damaged leader is often removed from any decision-making as the confidence in their abilities has moved from just waning to completely absent.

6.2 *Corrective Action*

At the discord stage, clear shifts in leadership power are evident. The shadow leaders are the ones who are given the clearance to call the shots. The influence both in and out of the organization are driven by the shadow leaders. In discord, the damaged leader is merely occupying their role, one might say via "renting," not "owning."

While corrective action is still possible, it is often met with even more resistance if the damaged leader attempts to rectify the situation on their own. The damaged leader must accept that their actions have brought discord, eliminating trust and confidence. If they wish to authentically recover, assuming they are given the chance, they must enlist the guidance of unbiased outside help to do so.

Outside help in this case could be represented in various forms. The damaged leader should seek private assistance from a trusted mentor or from someone who has been in a similar position before. The purpose of this is to help the damaged leader see previous mistakes and learn how ownership of those mistakes can lead to authentic recovery. The damaged leader must also develop the mindset of acceptance of blame, regardless of previous intentions. Vulnerability is key to authentic corrective action.

The other form of outside assistance would be through a mediator. This is more than a specialist in team building—it is too late for that. The outside mediator would be tasked with the role of hearing both sides and guiding them toward an acceptable solution that would benefit the organization moving forward.

While not always the case, the solution is not always beneficial to the damaged leader. When shadow leadership is allowed to reach the discord stage, even with thoughtful, transparent discussions, it may be too late to reestablish the damaged leader to the role of prominence. The goal of restoration is to stabilize the organization.

7 Destruction

Destruction in music is not a recognized element or form, but rather a metaphor for the sudden change to a musical piece. Destruction is a shift so far from the original melody that recovery is often not an option. The jarring tones signify that the original melody has collapsed and that the end is imminent.

Sadly, destruction in shadow leadership carries the same undertone. The destruction stage is not the destruction of the organization (though we recognize for some it could be), but instead, it is the destruction of the damaged leader. In simpler terms, destruction is a victory for the shadow leaders. They successfully navigated through the stages of shadow leadership to accomplish what they were originally seeking, a clear path divergent from the once-prominent leader.

7.1 *Identifying Destruction*

Destruction is the easiest to identify for the destructed leader. If still in their role, they understand they are just biding their time until the removal process can be executed. This could be in the form of being asked to step out of formal meetings for discussions that they were previously privy to. In a more drastic case, the destructed leader may be required to take a leave of absence or even be put on leave by the leadership peers tasked with that authority.

We intentionally do not include scenarios of destruction as we contend it looks the same regardless of the organization or situation. Destruction is end game for shadow leaders, and the lived reality for the destructed leader, regardless of the setting. Once destruction is identified, the destructed leader must look for solutions on how to navigate the destruction with grace and dignity intact.

7.2 *Corrective Action*

We use corrective action for destruction for consistency purposes with a clear understanding that corrective action is impossible at this stage in shadow leadership. It must be recognized that reverting to discord, let alone harmony, is unfeasible. So, without true corrective action, what must the destructed leader do?

To answer that question, the destructed leader must seek eloquent closure. This is to say, an exit that is full of acknowledgment and acceptance. It is a time to make amends, to concede wrongs without providing justification of previous actions.

A destructed leader should remain confident in their leadership abilities, while understanding that their failure within an organization is not a failure over their overall capabilities. Even at the most difficult times, a destructed leader should exemplify top leadership qualities such as keeping the focus on the health of the organization following their departure, being thankful for the opportunity to once be the prominent leader, and ensuring to leadership peers that no grudges are to be held and that they (the destructed leader) would do anything in the future to guide others to success.

8 Muted Mutiny

While we do not want to convolute shadow leadership, we feel it is important to point out that even those who remain in support of the prominent leader but remain silent are actually demonstrating more support, unintentionally, for shadow leadership. Muted mutiny is about taking a neutral position but ultimately supporting shadow leadership to prevail through silence. This could be from leadership peers who do not necessarily agree with what shadow leaders are doing, but they choose to stay silent believing their neutrality gives indication they do not support the dissent. However, this is what shadow leaders want because the silence also does not outwardly support the prominent leader. Muted mutiny is subtle but powerful when it comes to supporting dissent. This silence quietly endorses shadow leadership through inaction. It is a way that even the most well-intentioned distance themselves from actual conflict while inadvertently bolstering the influence of shadow leaders with their inaction. It is rebellion assistance amid a ruse of neutrality.

9 Conclusion

Shadow leadership on a whole is impossible to avoid within organizations. While everyone strives for constant harmony, it is futile to operate with that mentality. Constant harmony is impossible. Organizations regularly move in and out of descant as competing trains of thoughts provoke varying viewpoints on situations.

But even descants in the organization can be beautiful and pleasing. Allowing descants gives leadership peers the opportunity to be recognized and valued. It gives energy to the organization as a multitude of ideas are given credence. An organization cannot be led with a singular focus. It must allow everyone to operate independently with the intention; they are playing a consistent melody that moves forward without disruptions.

In musical parlance, when stray notes appear, it is the role of the conductor to guide peers back to the main melody. A prominent leader's focus should be in remaining aware of when descants lead down a path toward potential disruptions. Left unaddressed, shadow leadership threatens the overall health of an organization as well as the job security of the prominent leader.

It is also worth noting that a newly instated prominent leader selected to lead a company that previously experienced shadow leadership will not come into harmony. They are coming into a situation that cycled through the stages of shadow leadership to the destruction of the previous leader. The newly established prominent leader must make restoration of harmony a priority, even if this means remaining patient for an extended period of time before implementing new ideas.

Shadow leadership initially operates in stealth, but it is always hiding in plain sight. A keen understanding of how to guide an organization through the stages of shadow leadership is critical to the success of an organization and all leaders associated with the organization. Navigating shadow leadership is a test of one's true abilities to fully understand the dynamics of leadership and highlight their effectiveness in leading.

References

Apel, W. (1969). *Harvard dictionary of music*. Harvard University Press.

Gruenert, G., Donlan, R., & Frederick, M. A. (2018). Domains interrupted: The Guardian. In M. A. Frederick, P. A. Sasso, & J. M. Maldonado (Eds.), *The dynamic student development meta-theory: A new model for student success*. Peter Lang.

MacDonald, S. (2024). Descants are for life, not just for Christmas. American Organist, 58(12), 70–71.

Nobile, D. F. (2011). Form and voice leading in early Beatles songs. *Journal of the Society for Music Theory, 17*(3), 1–11.

Thriving in Industry 5.0: Leadership and Business Models for Value Creation

Jeffrey Moore, Everon Chenhall Maxey, and Josiah Hangen

Abstract Industry 5.0 redefines industrial success—placing human well-being, sustainability, and resilience at the center of value creation. This paper posits leadership as the decisive force in bringing that vision to reality. By integrating the Burke–Litwin Model with Complexity Leadership Theory, it outlines three critical roles—enabling, adaptive, and administrative—that transform top-down management into inclusive, agile, and innovation-driven cultures. Case studies from Walgreens and Sephora demonstrate how human-centric leadership not only enhances engagement, productivity, and retention but also unlocks the full potential of diverse teams. Leveraging assistive AI and technology in an ethical manner, leaders can amplify human strengths, empower employees with disabilities, and build trust in human–machine collaboration. The result: organizations that are not just prepared for the future, but actively shaping it, where technology serves humanity, and leadership drives enduring competitive advantage.

Keywords Human-centric · Leadership · Complexity · Sustainability · Industry 5.0

1 Introduction

Industry 5.0 represents a significant evolution from the digital age, shifting its focus from mere efficiency and productivity to the comprehensive integration of human-centricity, sustainability, and resilience across industrial operations (Breque et al., 2021; Nahavandi, 2022). This new paradigm requires leaders who can navigate complex environments, ethically integrate advanced technologies, and cultivate cultures that enable people to thrive. Leadership, not technology alone, will determine whether Industry 5.0 delivers on its promise of harmonizing human well-being with innovation and economic growth.

J. Moore (✉) · E. C. Maxey · J. Hangen
College of Business and Economics, Anderson University, Anderson, SC, USA
e-mail: jmoore@andersonuniversity.edu

A. M. Waite et al. (eds.), *Leading in Industry 5.0*, Management and Industrial Engineering, https://doi.org/10.1007/978-3-032-19207-3_10

2 Human-Centricity in Industry 5.0: Leadership as the Engine of Transformation

Industry 5.0 presents a vision that fundamentally rethinks the relationship between humans and technology, placing human well-being at the heart of industrial operations. This approach fosters continuous skill development, ensures the ethical integration of technology, and cultivates a supportive organizational culture that values every individual's contribution (European Commission, 2020). Realizing this vision requires a decisive shift from traditional, top-down management to adaptive, employee-centered leadership that can guide organizations through the increasing complexity and volatility of today's VUCA (volatile, uncertain, complex, and ambiguous) environments.

Achieving this transformation begins with understanding how leadership interacts with organizational systems and culture to drive change. The Burke–Litwin Model (1992) provides a comprehensive framework for analyzing these interactions by identifying 12 interconnected organizational components. Its transformational factors—external environment, mission, and strategy, leadership, and organizational culture—are particularly critical in shaping large-scale change. External pressures, such as rapid technological advancements and evolving customer expectations, necessitate that leaders establish a clear vision, align their strategy, and cultivate cultures that embrace adaptation.

While the Burke–Litwin Model maps the structural and cultural elements necessary for change, Complexity Leadership Theory (Uhl-Bien et al., 2007) offers a dynamic approach to leading in complex, adaptive environments. CLT identifies three interrelated leadership functions: adaptive, enabling, and administrative, which empower organizations to innovate, collaborate, and respond effectively to uncertainty. Together, these models position leadership as the central force in translating the human-centric vision of Industry 5.0 into practical, sustainable, and resilient business practices.

Table 1 outlines three critical leadership functions for navigating the future of work, how leaders can foster a human-centric and technologically integrated environment. The **Enabling** role focuses on building strong relationships, fostering transparent communication, increasing high connectivity throughout the organization, and ensuring employees feel motivated and involved in their own development. Furthermore, leaders are responsible for fostering increased interaction and information sharing among employees, which is essential for bottom-up, creative problem-solving involving informal leaders. The **Adaptive** role, conversely, drives innovation and a growth mindset, encouraging bottom-up solutions and embracing change. This involves championing new ideas, promoting continuous learning, and integrating human-machine collaboration while building trust in automation. Ultimately, the Administrative role seeks to streamline systems, eliminate obstacles, and empower frontline employees by minimizing restrictive policies and decentralizing decision-making. This role emphasizes supportive planning, viewing mistakes as learning opportunities, and ensuring ethical governance of AI and HR systems.

Together, these three roles provide a comprehensive approach for leaders to create agile, human-centric organizations that thrive in the evolving landscape of Industry 5.0.

Industry 5.0 calls for employee-centric leaders who foster a culture of continuous learning, ethical technology integration, and human-centered values. Leaders must prioritize employee well-being and inclusivity, creating supportive environments that enhance engagement, dignity, and autonomy. Continuous workforce development through reskilling, digital training, and leadership coaching is essential to adapt to evolving technologies and promote creativity and adaptability. Human-machine collaboration should augment rather than replace human roles, empowering workers to focus on meaningful, innovative tasks. Ultimately, incorporating ethical principles such as transparency, fairness, and accountability into technological systems is essential for fostering trust, mitigating bias, and promoting diverse, inclusive workplaces.

Technology in this framework is strategically employed to mitigate drudgery and reduce injury risks, transforming jobs into more engaging and less stressful experiences, rather than contributing to job insecurity. Furthermore, this vision extends to empowering workers through collaborative technologies. A cobot, or collaborative robot, is specifically designed to work safely alongside humans in a shared workspace, featuring features such as force limitations, rounded edges, and advanced sensors to enable direct and safe interaction (Fiveable, 2024). These advanced cognitive systems are designed to augment both cognitive and physical human abilities. These systems support human decision-making, allowing for human oversight and intervention when necessary, thereby fostering a true partnership between humans and machines, a key tenet of Industry 5.0.

Industry 5.0 embeds principles of transparency, accountability, fairness, and privacy into the design and deployment of collaborative systems. This is vital for addressing discrimination, particularly for vulnerable groups like immigrant workers, ensuring that automated HR systems do not perpetuate biases through prejudiced data inputs or by focusing on inappropriate social or political factors in assessments. Human oversight in automated systems is crucial for properly valuing diverse backgrounds and preventing the depersonalization of interactions that can lead to isolation and disconnection among employees.

By viewing employees as holistic individuals and transforming leaders into champions of employee success, they can reverse disengagement trends, fostering thriving workplaces with enhanced productivity, retention, and innovation. Organizations should invest in manager training, inclusive policies, and regular feedback to align with complexity leadership practices and employee-centric value driven Industry 5.0 philosophy.

Table 1 Industry 5.0 human-centricity and complexity leadership practices

Complexity leadership	Key responsibilities	Industry 5.0 themes	Examples and strategies
Enabling increases communication, facilitates collaboration, interdependence, and high connectivity	Invest in authentic relationships by learning how to communicate effectively with each employee. Learn what motivates each person. Involve employees with their development (ATP) increase interaction and communication across the organization. Build safe, encouraging and inclusive environments where diverse teams thrive	Employee well-being and engagement strategic emphasis on employee well-being workforce development and reskilling	Learn about employees in order to find out what motivates them ask the person (ATP) do not assume you know their perspective transparent communication job rotation models based on physical and emotional well-being

(continued)

Table 1 (continued)

Complexity leadership	Key responsibilities	Industry 5.0 themes	Examples and strategies
Adaptive fosters innovation, growth mindsets and bottom-up solutions	Committed to employee success. Adopts a change mindset and is willing to try new methods to improve individual and group performance champions new ideas from the lowest level, encourages creativity, continuous learning, supports informal leaders, and agile processes that support solving challenges	Human-machine collaboration human-centric culture Trust in automation	Produce shared ideas, information, products, resources (represents no one person but a collective emergence) company HR AI bots for streamlining information distribution and employee questions AI tools (Team's speaker coach) assisting disabled team members with real time feedback on communications during meetings and presentations
Administrative align systems, remove roadblocks and break down silos minimize rules and policies to support frontline staff	Supportive planning and coordination provide sense-making. Mistakes are not punished but seen as "learning." Minimize rules and policies, challenge norms and the status quo to support the frontline push down decision-making, power, and resource control. Build teams through improvement initiatives from team ideas	Digital leadership and competence management ethical governance of AI/HR systems Ethical tech integration	Coaching and mentoring programs. Employees become peer mentors for other team members with disabilities develop break/lunch areas to facilitate dialogue accountability in AI-driven HR tools AI Microsoft Copilot single platform to share data, improving workflow

3 Global Employee Engagement Crisis

The digital age, while revolutionary in efficiency and productivity, was primarily technology-centered and machine-oriented. This focus often overlooked the human factor and worker well-being, leading to concerns about job displacement, the erosion of the "human touch," and diminishing human relevance in increasingly automated workplaces. These shortcomings underscore the urgent need for a leadership-driven paradigm shift toward Industry 5.0, where human well-being, sustainability, and resilience are prioritized alongside innovation and productivity.

The *State of the Global Workplace: 2025 Report* by Gallup shows that employee engagement has dropped globally from 23 to 21%, with manager engagement falling from 30 to 27%. Europe has the lowest engagement rate at 12%, while the United States stands at 32% but has seen historic declines in well-being. Notably, managers, especially young and female leaders. Are experiencing the steepest drops in both engagement and overall thriving, signaling a leadership crisis rather than simply a workforce morale problem (Gallup, 2025) (Table 2).

Gallup concludes that engagement is the strongest driver of overall well-being, and that managers play a pivotal role in shaping it. Engaged leaders foster trust, reduce stress, and create cultures where employees thrive. The report identifies three proven leadership strategies that align with Industry 5.0 priorities: targeted leadership development, coaching skill training, and structured mentorship. Together, these approaches improve engagement, boost performance by up to 28%, and significantly increase leaders' sense of thriving.

AT&T's recent redefinition of its employer–employee relationship offers a stark contrast to the human-centric leadership model central to Industry 5.0. In a 2025 memo, CEO John Stankey openly stated that the company had "consciously shifted away" from an employment deal rooted in loyalty, replacing its long-standing

Table 2 Employee well-being and engagement in the United States and Europe

	United States (%)	Europe
Employee engagement	**32**	**13%**
Not engaged	52	73%
Actively disengaged	17	15%
Life evaluation—Thriving	**58**	**47%**
Struggling	44	48%
Suffering	4	5%
Daily stress (felt "a lot" yesterday)	**50**	**38%**
Daily anger	17	14%
Daily sadness	22	17%
Daily loneliness	15	1
Good time to find a job	**57**	**57%**
Intent to leave (actively seeking/watching)	50	30%

"familial" culture with a market-based approach that no longer promises long-term reciprocity between employer and employee (Ito, 2025). This approach reflects a broader corporate trend toward transactional employment, where workers are valued primarily for their immediate output rather than their long-term contributions and development.

In contrast, Industry 5.0 leadership emphasizes rebuilding the psychological contract around trust, mutual investment, and shared purpose. Rather than eroding loyalty, leaders in an Industry 5.0 framework actively cultivate it by aligning organizational strategy with employee well-being, fostering inclusive cultures, and leveraging technology to augment, not diminish, human potential. While AT&T's shift underscores the risks of disengagement and erosion of workplace meaning, the Industry 5.0 paradigm offers a pathway to sustainable value creation by treating people as long-term partners in innovation and growth.

4 Employee-Centric Models: Walgreens and Sephora's Inclusive Culture

The principles of human-centricity are not solely academic; they're being successfully implemented and recognized in leading organizations. The extensive case studies of Sephora and Walgreens provide compelling empirical validation and deeper insights into the practical operationalization of human-centricity. In recognition of their commitment, Disability:IN awarded Walgreens Boots Alliance the *Employer of the Year* Award for the 2023 Inclusion Award (Business Wire, 2023). Similarly, Sephora received the *Best Places to Work for Disability Inclusion* for 2023 from Disability:IN. Further underscoring the positive impact of this approach, during the *Human Flourishing Conference on Transforming Culture through Disability Employment*, Moore presented research findings demonstrating compelling 2021 Sephora metrics comparing their non-PWD (Persons With Disabilities) team members against their PWD team members. These metrics revealed that PWD team members were 480% less absent, 410% safer, and 7% more productive (Moore & Hanson, 2023).

Table 3 presents best practices in human-centric leadership, categorized within the CLT model, which boosts employee engagement and loyalty, improves well-being and safety, and cultivates positive attitudes through interaction with diverse teams. Effective leadership has been consistently linked to higher levels of employee engagement and lower turnover, underscoring the importance of investing in leadership development (Moore & Hanson, 2022). These findings highlight the shift from traditional top-down control to a supportive, collaborative, and empowering employee-centric leadership style of inclusive teams at Sephora and Walgreens.

The shift from rigid, top-down organizational structures to collaborative, flexible environments is profoundly demonstrated by the observed transformation in managerial styles. Managers in these organizations moved from autocratic, metric-focused

Table 3 Human centric manager practices at Walgreens and Sephora

Theme	Key elements	Descriptions
Enabling valuing and empowering employees	Promoting talents, diversity and empowerment	Embracing diverse talents, including individuals with disabilities, to expand workforce capabilities
	Courageous humility and empathy	Leaders demonstrate openness and concern for employee well-being, fostering emotional connections
	Building authentic relationships	Treating employees as respected individuals; cultivating trust and mutual support
	Focus on individual development	Prioritizing long-term personal growth over short-term metrics; adapting to individual learning styles
Administrative loosening and front-line support	Employee-centric management style	Shifting from top-down control to supportive, inclusive leadership that adapts to team needs
	Shared power and decision-making	Involving employees in goal-setting, decision-making, and innovation to promote ownership and engagement
	Minimizing controlling forces	Reducing rigid policies to increase flexibility and innovation using complexity leadership principles
	Addressing relational stigma and fear	Providing training to overcome discomfort or bias, especially in disability inclusion
Creating a supportive and adaptive culture	Disability inclusive culture	Embedding respect and equal opportunity into daily practices through leadership and policy
	Agility, resilience, and innovation	Encouraging problem-solving and adaptation, even at the risk of failure, to build organizational strength
	Higher engagement and loyalty	A sense of value and belonging boosts morale, retention, and organizational commitment
	Workplace learning and training	Promoting continuous development through real-time, job-embedded learning experiences

(continued)

Table 3 (continued)

Theme	Key elements	Descriptions
Prosocial and societal benefits	Improved employee well-being and safety	Inclusive practices lead to better safety and health outcomes for all employees
	Moral elevation and positive attitudes	Contact with diverse individuals fosters empathy, positivity, and aspirational behavior
	Community partnerships	Collaborating with local and disability organizations to enhance recruitment, training, and inclusion

approaches to inclusive, relationship-based leadership. This transformation was often initiated by an initial "fear of the unknown" associated with integrating individuals with disabilities. However, when this fear was overcome through structured support and executive commitment, it compelled managers to adopt higher levels of managerial competence and empathy, ultimately benefiting all employees. The ripple effects of disability-inclusive hiring extend beyond individual success stories and positively shape team dynamics (Moore et al., 2020a, Moore et al., 2022).

A critical attribute of successful inclusive managers, as revealed by the research, is "courageous humility." This denotes a willingness to acknowledge a lack of comprehensive knowledge, to be comfortable without having all the answers, and, crucially, to share power. This psychological shift from a control-oriented mindset to one that embraces vulnerability and shared authority is fundamental for building authentic relationships and fostering deep trust within teams. Human-centricity is further operationalized through managers building authentic relationships with employees. These relationships are characterized by individualized communication, a genuine understanding of employee motivations, and the provision of patient, empathetic coaching. This approach fosters a "circle of security" where employees feel psychologically safe to explore new ideas, take calculated risks, and effectively cope with challenges.

Employee empowerment and voice are direct outcomes of the "Participative (Adaptive Leadership)" role within Complexity Leadership Theory, as described in Sect. 3 of the paper. This translates into actively empowering frontline workers to generate solutions and contribute ideas. This is vividly exemplified by Sephora's "Ask The Person" (ATP) model, which encourages direct employee input, where volunteer hourly team members actively engage with management on critical operational issues (Hanson & Moore, 2023).

Table 4 highlights how 713 employees at Sephora with and without disabilities describe their "Best Boss Ever," based on statistically significant correlations. For employees with disabilities ($n = 63$), the strongest trait was a boss who investigates the root causes of performance issues. As a result of their "Best Boss Ever" they would recommend working at Sephora to their friends. Employees without disabilities ($n = 650$) strongly value bosses who care about them personally and strive to make every team member successful. Both groups highly appreciate leaders who give timely feedback, address problems promptly, and demonstrate personal care. These findings underscore the universal importance of supportive leadership, with a particular emphasis on fostering deeper understanding and building trust among employees with disabilities.

In Table 5, we examine 650 Sephora non-disabled employees and analyze the impact of their disabled team members by dividing them into three categories. The first category includes teams with employees who have disabilities. The second category comprises team members who have worked with disabled employees for over a year. The third category consists of team members without any disabled employees in their teams. These teams with and without employees with disabilities perceive the qualities of their "Best Boss Ever" based on statistically significant correlations. Teams that include employees with disabilities, especially those with over one year of experience, highly value bosses who focus on individual strengths, communicate expectations clearly, motivate their teams, provide training, and offer personal care. These traits consistently show strong correlations, often above 0.6, indicating a strong connection to positive leadership experiences. While all teams value supportive and effective leadership, teams without disabled members also highlight traits like humor and communication about upcoming changes. Overall, the findings suggest that although some leadership qualities are universally appreciated, inclusive teams particularly value clarity, encouragement, and strength-based leadership.

Sephora has effectively implemented human-centric business models that emphasize adaptation, enabling leadership, innovation, inclusion, and empathy—resulting in meaningful employee empowerment at both the individual and team levels. By

Table 4 Correlations for individuals with and without disability about their "Best Boss"

	Employees with disability	Employees without disability
	$n = 63$	$n = 650$
Best boss perspective—disabled team members		
When there is a drop in performance, the boss focuses on discovering and eliminating the underlying cause	0.552***	
I would recommend this as a great place to work to my friends	0.5195***	
Best boss perspective—Non-disabled team members		
Gives us timely and ongoing performance feedback	0.5242***	0.5055***
Does not let problems fester	0.5118***	0.5239***
Works to make every team member successful	0.5109***	0.5932***
Cares about me as a person	0.5272***	0.5907***

*** = p <.001 very strong statistically significant level

Table 5 Correlations for teams with and without employees with disability about their "Best Boss"

		TEAMS with employees with disabilities	TEAMS with employees with disabilities (over 1 year experience)	TEAMS without employees with disability
Best boss perspective—teams with disabled members				
	Is patient when dealing with team members	0.5298***	0.5313***	
	Gives us timely and ongoing performance feedback	0.5122***	0.5418***	
	Focuses on finding and using each person's strengths	0.5702***	0.5851***	
	Makes sure that everyone knows what is expected of them	0.5329***	0.5654***	
	Is competent and knowledgeable when it comes to the work	0.5052***	0.5518***	
	Is committed to achieving our goals	0.5007***	0.5366***	
Best boss perspective—experienced PWD teams (over 1 year)				
	Never loses his/her temper		0.5271***	
Best boss perspective—teams without disabled team members				
	Jokes with team members			0.6147***
	Keeps us informed about upcoming changes that might affect us			0.5839***
Best boss perspective—all Teams				
	Admits mistakes or when he/she does not have the answer		0.527***	0.5559***
	Is an excellent motivator	0.5889***	0.6057***	0.6342***

(continued)

Table 5 (continued)

		TEAMS with employees with disabilities	TEAMS with employees with disabilities (over 1 year experience)	TEAMS without employees with disability
	Works to make every team member successful	0.6052***	0.595***	0.5778***
	Does not let problems fester	0.5201***	0.5338***	0.5294***
	Cares about me as a person	0.5913***	0.6161***	0.5916***
	Is an excellent trainer	0.5927***	0.6192***	0.5093***
	Is generous with praise	0.5531***	0.5734***	0.5665***
	Looks for ways to improve things	0.5033***	0.5323***	0.5574***
	Maintains a positive outlook	0.5394***	0.5511***	0.5308***
	Helps me in exceeding my own expectations	0.604***	0.6789***	0.5722***

*** = p <.001 very strong statistically significant level

embracing diverse talent, including individuals with disabilities, and shifting from rigid, top-down structures to collaborative, relationship-based leadership, the organization has significantly enhanced employee well-being, engagement, and overall performance. Building inclusive organizations depends heavily on leader-employee trust and authenticity, which together create conditions for mutual development and belonging (Moore et al., 2020b). The research highlights the importance of "courageous humility" in managers, who build trust through openness, empathy, and shared decision-making. Survey data from over 700 participants reinforces that inclusive leadership, characterized by timely feedback, personal support, and a commitment to individual success, strengthens teams and improves workplace satisfaction. Sephora's case study exemplifies the practical application of the Industry 5.0 framework, demonstrating that a human-centric leadership model not only aligns with ethical values but also delivers measurable organizational value. Furthermore, these findings closely align with Gallup's recommendations for promoting employee flourishing through effective leadership and supportive workplace culture.

5 Leaders' Use of Inclusive AI Tools to Support Employees with Disabilities

According to the World Health Organization, (2023), an estimated 1.3 billion people, or about 16% of the global population, have a disability. Employees with certain disabilities may possess strengths in mathematical acumen yet may encounter challenges in managing cognitive load or articulating their thoughts. Assistive technologies provide employees with opportunities to perform new tasks with increased confidence and enhanced work experience, enabling them to thrive. In doing so, employees have the potential to augment employee engagement and performance. As highlighted by Barnett, (2024), companies like Microsoft are leveraging AI not only to empower employees with disabilities but also to model inclusive innovation across industries.

Industry 5.0 emphasizes synergistic collaboration between humans and technology, where machines are designed to augment human capabilities rather than replace them. According to Deloitte's, (2025) Human Capital Trends, AI can perform administrative and repetitive tasks, freeing up leaders to focus on areas such as employee development, work redesign, and strategic thinking. More time is spent optimizing human capital and participating in activities that add the most value-added activities. This means collaborative robots (cobots) take on repetitive and dangerous tasks, freeing human workers to focus on more creative, problem-solving, and critical thinking roles. This shift leads to enhanced productivity, creativity, and overall operational efficiency, reducing the fear of redundancy and making work more fulfilling. Employee flourishing managers seek tools and artificial intelligence to be more strategic in recruiting, training, coaching, and developing employees with disabilities.

6 Case for Inclusive AI Tools for Removing Barriers to Increase Employee Resilience

Employee flourishing managers are receptive to leveraging AI tools to increase their awareness and sensitivity toward barriers in the workplace, promoting an inclusive environment for employees. In the case of disability, managers utilize tools to identify and remove workplace barriers, enabling team members to develop increased resilience and potentially benefiting all employees. Depending on the nature of the disability, productivity challenges, and the type of work, employees with cognitive or physical impairments can utilize assistive technologies to optimize performance and engagement, with an increased emphasis on team member interactions. For example, employees with cognitive disabilities such as traumatic brain injuries, learning disabilities, and dementia can receive scheduled reminders of the tasks to be performed throughout the day using chatbots and virtual assistants. AI-powered tools can also assist with prioritizing tasks from emails. As a result, employees can spend less time on working out scheduling details, prioritizing tasks, and improving workflows, allowing them to focus more on strengthening work relationships. After all, the "health of interactions dictates the health of culture," as Halter, (2025) reinforced at a *Human Flourishing Conference on Transforming Culture through Disability Employment.*

Ernst and Young (EY, 2024) partnered with Microsoft EMEA to conduct a global study involving 300 employees from 17 organizations who self-identified as disabled or neurodivergent and used Microsoft Copilot 365. The study revealed that Copilot had a positive impact on barrier removals, productivity, and employee satisfaction. There was a range of responses regarding the results of using Microsoft CoPilot in various work contexts to address individual needs. Tools that simplify workflow, organization, and communication can reduce cognitive load, increase focus, and minimize distractions. Microsoft Copilot 365 is a tool designed to reduce the cognitive load through centralizing applications into a single platform for Word, Excel, Outlook, Loop, Power Platform, PowerPoint, and Teams (Barnett, 2024). Employees reported on the ease of accessing data, managing projects, and communicating using an organized, integrated approach. Some notable stress-reducing capabilities of Microsoft 365 Copilot include meeting transcriptions and task prioritization, which enable focusing on the most critical tasks currently and in the correct sequence. The GenAI for Accessibility Report indicates that 50% of surveyed employees "made meaningful improvements in creative or innovative tasks" (EY, 2024, p.29).

Whelpley and Perrault, (2021) discussed the challenges faced by autistic employees in the workplace and the need for improved communication with supervisors and team members. SAP, the German multinational software company, has developed meeting cards with inclusive meeting guidelines for Zoom and Teams sessions, enabling everyone to participate regardless of impairments or neurodiverse challenges (SAP Inc., 2021). Microsoft Teams offers a "speaker coach" feature that can be beneficial for autistic employees and all employees participating in virtual meetings. Employees are empowered to adapt their communication approaches by receiving real-time, one-on-one feedback based on tone, word choice, and pace, which enhances their communication skills during and after virtual meetings (Microsoft, n.d.). Microsoft utilizes CoPilot 365 to enhance verbal and written communication among employees, which is particularly useful for employees with visual, hearing, or motor impairments. The AI benefits include support for articulating messages effectively, utilizing voice-to-text and text-to-speech capabilities, and ensuring accurate grammar when communicating with team members, customers, and other stakeholders (EY, 2024).

Employees with hearing impairments often struggle to grasp important details and instructions during work conversations and to follow up accordingly. Providing effective feedback and responding to colleagues can be tough, especially in noisy environments and across different communication channels. AI-powered hearing aids can adjust to various surroundings, enhancing clarity and reducing background noise (Ahmad et al., 2025). These hearing aids can be connected to computers, phones, and smartwatches. Consequently, employees feel more confident and engaged when adapting to various modes of communication, whether in face-to-face or virtual meetings.

7 Using AI Tools to Monitor Employee Engagement and Well-Being in an Inclusive Culture

Sigstad et al., (2025) conducted a mixed-methods study across several industries in Norway to determine the characteristics of employers and cultures that support individuals with cognitive disabilities. One of the themes among the employers who successfully integrate employees with disabilities was monitoring employee well-being and sense of belonging in the work teams. Consistent with the Walgreens and Sephora case studies presented in Sect. 3, the sampled Norwegian organizations, characterized by inclusive work cultures, provide enabling leadership, where managers and peers support employees with disabilities. Job coaches are also provided. Another commonality was that the inclusive manager provides adaptive leadership by giving clear expectations, which is an essential support for successful job performance.

A core principle of Industry 5.0 is creating safe, inclusive, and motivating work environments that focus on workers' physical and mental health, dignity, autonomy, and privacy. This involves addressing issues such as ergonomic risks, noise, and vibration exposure, as well as boredom, through adaptable strategies like flexible job arrangements and multi-objective job rotation models that consider individual worker attributes, including experience, physical capacity, and perceived boredom. Industry 5.0 aims to foster a happy, motivated, and satisfied workforce, which can lead to reduced staff turnover and better retention. It also seeks to counteract psychosocial hazards, such as stress and job insecurity caused by automation, by offering proper training and clear communication about the benefits. Employers are encouraged to emphasize self-awareness, a positive work environment, and emotional support from family, which are often overlooked in current job postings.

Employee flourishing managers stay aware of employee needs, engagement levels, and sources of motivation. Therefore, sentiment analysis, an AI data collection technique, is used to measure employee engagement and well-being and to develop strategies for boosting engagement. There are AI tools available for managers and employees to create actionable steps and recommendations that enhance the physical and emotional well-being of staff. For example, AI-driven communication tracking devices monitor conversations to identify signs of bullying or harassment in the workplace. Additionally, AI tools can track stress and work patterns, spotting potential signs of burnout. Employees can utilize mobile apps and wearables to track their health and receive personalized recommendations (Sodales Solutions Inc., n.d.).

8 Examples of Inclusive AI Tools for Value Creation

Goodhue and Thompson's Task-Technology Fit (TTF) Model (1995) emphasizes aligning technology with specific job tasks based on desired outcomes. Hamideh et al., (2025) outlined some potential implications of the TTF Model for improving employee learning capacity and motivation through AI apps and chatbots. For example, chatbots can be customized to meet the learning needs of companies for various tasks. If an employee transitions to a new role, the chatbot can help capture knowledge of their previous role to share with their successor. Hamidah's, (2025) study in Germany showed how employees with disabilities developed chatbots tailored to the technology's compatibility with specific tasks and employee needs. Like job aids, chatbots or AI apps can be used to learn new tasks and roles or to enhance the effectiveness of current jobs. In the study, employees with disabilities collaborated to develop functionalities and use cases for pre-coded chatbots, focusing on helping employees request sick leave, vacation time, and utilize the Suggestion Box chatbot. Employees interacted and communicated during training, development, and after implementation through ongoing feedback and evaluation. The study highlights how AI tools can help employees with disabilities improve their skills, including digital and communication skills. Furthermore, people with disabilities can be empowered to create their own chatbots, meeting their needs and fostering an inclusive workplace culture.

Managers should consider the problems they intend to solve and how employees can collaborate to provide effective solutions. Applying the Burke-Litwin Model (Burke & Litwin, 1992), as referenced in Sect. 2, managers can understand the "what" and "why" questions related to the needed changes in culture, leadership, behaviors, processes, and systems that support disability inclusion. In doing so, managers should consider the role of the employees and potential uses of AI in determining which tools are most appropriate in an inclusive environment. When piloting and selecting AI tools, questions should be asked about the types of data that will be produced, how it will be used, and what security or privacy concerns may arise. Consideration should be given to the applications of the tools in various geographic regions of global companies (Farrar et al., 2025).

The successful integration of human-centricity within Industry 5.0 also mandates a rigorous focus on ethical considerations throughout the technological deployment and development lifecycle. Key ethical principles include ensuring transparency and explainability in AI systems, maintaining strict privacy and robust data governance, upholding human autonomy and agency by preserving human control over machine actions, clearly defining accountability in shared human-AI decision-making scenarios, and actively preventing algorithmic and data biases to ensure fairness and non-discrimination. Paramount to this approach is also the commitment to the safety and reliability of collaborative robots and AI, guaranteeing both physical and psychological well-being in these collaborative environments.

Ultimately, human-centricity within Industry 5.0 aims to create an industrial future where technological progress is harmonized with societal well-being, providing prosperity while respecting planetary boundaries and prioritizing the well-being of workers at the heart of every production process. The transition to Industry 5.0 requires digital leaders who can guide their teams through technological change, fostering a culture of continuous learning and adaptability. Leaders must champion the ethical integration of technology and demonstrate accountability for their actions. A profound cultural shift is needed within organizations to elevate human values, foster unity, and empower individuals to strive for excellence together. Employee engagement is recognized as a key outcome of effective competence management, necessitating management to prioritize well-being in their strategic decisions.

Hiring

Companies across various industries, including retail, banking, healthcare, pharmaceuticals, and technology, have adopted autism hiring initiatives. SAP (2024) has been recognized for its best practices in hiring employees with autism and Parkinson's disease. The organization has gone beyond simply accommodating autistic employees to realizing the benefits of innovating, including modifying hiring practices (Khan et al., 2023). SAP, (2024) offers an Autism *at Work* initiative that has been in place since 2013 and now has more than 240 employees in 16 countries, with roles in AI, cybersecurity, data analytics, software development, Finance, HR, and other areas across the organization. The retention rate is approximately 90% for SAP employees in the Autism at Work program (BluewaveSELECT, 2024). Another example is Dell, which was the 2023 recipient of the *Autism Inclusion Company of the Year* for innovation in hiring and retaining autistic employees at a 96% rate (Business Wire, 2023). For example, neurodiverse candidates submit a professional portfolio, which they share with the interview team, and then participate in skills-based interviews to demonstrate their ability to perform specific tasks required for the job (McGlauflin, 2023).

Utilizing the right data for model building, combined with algorithmic training and testing, managers can leverage HR technology to develop models that reduce bias in sourcing candidates and screening resumes (Novartis, 2020; SAP, 2025).

Managers must continue to evaluate technology for potential biases to ensure that selection tools, for example, do not disproportionately rule out qualified candidates from protected classes. HR departments utilize AI tools to match applicants with jobs based on their strengths, enabling high-impact hiring practices that improve both the quality and efficiency of matching top applicants with jobs in an expedited manner. For example, SAP, (2025) has developed AI-powered solutions for assisting with creating job descriptions, screening candidates, and scheduling interviews, among other talent management activities. Beginning with the application stage, organizations that offer AI chatbots help employees with disabilities in completing pre-hire documents, scheduling interviews, and answering questions (Nacheva, 2024). HR managers can consider utilizing AI tools to assist in developing interview guides and some of the hiring activities, thereby allowing more time to train hiring managers on interviewing and selecting the best candidates (Farrar et al., 2025). Additionally, companies can offer manager training on how to interview applicants with disabilities and provide support to individuals with disabilities throughout the hiring and onboarding processes.

Onboarding

Onboarding ensures that new hires are well-integrated, connected with other team members, and aligned with performance expectations. Typically, there is close coordination between HR and hiring managers. AI tools have the capability of prompting hiring managers when onboarding activities and actions are overdue, ensuring they provide employees with resources and documents on a timely basis (Farrar et al., 2025). Smaller organizations have integrated AI tools with Google Docs to simplify access to onboarding materials, enabling the quick location of relevant information, such as standard operating procedures in employee handbooks (Korba et al., 2025). Larger companies, such as Walmart, use apps for new hires to access orientation documents and annual enrollment information. Augmented reality assists with training employees. Employees can also use the "Ask Sam" app, based on generative AI, to ask questions about work schedules, product locations, or instructions for carrying out work tasks. Associates at the corporate office are experimenting with innovative ways to work using generative AI tools, which is crucial for employee engagement while also leveraging technology to enhance customer experience (Marr, 2024; Walmart, 2025).

Dell introduces employees to a network and provides professional development resources for neurodiverse employees. Other companies have also become innovative in their inclusive hiring practices to attract workers in the community who meet business requirements. An automotive company sought to hire individuals experiencing homelessness and provide them with resources, including shoes, to support their employees. Other companies that hire workers with disabilities have learned about the transportation needs of their employees, including getting to and from work, and have partnered with local organizations or agencies to make transportation possible (SAP, 2024).

Training and Continuous Learning
Industry 5.0 necessitates continuous learning and reskilling to keep pace with evolving technologies. The Gallup State of the Global Workplace: 2025 Report strongly supports continuous learning, emphasizing the crucial role of manager development in cultivating an engaged and adaptable workforce. The report indicates that managers who receive training in best management practices are 22% more engaged, and their teams, in turn, experience an 18% increase in engagement. Gallup explicitly identifies training every manager and teaching coaching skills as concrete actions to reverse the current engagement crisis. They emphasize that leadership is not about command-and-control, but rather about coaching, with managers trained in these techniques seeing 20–28% improvements in performance and their teams achieving up to 18% higher engagement. Human resource management (HRM) practices play a crucial role in digital skills training, digital leadership development, and programs designed to enhance employee engagement. The development of skills that emphasize creativity, flexibility, and open-minded thinking, alongside digital and technological abilities, is essential. For instance, engineering education is being rethought to train future engineers in technology, data, and knowledge fluency, to create more human-centric industries.

Managers should carefully discern how AI can expand the learning capacity of the workforce. AI tools can personalize and customize training content, evaluate it, and even translate it into different languages for use across various geographic locations. Wireless technology can provide real-time performance feedback, along with automated instructions, as employees perform tasks. Some chatbot features generate text chats, provide task reminders, and track the progress of learning and applying new skills. Human flourishing managers can use AI as a tool to identify when additional employee support is required. Adaptive leaders can utilize software results to provide on-the-spot coaching and adjust training as needed, based on employee engagement levels with the training content (Noe, 2024).

9 Conclusion

In conclusion, Industry 5.0 represents more than a technological revolution, it is a cultural and leadership transformation that redefines value creation around human well-being, inclusivity, and resilience. By integrating frameworks such as the Burke–Litwin Model and Complexity Leadership Theory, leaders are positioned not only to navigate complexity but also to create environments where people thrive and technology serves as an enabler rather than a replacement. The evidence from organizations like Walgreens and Sephora demonstrates that human-centric leadership yields measurable gains in engagement, productivity, and innovation, while inclusive AI tools expand opportunities for employees with disabilities. Ultimately, the realization of Industry 5.0 lies in leaders who embrace courageous humility, ethical responsibility, and collaborative vision, driving organizations toward sustainable success where human flourishing and technology are integrated.

References

Ahmmad, J., Al-Dayel, O. A., & Khan, M. A. (2025). AI-assisted technology optimization in disability support using fuzzy rough MABAC decision-making. *Scientific Reports, 15*, 18335. https://doi.org/10.1038/s41598-025-02362-8

Barnett, N. (2024, December 3). Empowering potential: AI's role in disability inclusion. Microsoft Accessibility Blog. https://blogs.microsoft.com/accessibility/empowering-potential-ais-role-in-disability-inclusion/.

bluewaveSELECT. (2024, October). The neurodiversity in SAP report 2024. https://bluewaveselect.com/wp-content/uploads/2024/10/The_Neurodiversity_in_SAP_Report_2024.pdf

Breque, M., De Nul, L., & Petridis, A. (2021). Industry 5.0: Towards a sustainable, human-centric and resilient European industry. *European Commission, Directorate-General for Research and Innovation*. https://doi.org/10.2777/308407

Burke, W. W., & Litwin, G. H. (1992). A causal model of organizational performance and change. *Journal of Management, 18*(3), 523–545.

Business Wire. (2023, July 13). Disability: IN announces 2023 Inclusion Awards with winners in 12 categories: Walgreens Boots Alliance is named Employer of the Year, and Dell Technologies is named Autism Inclusion Company of the Year. *Business Wire*. https://www.businesswire.com/news/home/20230713666878/en/DisabilityIN-Announces-2023-Inclusion-Awards-with-Winners-in-12-Categories-Walgreens-Boots-Alliance-Named-Employer-of-the-Year-Dell-Technologies-Named-Autism-Inclusion-Company-of-the-Year

Deloitte. (2025). Global Human Capital Trends. Is there still value in the Role of Managers? https://www.deloitte.com/us/en/insights/topics/talent/human-capital-trends.html#role-of-managers

European Commission. (2020). *Directorate-General for Research and Innovation, industry 5.0—Human-centric, sustainable and resilient*. Publications Office. https://data.europa.eu/doi/10.2777/073781

EY. (2024). *GenAI for accessibility: more human, not less: How does Microsoft 365 Copilot impact the working experience of people living with disability and/or neurodivergence?* https://www.ey.com/content/dam/ey-unified-site/ey-com/en-uk/services/ai/documents/ey-gen-ai-for-accessibility-more-human-not-less.pdf

Farrar, P., Johnson, Z., Mondesir, M., & Hayes, M. (2025, June 25). *Agentic AI for HR: From automation to true workforce transformation* [webinar]. HRCI. https://events.zoom.us/ev/AqvfVzuVulSFHe73JcCucUkCFcji2nCKtvJkeqsWhYZJkzMLeciS~AoKLwJMXqL5Jsq-tOnsl6_u4IKvtrsYcnxAjjcHTpoENbHMfd5_CWzasnw

Fiveable. (2024, July 31). Collaborative Robots (Cobots)—Intro to Industrial Engineering. Accessed June 23rd 2025, from https://fiveable.me/key-terms/introduction-industrial-engineering/collaborative-robots-cobots

Gallup, Inc. (2025, April 23). State of the global workplace: 2025 report. Gallup. https://www.gallup.com/workplace/349484/state-of-the-global-workplace.asp

Goodhue, D. L., & Thompson, R. L. (1995). Task-technology fit and individual performance. *MIS Quarterly, 19*(2), 213–236. https://doi.org/10.2307/249689

Halter, C. (2025, June 2). Employee Discretionary Effort & Transformation. [conference presentation]. Human flourishing conference, Anderson University, Anderson, SC. https://www.canva.com/design/DAGeKPm-sB4/n9lFjRFRZLttfQdoWff6Qw/edit

Hamideh, K. S., Kirchhoff, B. M., Adolph, L., & Bächler, L. (2025). A study on Chatbot development using no-code platforms by people with disabilities for their peers at a sheltered workshop. *Technologies, 13*(4). https://doi.org/10.3390/technologies13040146

Hanson, W., & Moore, J. (2023). Conditions and pressures promoting inclusive organizations (chapter 12). In J. Ingold & P. McGurk (Eds.), *Employer engagement: Making active labour market policies work*. Bristol University Press.

Ito, A. (2025, August 7). *AT&T CEO's memo just made it official: Workplace loyalty is dead. Business Insider*. https://www.businessinsider.com/at-t-ceo-memo-workplace-loyalty-is-dead-2025-8

Khan, S. M., Fowler, B., Adams, R., & Vainauskas, J. (2023). Neurodiversity in the workplace: Lessons learned from 10 years of employer initiatives. *Journal of Organizational Psychology, 13*(2), 28–37.

Korba, S., Hamirani, Q., & Hayes, M. (2025, June 20). *Hype to high-impact AI* [webinar]. HRCI. https://events.zoom.us/ev/Au0kerzEtqoZ2H-OI_fCvwI4pzXmPxiP1aqsEefNzkOVFek7f97F~As6kqIclVFgCkgTGExWSVup8T4L57PPLkSYsUr7WnglNW3_9j6c9M7lDjQ

Marr, B. (2024, February 15). The amazing ways Walmart is using generative AI. *Forbes*. https://www.forbes.com/sites/bernardmarr/2024/02/15/the-amazing-ways-walmart-is-using-generative-ai/

McGlauflin, P. (2023, October 27). How companies are revamping their interview process for neurodivergent talent. *Fortune*. https://fortune.com/2023/10/27/ey-dell-ubisoft-neurodivergent-employees-hiring/

Microsoft. (n.d.). *Speaker coach in Microsoft teams meetings*. Microsoft Support. https://support.microsoft.com/en-us/office/speaker-coach-in-microsoft-teams-meetings-30f50d15-5f62-4e09-b3bf-cadeb806386a

Moore, J., Maxey, E. C., Waite, A. M., & Wendover, J. D. (2020a). Inclusive organizations: Developmental reciprocity through authentic leader-employee relationships. *Journal of Management Development, 39*(9/10), 1029–1039. https://doi.org/10.1108/JMD-05-2019-0211

Moore, J., Hankins, S., & Doughty, S. E. (2020b). Successful employees with disabilities through the lens of Bronfenbrenner's ecological systems theory: A case study at Sephora. *Journal of Business Diversity, 20*(5), 10–19. https://doi.org/10.33423/jbd.v20i5.3924

Moore, J., & Hanson, W. (2023). Sephora—Journey to inclusive workplace—Let us belong philosophy (chapter 13). In J. Ingold & P. McGurk (Eds.), *Employer engagement: Making active labour market policies work*. Bristol University Press.

Moore, J., & Hanson, W. (2022). Improving leader effectiveness: Impact on employee engagement and retention. *Journal of Management Development*.

Moore, J., Hanson, W., & Maxey, E. C. (2022). Beyond inclusion: Team impact of hiring people with disabilities at Sephora and Walgreens. *Organization Development Journal, 40*(3), 66–76.

Nacheva, R. (2024). Hired by artificial intelligence: Digital inclusion practices for people with disabilities. *Journal of Underrepresented & Minority Progress, 8*(2), 10+. https://link.gale.com/apps/doc/A816649484/OVIC?u=anderson_cl&sid=summon&xid=6b6f6e00

Nahavandi, S. (2022). Industry 5.0—A human-centric solution. *Sustainability, 14*(3), 1287. https://doi.org/10.3390/su14031287

Noe, R. A. (2024). *Employee training and development* (8th ed.). McGraw-Hill Education.

Novartis, Inc. (2020). Novartis' commitment to the ethical and responsible use of artificial intelligence (AI) systems: A human-centered approach in using artificial intelligence to reimagine medicine. https://www.novartis.com/sites/novartis_com/files/novartis-responsible-use-of-ai-systems.pdf

SAP Inc. (2021, October 20). Inclusive meeting cards: Free to download! *SAP Community*. https://community.sap.com/t5/technology-blogs-by-sap/inclusive-meeting-cards-free-to-download/ba-p/13507139

SAP Inc. (2024, May 20). Close the labor gap with socially conscious hiring. *SAP Blog*. https://www.sap.com/croatia/blogs/close-the-labor-gap-with-socially-conscious-hiring.

SAP Inc. (2025). AI in human capital management. SAP. https://www.sap.com/hk/products/hcm/ai.html.

Sigstad, H. M. H., Garrels, V., Dean, E. E., & Wendelborg, C. (2025). Characteristics of workplaces and employers that embrace cognitive diversity on the work floor. *Intellectual and Developmental Disabilities, 63*(2), 105–119. https://doi.org/10.1352/1934-9556-63.2.105

Sodales Solutions Inc. (n.d.). *The role of AI in creating a more inclusive and supportive work environment. Sodales Solutions*. https://www.sodalessolutions.com/the-role-of-ai-in-creating-a-more-inclusive-and-supportive-work-environment/

Uhl-Bien, M., Marion, R., & McKelvey, B. (2007). Complexity leadership theory: *Shifting leadership from the industrial age to the knowledge era. The Leadership Quarterly, 18*(4), 298–318. https://doi.org/10.1016/j.leaqua.2007.04.002

Walmart. (2025, June 24). Walmart unveils new AI-powered t[...]. *Walmart Corporate News*. https://corporate.walmart.com/news/2025/06/24/walmart-unveils-new-ai-powered-t

Whelpley, C. E., & Perrault, E. (2021). Autism at work: How internal and external factors influence employee outcomes and firm performance. *Journal of General Management*, 46(3), 210–219. doi:https://doi.org/10.1177/0306307020961987 (Original work published 2021)

World Health Organization. (2023). *Disability and health*. WHO Fact Sheet. Retrieved from https://www.who.int/news-room/fact-sheets/detail/disability-and-health

AI-Driven Leadership: Strategies and Tactics

Seth-Aaron Martinez

Abstract This chapter explores the critical role of leadership in guiding organizations through the integration of artificial intelligence (AI), with emphasis on both analytical AI and generative AI (GenAI). Drawing upon professional practice and emerging scholarship, the discussion highlights the conditions leaders must establish to ensure effective adoption of AI technologies, including the development of new work practices, role relations, and organizational networks. A central argument of this chapter is that successful AI integration requires a culture of data excellence, where governance, ethics, and cross-functional collaboration underpin sustainable value creation. Leadership responsibilities are framed through three key activities—framing, structuring, and evaluating—that align technological adoption with organizational identity, values, and strategic direction. I further argue that leaders who intentionally design conditions for learning, collaboration, and ethical stewardship are best positioned to navigate AI's transformative impact. By fostering readiness across data, governance, and culture, leaders enable enterprise-scale deployment of AI. The chapter's synthesis of practice-based insights and scholarly research provides a framework for understanding how effective leadership can proactively shape organizational futures in the age of intelligent technologies.

Keywords Leadership · Artificial intelligence · Leading AI

1 Introduction

Several years ago, while managing Adobe's global artificial intelligence (AI) and machine learning (ML) training program, I witnessed firsthand the far-reaching implications of AI capability building. The AI/ML training program primarily targeted 5500 Adobe technical professionals (i.e., software engineers, computer scientists, data scientists) but quickly expanded to include a secondary audience of product managers, sales staff, customer/client support, and HR—multiplying the program's

S.-A. Martinez (✉)
College of Education, Boise State University, Boise, ID, USA
e-mail: SethAMartinez@boisestate.edu

A. M. Waite et al. (eds.), *Leading in Industry 5.0*, Management and Industrial Engineering, https://doi.org/10.1007/978-3-032-19207-3_11

impact across the organization by more than two-fold. Across my three years in this role, the experience revealed to me both the technical affordances of AI and the leadership practices essential for its adoption.

Today, nearly every organization faces the challenge of adapting to the disruptive realities of ML and AI (Ransbotham et al., 2022). This chapter synthesizes lessons from my industry practice and scholarly research to examine how leaders can enable effective AI integration. Specifically, I explore how leaders create the right organizational conditions, foster a culture of data excellence, and make informed strategic decisions. The discussion concludes with how leaders can anticipate and guide systemic change as AI continues to transform work at scale.

2 Creating the Right Conditions

2.1 Analytical AI Versus Generative AI

It is important to first distinguish the characteristics of the two types of algorithms: analytical AI and generative AI (GenAI). Analytical AI has generally been used for quantitative tasks. Analytical AI tools rely on specifications and algorithms that measure precise points of interest. Analytical AI models make precise measurements and work through algorithmic structures, such as decision trees, to quantify the likelihood of a future outcome based on given inputs. One familiar use case is the presentation of content on social media and other digital applications. Based on user behavior, analytical AI models are trained to predict what type of content is presented to a specific user.

Recently, since the release of ChatGPT, interest in large language models (LLMs) has grown (McKinsey, 2025; MohanaSundaram et al., 2024). LLMs, which is a subset of GenAI models, aim to generate human-like text and content. While both LLMs and other machine learning models leverage data and algorithms, GenAI models like LLMs focus on creating new content, whereas many traditional analytical AI models focus on making predictions or classifications based on existing data. As Table 1 highlights, organizations would be wise to keep these distinctions in mind as they plan to build internal AI capability.

In some ways, one of the clearest contrasts between analytical AI and generative AI (GenAI) lies in the data and analyses that each use. Analytical AI typically works with your organization's internal, structured data repositories: the tables, logs, and rows of information accumulated over time. For instance, you might use analytical

Table 1 Distinction between analytical AI versus generative AI

	Purpose
Analytical AI	To analyze, measure, and predict
Generative AI	To generate and create

AI to predict customer churn based on past purchases, support tickets, or account activity. These models are built to predict specific outcomes from historical data. A doctor might use analytic AI to predict a cancer diagnosis based on radiologic images. Aviation experts at GE might detect single engine problems across their fleet of planes. Each of these examples use analytical AI to predict specific instances of important outcomes, and in some cases automate the organizational response based on the prediction.

By contrast, GenAI refers to the large language models (LLMs) that are trained on massive, unstructured datasets. These models are optimized not to predict an outcome, but to generate new data or new content: a blog post, a code snippet, a product description, a legal clause, a strategic summary. The analyses underpinning these models also differ somewhat. These differences in data and algorithm are real and important. But I do not suggest one is better than the other. Rather, each plays a different role, and increasingly, hybrid use cases are emerging where the output of analytical AI becomes the input to GenAI, and vice versa. Increasingly, organizations are learning to use both together (Pecan, n.d.). When these different functions are considered, it becomes clear that the structures leaders create within an organization as it builds algorithmic capabilities must be tailored to the desired outputs.

2.2 *Algorithmic Capabilities and Change*

Building *algorithmic capabilities* refers to an organizations' implementation and adoption of analytical AI and GenAI tools (Ghani et al., 2022; Horvat et al., 2023). This means ensuring that stakeholders throughout the organization understand, contribute to, and utilize new tools. While the specific objectives organizations want to achieve by building algorithmic capabilities differ, to ensure the successful adoption of these algorithmic capabilities, there are broader change management considerations that must be understood by all. However, before exploring analytical AI and GenAI in greater depth, I will briefly consider the foundation that should be established by leaders when building algorithmic capabilities.

To be clear, building algorithmic capabilities is a transformational change. Leaders and decision-makers need to understand that the first step in a truly transformational change is the development of new skills and/or work practices throughout the organization (Hossain et al., 2025). As individuals and teams begin to utilize new skills, they rely on and develop new organizational networks. Teams that were once disparate might find themselves working closely together or vice versa. It is up to leaders to frame these changes and build structures that allow for the development of new skills and collaborative networks across the organization (i.e., organizational networks). To understand a process such as building algorithmic capabilities, leaders and stakeholders need to understand that a successful development of algorithmic capabilities requires changes to three spheres: work practices, role relationships, and organizational networks (Alshdaifat et al., 2025).

2.2.1 Work Practices

Work practices refer to the tasks, jobs, and social relationships that constitute work within organizations (Aroles et al., 2019). Barley, (2020) emphasized the importance of empirically studying how technological change impacts and transforms these practices. Building algorithmic capabilities will result in changes in work practices, the use of new data, new measurements, and new analyses. Individuals and teams will require new skills, and the organization itself might require new professions.

2.2.2 Role Relations

Role relations refer to the social relationships and interactions between different jobs, tasks, and roles that constitute work practices within organizations (Marwell & Hage, 1970). In the transforming workplace, organization-wide change and the adoption of new skills require close collaboration across teams and strong role relations (Gupta et al., 2010). To build algorithmic capabilities, this collaboration often occurs between teams that have had little prior contact. It is important for leaders to ensure that data teams are not working in silos but rather have access to the domain experts they need to identify essential features and tags.

2.2.3 Organizational Networks

Barley, (2020) described organizational networks as patterns of relationships between roles and positions which constitute an organization's social structure. In any analytical AI or GenAI integration, leaders should maintain a clear view of their team's objectives and priorities and understand how the organizational networks function within their organization. They can also benefit from the use of these technologies. Regardless of how carefully planned such changes are, decisions will need to be continuously made, and relying on clear objectives can help leaders make the most beneficial trade-offs throughout the process.

2.3 *Organizational Readiness*

It is necessary to next outline what *organizational readiness* for use at scale really means: organizational alignment across the enterprise (Hradecky et al., 2022). Leaders who assess and prepare across these five dimensions will be far better positioned to move from scattered pilots to enterprise-scale, high-value deployments (Nasrollahi & Ramezani, 2020). Readiness requires creating the conditions in which GenAI can work at scale, reliably, and safely, over time. It is most helpful to consider readiness as a layered system (Hossain et al., 2024). While there are likely more,

I will discuss four distinct layers of readiness: data, collaboration, governance and incentives, and implementation.

2.3.1 Data Readiness

GenAI is entirely dependent on the data it draws from. If an organization's knowledge management or internal documentation are disorganized, your processes undocumented, or your systems siloed, your GenAI tools will reflect that (Baten, 2024; Seamans & Raj, 2018). Foundational readiness begins with making internal work and knowledge visible. That might mean restructuring repositories, reformatting documents, or mapping workflows that have never been clearly diagrammed.

2.3.2 Collaboration Readiness

GenAI blurs traditional boundaries between data science, domain expertise, and operations. Therefore, success depends on true cross-functional collaboration. Data teams should shift from considering themselves only as model builders and begin seeing themselves as infrastructure architects, tooling designers, and feedback integrators (Vaittinen & Martinsuo, 2019). Meanwhile, domain experts must take ownership of prompts, data work, ongoing feedback and evaluation, and continuous improvement.

2.3.3 Governance and Incentives

Governance refers to who is doing what, where, and with what constraints. Organizations scaling GenAI must define access rights, escalation paths, and monitoring policies (Harvard Law School Forum, 2025). They must also reward the often-invisible work that makes GenAI successful: curating datasets, documenting workflows, correcting failures, and refining prompts.

2.3.4 Implementation

There is a level of discipline required to successfully implement digital change at scale. Implementation must become an iterative, observable process: phased rollouts, instrumentation for tracking failure modes, and infrastructure for learning from real-world use. However, organizational readiness for generative AI is about more than infrastructure; rather, it centers on the clarity of process, shared responsibility, and institutional memory. Effective leaders treat readiness as the system that makes innovation sustainable.

2.4 The Role of Leadership

When approaching organizational changes, such as implementing AI models, leaders need to be aware of their role in three key activities. First, leaders must *frame* the change in ways that are specific to both the organization and the individuals that comprise it. Second, leaders need to *structure* the change in terms of the steps that will be taken during implementation and describe what the organization will look like once the change has been accomplished. Finally, leaders must be prepared to *evaluate* the changes through the development and deployment processes. These activities help leaders align AI integration with their organization's identity, ensure workflows and governance support adoption, and build feedback systems that promote learning and improvement. While these activities matter for both predictive analytical AI and GenAI, they often look quite different in practice. Predictive AI use cases typically involve clear problem definitions, structured data pipelines, and measurable outputs, making them easier to scope, centralize, and evaluate with traditional business metrics. By contrast, GenAI introduces open-ended workflows, fuzzy output boundaries, and faster diffusion across creative, customer-facing, and operational roles. These differences push leaders to develop more holistic framing that helps people understand leaders' vision for the future, organizational structures that support federated governance of AI agents and vast kinds of data, and sustainable, interactive evaluation of AI model outputs. I discuss each of the three key leadership activities of framing, structuring, and evaluating next.

2.4.1 Framing

Over the past two decades, organizations have undergone several changes under what has broadly been called *digital transformation* (AWS, n.d.). The move from brick-and-mortar stores to online markets and the migration of services to the cloud are two common examples of digital transformations. The process of digital transformation has shown that organizations can approach such changes in a variety of ways; however, success often depends on how prepared individuals and teams throughout the organization are before such changes are implemented. Framing refers to how leaders define the nature of a problem or opportunity. Such a lens largely determines how the team will respond to it. Research shows that when leaders frame challenges clearly and consistently, people will align their workflows, their priorities become more visible, and energy focuses (Carreño, 2025). Conversely, when framing is vague, people will default to their own interpretations: they become misaligned, they duplicate effort, or they compete over who is right.

Effective framing helps teams understand why GenAI matters, where it fits in the organization's identity and goals, and how to engage with it productively. This is especially important given the speed and ambiguity surrounding GenAI adoption. A strong framing narrative can help reduce anxiety, channel experimentation, and align stakeholders around shared ambitions.

So, what does effective AI framing consist of? Effective framing sets the foundation for collective sense making. Strong framing defines the organization's shared reality, thereby providing clarity and unity of focus. Additionally, leaders' framing activities need to be responsive to the fact that GenAI can be seen as a threat. With GenAI, leaders might need to be more flexible, and they should frame this flexibility. The overall frame can be one of trying and failing, and when something does succeed, being ready to take steps to scale it.

It is perhaps unsurprising, then, that organizations looking to implement GenAI tools are drawing on leadership with diverse backgrounds. Technical leaders, such as the chief data strategy officer at the advertising agency M&C Saatchi, as well as creative leaders, such as head of global creative strategy at Coca-Cola, are being tapped to lead GenAI integration efforts (Baskin, 2024). Innovation is likely to be an essential element of framing GenAI adoption for the foreseeable future.

One of the most difficult aspects of GenAI integration is how rapidly the field is moving. Another executive likened GenAI implementation to putting a spare tire on a car. One nut could not be overly tightened, instead it was a gradual process of trying a GenAI solution here, another there, and adjusting until these GenAI tools were working across the organization (Eastwood, 2023).

The rapid pace of development and the ever-increasing number of tools on offer leads to one of the main challenges for leaders: simply getting started. The initial exploratory steps into GenAI implementation can seem daunting and for many current knowledge workers it will take dedication to learn what these GenAI tools do and how to use them. (This relates to the structural aspect of time discussed in the previous module.) Leaders can frame the process in a way that helps overcome initial concerns and the paralysis of choice by establishing an AI-first team culture and fostering a "*we'll learn together*" attitude.

Building this team culture supports tentative first steps and encourages the gradual adoption suggested above. Leaders can frame such a culture as experimentational and messy where mistakes are expected as the team builds competence. In this way, the transformative change itself can be framed as a relearning experience rather than a disruptive overhaul. For example, professionals now view platforms such as Google Drive or tools such as Microsoft Teams or Slack as necessary to get their work done. Anchoring GenAI as another tool to learn just as "everyone needs to know how to browse the internet" can help stakeholders meet the challenge of GenAI implementation.

However, each organization has its own culture and ways of communicating. So the work of framing provides a cultural container and supports the emotional management for change. To appropriately frame the upcoming change, leaders should consider addressing three basic issues: values, identity, and technology.

Values. Every organization embodies a unique set of values. These values guide decision-making processes and generally guide different teams' approach toward their assigned tasks. For instance, some organizations value innovation and thinking outside the box and teams will approach problems with the expectation that innovative solutions will be considered and possibly implemented. Other organizations value structure, predictability, and minimization of risk. Before an organization begins taking steps toward transformative change, leaders must clearly communicate to stakeholders how this change is tied to organizational values. That is, for organizations that value customer or client interaction, the reason for building algorithmic capabilities must reflect how this will contribute to and improve customer or client relationships or experience. Similarly, for an organization that values quality in manufacturing, leaders must explain how the analytical AI or GenAI integration will address pain points in manufacturing processes or improve employee operations in some way.

Identity. Leaders likely know that change can be uncomfortable and even resisted. When planning to build algorithmic capabilities, stakeholders throughout the organization must be given a compelling reason for the proposed changes. The "why" should be grounded in organizational culture and identities. Additionally, leaders need to communicate a vision of what the organization will become if it succeeds in the transformation. Individuals and teams within an organization share a collective identity. When undergoing transformation, it is helpful to anchor oneself on the intended result, which is the new identity. Leaders can help ensure the success of transformational change by providing a vision of the collective identity that will emerge. Transformational change can also often result in users reimagining their professional identities. By framing potential professional identities, leaders can make this reimagining an anticipated and empowering change for users.

Technology. There are many ways that the technology of analytical AI and GenAI can be viewed. These tools can be seen as a threat, reflected in headlines stating that GenAI models will replace jobs. They may be seen as a tool available for workers to address a task or problem. Or they may even be seen as colleagues. Ultimately, however, it is up to leaders to frame the organization's vision of these technologies.

2.4.2 Structuring

While framing is the communicative responsibility of leaders, structuring is the essential operational planning they perform. Structuring is about the practicalities of realizing each step of implementation. Leaders and decision-makers are constantly structuring organizations in a variety of ways. Even the basic seating floor plan is an important structuring practice that leaders undertake; choosing where each team sits in relation to other teams is a key consideration for how an organization will operate. When an organization is preparing for a transformation, such as building algorithmic capabilities, leaders should consider several structural elements. By anticipating these structural elements, leaders enable their teams to meet the challenge and realize the ultimate vision.

Structuring for GenAI demands a fundamental shift in digital governance. Unlike analytical AI, which is typically built by data scientists to answer well-scoped business questions, GenAI is often user-driven, exploratory, and embedded directly into day-to-day workflows. This ambiguity and flexibility make it difficult to anticipate how tools will be used, and by whom, which in turn elevates the importance of robust digital governance. Leaders must design systems that go beyond one-time model deployment to continuously audit access and usage, monitor inputs and outputs, track user interactions, and ensure data quality over time.

Leaders who set up their tech developers for success and invest in these digital governance capabilities are far more likely to succeed. As you reflect on GenAI in your organization, you can consider: "*Have we structured this environment to make safe, consistent, and reliable enterprise AI use possible?*" Because without the right structure, even the best AI models are not going to go very far.

Time. It is difficult to pursue change if people lack the necessary time. Leaders must set aside time for training, workshops, and opportunities to hear employee feedback. When building out algorithmic capabilities, for example, allowing individuals the time to play around with analytical AI and GenAI tools to see how they can configure them to their personal workflows is key. Change management research highlights the tendency of falling back into familiar habits and routines when stakeholders are not given the time to grow accustomed to new operational mechanisms (Zhang et al., 2025).

Accountability. There are several potentially confusing aspects of accountability as an organization undergoes transformational change. As changes are being implemented, questions can arise as to who is responsible and ultimately accountable for each step of implementation. As algorithmic capabilities are being built, there are essential contributions from multiple teams, e.g., data scientists, domain experts, etc. While several teams may be responsible for contributions related to the change, accountability must rest with only one person. Leaders must make clear what aspects each team and team member is accountable for. Furthermore, individual and team roles may change, sometimes significantly, pre- and post-transformation. As leaders communicate "who we will become," it is important that they structure the landscape of accountability post-change so that stakeholders can have that collective identity clearly in mind.

Collaboration. Leaders also need to structure the nature of collaboration. That is, they should plan out what each team is doing during each step in the process and how complementary tasks can be divided or shared across teams. When building algorithmic capabilities, it is particularly important that the processes of collaboration are structured clearly and are easy to implement. To do so, engineers and data scientists need to work with domain experts to develop useful models. In fact, the working relationship between users and developers can be an important predictor of success (Vishwanatha et al., 2023).

2.4.3 Evaluating

The purpose of evaluation is to determine and focus on what is working and adjust what is not. Evaluation, done correctly, is what helps make an organization's GenAI safer and more reliable over time. Evaluating analytical AI and GenAI capabilities requires a new skill set that leaders and their teams need to develop as part of the implementation process. Building algorithmic capabilities is a transformative change, one that includes not only leaders but also input from other stakeholders throughout the organization. For a brief overview, I consider some of the skills in the evaluation toolkit.

1. **Articulate the strategy early**. The tools for measuring and evaluating analytical AI and GenAI models will be developed with the models themselves. However, to ensure these are comprehensive, leaders and other stakeholders must be able to articulate the desired strategy fully upfront.
2. **Identify needed data/evidence**. Predictive analytical AI models are only as useful as their users perceive them to be. For data science teams to develop useful models, they need the users' domain expertise. For users to rely on the models, both the data teams and the domain experts need to learn what data is available and what data is ultimately useful. Leaders must also develop their ability to identify needed evidence so they can adapt structures to facilitate collaboration between relevant teams.
3. **Explicitly narrate trade-offs**. Leaders, data scientists, and domain experts all need to learn how to communicate their metrics across teams. Teams in siloed organizations often develop specific metrics and jargon. However, as algorithmic capabilities are built throughout an organization, teams need to be able to express how decisions are likely to affect them. Communicating how trade-offs are made for a team will aid both the data team in creating a more useful model and the users by clarifying potential trade-offs.
4. **Decide on relevant ROIs**. Leaders can demonstrate and evaluate the success of their initiatives through return on investment (ROI) calculations. ROIs are particularly useful in the learning and development (L&D) stages, as they can help maximize the effectiveness of L&D programs (Rudy, 2023). In the context of analytical AI implementation, ROIs can help leaders demonstrate the model's effectiveness to relevant stakeholders within the organization.

3 A Culture of Data Excellence

A culture of data excellence refers to the collective values, beliefs, practices, and organizational systems that support the effective governance, management, and utilization of data. Establishing such a culture is a prerequisite for the development of reliable machine learning and generative artificial intelligence models, as these depend upon the integrity and accessibility of organizational data. This next section will examine how digital management practices, and the creation of a robust data infrastructure serve as essential foundations for cultivating a culture of data excellence. I situate data within a business context and highlights common challenges organizations encounter in their efforts to leverage data effectively. I then introduce the concept of data as a product (DaaP) as a strategic framework and outline practical steps leaders can adopt to operationalize DaaP.

3.1 *Mapping the Problems*

Leaders of all types should be focused on fostering a strong data culture within their function and the overall organization. Commitment from data leaders and an open-ended, informed dialogue with decision-makers is imperative for cultivating a good data culture. Yet, many companies still struggle to manage and leverage data effectively. In the following sections, I will explore the common issues that companies experience related to data quality, data misuse, data silos and integration, and data governance.

3.1.1 Data Quality

Data quality has been a long-standing issue in the industry, and while it remains a persistent challenge, the landscape has evolved with the rise of GenAI. Despite the increased pressure on data teams to deliver AI products, many data leaders feel their data is not yet ready for such initiatives (Nemhe, 2024). Data quality is imperative to ensure the accuracy of the data used by AI systems.

Incomplete Data. Missing values that can lead to inaccurate analysis and misinformed decisions. For example, in health care, missing patient records can lead to incorrect treatment plans and adverse health outcomes. If critical lab results are not recorded in the system, a physician might make a treatment decision based on incomplete information. As another example, in manufacturing settings, incomplete maintenance records for machinery can lead to equipment failures. If scheduled maintenance activities are not recorded, critical components may not be serviced on time, leading to unexpected breakdowns and production delays.

Inconsistent Data. Discrepancies in data formats or definitions across systems can create confusion and lead to errors. For example, discrepancies in data formats (e.g., different date formats or inconsistent naming conventions for financial instruments) can lead to errors in financial reporting and analysis. For instance, inconsistencies in transaction dates can result in inaccurate financial statements.

Outdated Data. Using outdated data can result in decisions that do not reflect current realities. For example, using outdated inventory data can result in stockouts or overstocking. If the inventory system is not updated in real-time, a retailer might believe they have more stock of a popular item than they actually do, leading to missed sales opportunities.

So, what is causing these issues? According to Sambasivan et al., (2021), ensuring data quality is often not simply a technical problem, but rather a social problem. Sambasivan et al., (2021) found that data scientists are not incentivized to do data work given an "overall lack of recognition for the invisible, arduous, and taken-for-granted data work in AI [leading] to poor data practices." Participants noted that data work is often expected to be done within tight deadlines and under great constraints. Despite the effort involved in data work, it is often not rewarded in the same way that developing algorithmic models is, therefore many data scientists choose to focus on models to advance in the field.

When managing data for AI-driven systems, it is crucial to move beyond just detecting data issues and focus on understanding and resolving the root causes of these problems. As data landscapes become increasingly complex, data issues can arise from multiple sources, including the data itself, the code used to process it, and the underlying systems or infrastructure. Effective data management requires visibility into all these components and the ability to correlate issues with specific changes in the system. This complexity means that simply cleaning data is no longer sufficient; businesses must adopt more sophisticated approaches to ensure reliable and trustworthy data for AI applications.

3.1.2 Data Silos and Integration

Data Silos. Isolated data systems prevent a comprehensive view of information, leading to fragmented insights and suboptimal decisions. For example, a large retail chain might operate both physical stores and an online store, but the data from these two channels might be stored in separate systems. The physical stores use a traditional point-of-sale (POS) system, while the online store uses an e-commerce platform. This separation prevents the company from having a unified view of customer behavior and inventory levels. As a result, the retailer struggles with inventory management, leading to stockouts in some locations and overstock in others, and misses opportunities for targeted marketing campaigns.

Integration Challenges. Difficulties in integrating data from multiple sources can hinder the ability to derive accurate insights. For example, a software company might develop an enterprise resource planning (ERP) system but face difficulties integrating data from its CRM, accounting software, and third-party applications. Each system uses different data formats and structures, leading to challenges in creating a unified data view. As a result, the company struggles to generate accurate and comprehensive business insights, impacting strategic decision-making.

Decision-making is an increasingly important skill for employees, with 34 percent of jobs requiring decision-making skills—such as problem-solving and prioritizing—in 2018 compared with just six percent in 1960 (Deming, 2021). However, when teams or individuals only have a limited view of the available data, decision making is likely to remain suboptimal. Therefore, the presence of data silos or a lack of data integration can present major obstacles to successful data and AI adoption.

Agrawal et al., (2022) described how decision making can be strengthened in nearly every field from transportation to finance to competitive sailing. A key point is that successful data integration can shift decision making ability, decentralizing it in some cases such as ride sharing apps and centralizing it in others such as human resource departments. Whether decision making becomes centralized or decentralized, the two main functions of synchronizing work and assigning resources require differing levels of access to available data for human-only decision making and algorithm-enhanced decision making. Algorithms produce more effective recommendations when given access to data from across traditional team or departmental boundaries and Agrawal et al. even suggest that new systems of communication will be needed to ensure data is not siloed and fully integrated, offering this insight:

3.1.3 Lack of Data Governance

Without proper governance, data can be mishandled, leading to issues like data breaches or non-compliance with regulations. This is especially true for multinational organizations. Differing regulatory landscapes across borders requires precise data governance to ensure that regulations are met for every market an organization operates in. Additionally, mishandling sensitive data can result in privacy violations and legal consequences. Lastly, if the data contains biases, it can perpetuate or even exacerbate inequalities, which can promote unfair or discriminatory decisions. Ensuring that data used for training GenAI or analytical AI models is as free of bias as possible requires careful planning at the data collection stage. For example, sample sets must be scrutinized for under- or overrepresentation of demographic groups.

Data Breaches. Cyber-attacks on large corporations have become commonplace. According to Jain and Ropple, (2018), cyber-attacks are "often state-sponsored [and] these breaches are insidious, difficult to detect, and may implicate personal information relating to millions of individuals." Current methods for protecting sensitive data fall short. Many different laws at the national and state levels require companies to have "reasonable" security, but this term is vague and hard to define given the ever-evolving nature of cyber threats. A change at the policy level is needed to encourage companies to collaborate with the government on reducing cyber risks. Identity and access defense should be prioritized given companies are increasingly migrating to the cloud, in turn placing more significance on security performance measurement (Chertoff, 2023).

3.2 Robust Data Governance

Data quality fundamentally centers on priorities and governance. Leaders can focus on data quality through the lens of priorities and governance to ensure data integrity, security, and compliance. These practices enable businesses to leverage data effectively while mitigating risks (Pahwa et al., 2025).

Data quality practices:

- Catalog and organize data: Maintain a comprehensive data catalog to ensure all data assets are well-documented and easily accessible. Implement metadata management tools to enhance the catalog with detailed metadata, lineage, and governance information. Regularly update and audit the catalog to ensure accuracy and completeness.

- AI Assistance: Use AI tools to automate and enhance data quality checks, speeding up the evaluation process.

  Data access policies:

- Define access controls: Clearly specify what data AI systems can access and protect sensitive information by redacting personal identifiers. Implement role-based access control (RBAC), where access rights are assigned based on user roles. Regularly review and update these roles to adapt to organizational changes.
- Centralized policy setting: Set data access policies at a central level but enforce them locally. Improve this by incorporating automated policy enforcement tools that ensure real-time compliance and monitoring. Additionally, provide training for local administrators to understand and implement these policies effectively.

  Monitoring and enforcement:

- Continuous monitoring: Actively monitor data inputs and outputs to ensure compliance with governance policies.
- Track changes: Keep track of how models evolve and adapt to real-world data, ensuring they remain compliant and effective.

  Integrating governance from the start:

- Frameworks and platforms: Implement governance frameworks and platforms from the beginning to manage data lifecycle efficiently. Choose scalable and flexible governance frameworks that can evolve with your organization. Incorporate industry best practices and standards (e.g., DAMA-DMBOK, GDPR compliance) to ensure robustness. Regularly review and update the governance frameworks to adapt to new technologies and regulations.
- Lifecycle management: Source, catalog, filter, and transform data systematically, integrating governance throughout the data lifecycle. Implement automated data lineage tracking and version control to ensure transparency and traceability throughout the data lifecycle. Use data quality dashboards and KPIs to monitor and manage data health continuously. Promote a data stewardship culture where data ownership and accountability are clearly defined.

3.2.1 GenAI-Specific Considerations

A recent McKinsey study found that 72 percent of those surveyed have used GenAI in their business over the last year (Singla et al., 2024). The emergence of GenAI and its use in the workplace has presented numerous novel hurdles that organizations must overcome before successfully implementing GenAI into their processes. At the forefront of these challenges is preparing the data that the GenAI models will use.

The same McKinsey survey found that 70 percent experienced data issues, including definition, governance, training, and integrating data into GenAI models (Singla et al., 2024). This research highlights the importance of data and the pivotal role business leaders and chief data officers (CDOs) will play in the future, since they will determine which data is relevant for successful GenAI implementation (Davenport & Tiwari, 2024). Successfully determining the data to be used and how to leverage it best can make or break the success of GenAI implementation in a business.

3.2.2 Ethical Data Handling

Segalla and Rouziès, (2023) recommended creating a special taskforce in charge of assessing projects that involve handling data. These taskforces should focus on five areas of data safety which they refer to as "the five P's:" provenance, purpose, protection, privacy, and preparation.

Provenance. It is important to consider the source or origin of data used by different companies and applications. Even when companies are transparent about their intentions, this is not sufficient to ensure ethical data collection. Companies should scrutinize the origins of both new and existing data which often possess *dark data*—unused, forgotten information like customer logs, emails, and machine-generated logs (Segalla & Rouziès, 2023). Despite being unstructured and hard to integrate, dark data holds significant value, prompting the development of tools to harness its potential.

Purpose. In corporate contexts, data collected with consent for a specific purpose is often repurposed without informing the data providers. This necessitates obtaining additional consent for new purposes. Customer data is often treated less carefully, viewed as a revenue source, and sold to third parties. Therefore, this practice is increasingly scrutinized. Leaders are wise to ensure that the data is not being used for a function different from its initial intent.

Protection. There were approximately 4000 data breaches in the United States in 2024, exposing the personal details of billions (Statista, 2025). Europe faces similar issues despite having strong protective laws. Storing data with expert third parties does not always ensure better protection. Although encrypted, data are sometimes not anonymized, risking linkage to personal profiles for commercial or political use.

Privacy. Companies face a dilemma in balancing data anonymization (Segalla & Rouziès, 2023). Too little anonymization fails to meet government regulations, while too much renders data useless for marketing. Various anonymization techniques exist, such as aggregating data, using approximate values, making slight changes to variables, or pseudonymizing data. However, even minimal data like gender, birth date, and postal code can lead to individuals' identities being uncovered when combined with other datasets.

To address privacy concerns, corporate institutional review boards (IRBs) must evaluate the effectiveness of anonymization techniques, considering the power of data analytics (Segalla & Rouziès, 2023). Differential privacy adds an extra layer of protection by preventing algorithms from disclosing specific information about individuals. However, privacy can still be compromised by how data is collected and processed.

Table 2 Critical questions related to the five P's

Principle	Questions to consider
Provenance	What is the original source of the data? Was the data obtained in compliance with relevant laws and regulations? Did contributors provide the necessary permissions or informed consent for their data to be used?
Purpose	Is the data being used for a function different from its initial intent? Would the individuals or organizations that provided the data support its use for this new purpose?
Protection	What safeguards are in place to ensure data security? Who holds the responsibility for securely disposing of the data once it is no longer needed?
Privacy	Which personnel or groups will be allowed to access personally identifiable information? What methods will be used to de-identify individual records within the dataset?
Preparation	What procedures were followed to clean and preprocess the data? Are merged datasets being handled in ways that maintain participant anonymity? What steps are being taken to validate the data's accuracy and make corrections where required?

Preparation. Preparing data for analysis involves several key steps to ensure its accuracy and completeness. Errors and missing data can significantly impact statistical analysis quality. Cleaning data, especially from diverse sources, can present challenges. Therefore, leaders should ensure that proper procedures are followed to clean, process, and validate the data (Table 2).

3.3 *Creating the Culture of Data Excellence*

When thought of as a "culture," data excellence then encompasses not only practices and structures, but also the values and beliefs that guide all jobs and roles throughout the organization. As the problems mentioned in the previous page illustrate, it is a mistake to think that data quality is strictly the purview of data science. To be truly "data-driven," companies need everyone's help to ensure that their data are accurate, timely, high quality, unbiased—and that their decision-making relies on data in robust, systematic, evidence-based ways.

3.3.1 Reducing Discovery Time

By improving data discovery and quality checks, Lyft aims to reduce the time data scientists spend on finding, understanding, and validating data. This allows them to focus more on analysis and generating insights, enhancing overall productivity. Democratization of data means making data accessible to everyone within the organization while ensuring compliance and security. Lyft uses tools like Amundsen for data discovery, enabling employees to find and use data efficiently.

3.3.2 Network and Collaboration

Lyft emphasizes creating a network where employees help each other with data issues through channels like Slack. This community approach enhances knowledge sharing and reduces the burden on data owners.

3.3.3 Cross-Functional Collaboration

Cross-functional teams, including data engineers, data scientists, and product leaders, collaborate to ensure that data systems meet the diverse needs of the organization. This collaboration is facilitated by the shared tools and platforms provided by Lyft's data infrastructure.

3.3.4 Continuous Improvement

Lyft acknowledges that they are still "scratching the surface" of data discovery and quality management. Ongoing improvements and research are necessary to fully address these challenges and optimize their data systems.

3.3.5 Balancing Cost and Performance

Managing the trade-offs between cost and performance is an ongoing challenge. Lyft aims to make costs transparent and allow teams to make informed decisions about their data usage based on their budgets and needs.

3.3.6 *Toward a Positive Data* Culture

Many leaders try to change company culture through formal, analytical methods, such as workshops to define values or updating HR protocols (Day, n.d.). However, this approach is rarely successful. Fostering change for large-scale issues like cultural change requires a deeper dive into challenging employees' values, ideas, and frames. Barney et al., (2023) argued that organizational culture is rooted in the stories shared between employees. Therefore, when it comes to initiating culture change, leaders should begin by creating new narratives that highlight and encourage the desired culture shift.

1. Consider using stories to create change.
2. Consider using peer learning to create culture change.
3. Consider structuring collective responsibility.

3.4 Data as a Product (DaaP)

Data as a product (DaaP) is an approach that involves creating data products and curated collections of data designed with specific purposes and end-users in mind (Huang et al., 2015). These data products provide valuable insights and support informed decision-making (Mucci & Stryker, 2024). By focusing on usability, quality, and integration, data products transform raw data into actionable business intelligence that can be applied broadly across an organization (Desai et al., 2022). This ensures that data is not only accessible but also reliable and tailored to meet the unique needs of different users. Examples of data products include dashboards, chatbots, and recommendation algorithms (Mucci & Stryker, 2024).

By assuming a DaaP position, companies can create targeted, valuable data resources that different members of the organization can use widely. This approach transforms data into a product-like entity that is discoverable, understandable, and usable by various teams within the organization, leading to enhanced decision-making and innovation (Krantz & Jonker, 2025).

A robust data management system is essential for achieving better results when implementing AI models within a business. Effective data management ensures that the data used for training and deploying AI models is of high quality, reliable, and relevant. This, in turn, enhances the accuracy and performance of AI models, driving better business outcomes.

3.4.1 The DaaP Framework

Desai et al., (2022) outlined various types of consumption that can benefit from a robust DaaP framework. These archetypes include:

- Creating applications: Developing applications that rely on the available data to provide functionalities and insights. This includes customer-facing apps, internal tools, and automated processes that leverage data for improved efficiency and user experience.
- Advanced analytics: Utilizing data for advanced analytical purposes, such as predictive modeling, machine learning, and data mining. This allows organizations to uncover patterns, trends, and insights that can inform strategic decisions.
- Reporting: Generating comprehensive reports that provide a detailed view of business operations, performance metrics, and key indicators. Effective reporting helps stakeholders understand the current state of the business and make informed decisions.
- External data sharing: Sharing data with external partners, customers, or regulatory bodies in a secure and controlled manner. This can enhance collaboration, transparency, and compliance with industry standards and regulations.
- Discovery sandboxes: Providing experts with sandbox environments where they can experiment with data, test hypotheses, and discover new use cases.

These sandboxes enable innovation and the development of novel solutions and applications.

Implementing a robust DaaP framework allows organizations to maximize the value of their data assets, fostering a data-driven culture and enabling various teams to leverage data for a wide range of purposes. This holistic approach to data management and utilization is crucial for staying competitive in today's data-centric business landscape.

3.4.2 Crafting the Plan

Business leaders might think implementing the DaaP framework is a good option for their organization. However, Desai et al., (2022) warned that they must consider the feasibility and potential value of DaaP. The authors suggest leaders categorize the use cases by data products and consumption archetypes for more efficient implementation and a faster return on investment. Below is a roadmap for implementing the DaaP framework.

1. **Establish a clear vision and objectives**: The vision outlines the plans for the data product and how it serves the organization's strategic goals. The objectives must be measurable and clearly defined, such as improving the UX experience or providing data scientists with better resources to conduct the best possible analysis. Decide if this product will be released internally or to a wider audience.
2. **Conduct user research**: Involve key stakeholders, whether internal or external, and ask them what data products best suit their needs. These interviews will help you identify the key features and functions the data product should include.
3. **Define the product requirements**: Once you have interviewed stakeholders, you can determine the criteria that your data product will meet. These requirements should be aligned with the vision and objectives you identified above. Prioritize the data products that are deemed most needed.
4. **Identify key milestones**: These are critical during the process because they allow your organization to track the success of its strategy. These milestones should be aligned with the requirements and achievable within a given timeframe.
5. **Prioritize features**: Start by developing the features your previous research identified as the most critical.
6. **Define a release plan**: Based on the priorities you discovered in the previous step, create a plan for your product release to your team members or wider audience. This plan should include the features included in each release and a timeline. It should also align with the key milestones and the product vision and objectives.
7. **Monitor and evaluate progress**: Once the roadmap has been defined, monitor and evaluate the progress regularly. These processes will help you identify any issues, so you can adjust when needed.

4 Configuring GenAI into Workflows and Decisions

4.1 GenAI Tools

It has been nearly three years since generative AI (GenAI) blasted into the mainstream conversation. In that time, we have moved from everyone wowed by the first time they asked for a parody song or a deep fake video creation to the more mature, transformative tools used in the workplace. In this section, I will explore three major ways through which individuals interact with GenAI: prompts, agents, and developers.

4.1.1 Prompts

When you are using ChatGPT, Claude, Gemini, or a custom tool, everything begins with a prompt. Many Instagram accounts offer advice on exact prompts and specific prompt engineering techniques if you are looking for specifics. However, most of the improvement in AI has not come from better prompts, but from better models. For example, previously, there were no prompts that could produce the level and quality of writing needed for academic writing; GenAI was used for other tasks. But, with one update, the quality of outputs became so much better that the need for highly engineered prompts in some cases decreased.

4.1.2 Agents

Another interface that is becoming more prevalent, even for individuals, are AI agents that can execute multiple actions on your behalf.

4.1.3 Developers

A third and powerful way that professionals are interacting with GenAI is creating their own custom technology. People across all industries are using AI, low-code, and no-code tools to build their own apps, models, and automations.

4.2 Configuring an Individual Workflow

AI adoption often begins at the level of individual workflows, where employees experiment with integrating AI tools into specific tasks. Empowering workers as domain experts to explore how AI can add value to their roles fosters innovation and creates practical use cases for the technology. Yet, organizations must balance experimentation with governance by building environments that allow employees to test AI tools securely, especially in regulated industries. The lack of clear guidance can lead to struggles in determining the most effective use cases for AI, making the integration process somewhat fragmented and inconsistent.

4.3 GenAI

One of the primary challenges leaders face when integrating AI at the team level is managing the change process itself. Unlike organizational-level changes that are typically guided by strategic initiatives, these team-level changes often occur in a more ad-hoc, bottom-up manner. These leaders found themselves in a position where they must balance top-down expectations with the practical realities of their teams' varying levels of AI adoption. Some team members might quickly embrace new technologies, while others may struggle, leading to disparities in productivity and workflow efficiency. Leaders must navigate these differences while ensuring that the overall team workload remains balanced. This process, known as level loading, involves adjusting tasks and responsibilities to match each team member's capacity and expertise, a task made more complex by the unpredictable nature of AI tool adoption. Furthermore, leaders must do this while continuing to meet existing operational demands, adding another layer of difficulty to an already challenging process.

4.4 The Problem of Working with Immature or Generic Tools

Developed tools are often not mature enough to meet the needs of all users. Leaders report that some tools frequently produce errors or hallucinations, which undermines their reliability and value. This presents a dilemma: leaders are encouraged to promote the use of AI within their teams, but when the tools fail to deliver, it can lead to frustration and wasted effort.

Leaders and teams face significant challenges in configuring AI into workflows due to the general-purpose nature of these tools versus the need for tailored, specific applications. Generic tools like ChatGPT or other AI systems are versatile and can be applied broadly, but often lack the specificity needed for particular tasks. Customizing these tools for specific workflows—such as automating industry-specific tasks or integrating them into proprietary systems—can be complex and time-consuming. It requires deep domain knowledge, precise data input, and careful alignment with business goals, often involving substantial training, tuning, and trial-and-error processes.

In some cases, and for some workflows, generic tools can be faster to deploy, but for many workflows, achieving a high return on investment (ROI) requires iterative refinement. Teams often struggle with balancing the flexibility (but lack of real impact) of more generic AI tools and the effort required for customization.

Leaders must not only facilitate experimentation with AI tools but also recognize when these tools are not providing sufficient value. This discernment is crucial in preventing teams from abandoning tools that are proving useful or becoming bogged down by inefficient tools. Notably, this finding highlights the complicated role of leaders who need to be both advocates for and discerning critics of this emerging technology. Furthermore, while tools do not always work exactly as desired or even designed, the speed with which capability gaps are being closed by GenAI toolmakers is increasing.

4.5 Long-Term Implications for Organizations

In considering AI at the level of individual workflows, early evidence shows that it may impact workers differently based on their experience and expertise (Butler et al., 2023). A few studies have now shown that AI tools can significantly enhance the efficiency and effectiveness of new or novice employees by providing them with rapid access to knowledge and automated support that accelerates their learning curve (Butler et al., 2023). For these individuals, AI serves as a crucial enabler, offering guidance and resources that help bridge gaps in their skill sets. Conversely, studies note that more experienced workers, who already possess a high level of proficiency and domain knowledge, experience less dramatic improvements with AI tools. Instead, their workflows may see only incremental benefits from AI, reflecting the tool's role in optimizing rather than transforming established practices. This dichotomy illustrates how AI's impact on individual workflows varies widely, depending on the worker's existing skills and the nature of the tasks they perform.

Similar patterns have been observed in both writing and software development. Noy and Zhang, (2023) demonstrated that AI tools significantly enhanced the performance of individuals who initially performed poorly on a writing task by providing them with targeted guidance and feedback. Similarly, AI tools were found to have a more pronounced effect on less experienced developers compared to their more seasoned peers. However, research by Haslberger et al., (2024) indicated that AI can sometimes exacerbate performance disparities among workers. This research revealed that certain demographic groups are taking greater advantage of AI tools, resulting in a growing skills gap between typically young, educated, professional males and other groups. Haslberger et al. noted the potentially disruptive results of only these "usual suspects" benefiting from the adoption of AI tools.

On the other hand, overreliance on AI can be a problem, just as with analytical AI tools. Some of the sources of overreliance can be a lack of expertise in the subject or a lack of technical comfort with the AI tools. New workers may not have the domain knowledge to accurately scrutinize the output, or interrogate the AI content, and may therefore simply trust the AI and over-rely on it. There is some software that can help with such overreliance by highlighting or co-auditing. Balancing the trade-offs between aiding new workers and leading to overreliance is crucial in the integration of AI into individual workflows. Leaders can help navigate this by ensuring proper onboarding and reskilling for the use of AI tools.

5 How Might AI Change the Work of Leaders?

5.1 Management of Algorithms

Leaders would be wise to consider ways that algorithms can be used as more exploitative and/or empowering leadership functions. Leaders perform a variety of tasks, including exerting control over workers. Algorithms are increasingly integrated into leadership roles and tasks, and like their integration into other workflows there are varying levels of success. Some of these integrations leave employees feeling frustrated and helpless. However, there are examples of successful algorithmic integrations that leave employees feeling empowered and motivated.

As stated above, one way of thinking about organizational control is through the activities leaders perform: directing, evaluating, and disciplining. In the field of labor process theory, the relationship between management and employee is understood dialectically. On one hand, leaders push to maximize the value of employees' labor. On the other hand, employees attempt to maintain autonomy and dignity by resisting increased leadership control. This dynamic tension shapes the organizational landscape, influencing policies, practices, and the overall work environment.

Many workplaces today have already implemented algorithmic management systems to perform the tasks of directing, evaluating, and disciplining. These new data-intensive algorithmic management systems involve four key characteristics:

- Comprehensive: Algorithms can incorporate a wide variety of data sources and inputs from sensors, cameras, smartphone and GPS data, among others, to learn about and understand employee behavior.
- Instantaneous: Algorithms allow for real-time data analysis and feedback. This can enable continuous employee performance evaluation and improvement.
- Interactive: Algorithms can help leaders gather important feedback from employees and customers. They can be used to alter team hierarchies or workflows as the need arises.
- Opaque: Algorithms can ensure that sensitive company data remains secret. In addition to being able to deliberately hide company data, algorithms are opaque by nature, requiring technical knowledge to understand their operations.

5.2 *Algorithmic Organizational Design and Coordination*

Organizational structures function as critical mechanisms for coordinating adaptation, enabling teams to pursue collective goals that exceed individual capacity. Central to this function are information flows, which determine how data is collected, processed, and disseminated to support decision-making, alignment, and coordination (Galbraith, 1973; Tushman & Nadler, 1978). Teams serve as the fundamental units of these flows, interpreting task-relevant information, sharing insights across groups, and integrating organizational feedback (Kozlowski & Ilgen, 2006).

The design of team structures significantly shapes patterns of information exchange. Hierarchical arrangements channel communication vertically, while flat and cross-functional designs promote lateral flows that accelerate collaboration (Mintzberg, 1979; Hackman, 2002). Interaction norms further influence whether communication is open and continuous or formal and episodic (Edmondson, 1999). Importantly, no single team structure is universally effective; optimal configurations depend on task demands, member characteristics, and organizational context. Innovative initiatives often require cross-functional collaboration to foster idea generation, whereas routine operations may be better served by hierarchical clarity (Lawrence & Lorsch, 1967).

Leaders play a decisive role in aligning team design with organizational objectives. By understanding the strengths and limitations of structural alternatives, leaders can configure teams to optimize information flows, enhance decision quality, and balance the dual imperatives of efficiency and creativity (Ancona & Caldwell, 1992; Yukl, 2013).

5.2.1 The Tension Between Algorithms and Organizational Charts

The impact of AI on organizations is still relatively unknown. Traditionally, organizational design has relied on human decision-makers and organizational charts to divide tasks, manage information flows, and coordinate teams. However, the integration of analytical AI and GenAI tools is reshaping these decision spaces by expanding the scope and complexity of tasks that algorithms can handle.

Key takeaways include:

- Decisions: AI can manage and analyze larger sets of data more efficiently than humans, enabling broader decision spaces. While traditional org charts divide roles based on what individuals can handle, AI tools can optimize decisions across teams, not just within specific job roles, leading to more effective decision-making.
- Organizational charts: As AI takes on more decision-making responsibilities, the traditional hierarchical structure of organizational charts should be revisited. AI can enable more flexible and dynamic team structures, allowing organizations to adapt faster and restructure workflows for efficiency.
- Avoiding constrained bins: The module teaches that limiting AI to the existing human-created decision spaces can hinder its potential. By freeing AI from these

constraints and allowing it to explore data and decisions holistically, organizations can improve performance across teams.

- People and algorithmic structures: AI encourages a separation between human and algorithmic roles. While algorithms handle large-scale data analysis, human team members can focus on balancing the algorithm's output with their expertise, fostering collaboration between people and machines.

6 Conclusion

The research shows that leaders who intentionally cultivate a strong data culture position their organizations to maximize the benefits of AI integration (Hossain et al., 2025). As processes shift toward algorithmic enhancement, the fundamentals of data—its sources, storage, and use—become organizational imperatives. Yet technical expertise alone is insufficient. The real momentum emerges when leaders create conditions for experimentation and foster collaboration across diverse teams, particularly among knowledge workers and data scientists. This intersection of people, process, and technology provides the surest path forward.

In sum, the following key takeaways are essential for leaders navigating analytical AI and GenAI integration:

1. **Experimentation matters**. Leaders must normalize experimentation by encouraging teams to "play" with analytical AI and GenAI tools. Even when such exploration brings short-term disruption, it builds the adaptive capacity necessary for long-term gains. Often, integration is most effective when it arises through bottom-up initiatives rather than solely top-down mandates.
2. **Data excellence is both mindset and practice**. AI integration requires more than technical upgrades; it calls for a cultural shift toward data-centered ways of working. If employees across the organization lack fluency in the collection, handling, storage, and use of data, the potential of algorithmic tools will be undermined. Leaders are responsible for embedding a culture where data consciousness permeates every workflow.
3. **Keep business objectives in focus**. Data science must remain tethered to organizational purpose. Leaders set the strategic aims, and data teams translate those aims into algorithms that drive impact. Without this clarity, AI adoption risks becoming a technical exercise disconnected from meaningful business outcomes.
4. **Engage knowledge workers in design**. Successful integration depends on the participation of knowledge workers, who both shape the tasks that algorithms automate and serve as their primary users. Their resistance can stall adoption; their buy-in accelerates it. Leaders must ensure these workers have voice and agency in the design process.
5. **Reimagine organizational design**. Analytical AI and GenAI tools not only alter workflows—they can reshape the very structure of organizations. Unlike traditional hierarchies built around human-centered decision-making, algorithmic

systems enable new forms of coordination, information flow, and decision design. Leaders must be willing to question established hierarchies and embrace more flexible, adaptive forms of organizing. At the same time, they must safeguard the human element as integral to organizational life.

Leading in the era of AI requires a healthy combination of technical understanding and cultural foresight. The most effective leaders are those who balance experimentation with purpose, align data practices with business needs, and embrace organizational transformation while centering human values.

References

Agrawal, A., Gans, J., & Goldfarb, A. (2022, November). From Prediction to Transformation. https://hbr.org/2022/11/from-prediction-to-transformation.

Alshdaifat, S. M., Al-Hazaima, H., Alhasnawi, M. Y., Hu, G., Al-Shbiel, S. O., & Alshdaifat, K. (2025). Factors determining the adoption of artificial intelligence in higher education: Insights from accounting and finance education in Iraq.

Ancona, D. G., & Caldwell, D. F. (1992). Bridging the boundary: External activity and performance in organizational teams. *Administrative Science Quarterly, 37*(4), 634–665.

Aroles, J., Mitev, N., & de Vaujany, F. X. (2019). Mapping themes in the study of new work practices. *New Technology, Work and Employment, 34*(3), 285–299.

AWS. (n.d.). What is digital transformation? https://aws.amazon.com/what-is/digital-transformation/#:~:text=Digital%20transformation%20is%20the%20process%20that%20an,new%20technology%20adoption%20initiatives%20in%20the%20future.

Barley, S. R. (2020). *Work and technological change*. Oxford University Press.

Baskin, K. (2024, March 28). Who should be the head of generative AI - and what they should do. *MIT Management Sloan School.*. https://mitsloan.mit.edu/ideas-made-to-matter/who-should-be-head-generative-ai-and-what-they-should-do

Baten, R. B. A. (2024). How are US hospitals adopting artificial intelligence? Early evidence from 2022. *Health Affairs Scholar, 2*(10).

Butler, J., Jaffe, S., Baym, N., Czerwinski, M., Iqbal, S., Nowak, K., Rintel, R., Sellen, A., Vorvoreanu, M., Hecht, B., & Teevan, J. (2023). *Microsoft new future of work report 2023 (Microsoft Research tech report MSR-TR-2023-34)*. Microsoft. https://www.microsoft.com/en-us/research/publication/microsoft-new-future-of-work-report-2023/

Carreño, A. (2025). Aligning change initiatives with executive leadership priorities: The bridge between strategy and continuous transformation. https://adolfocarreno.com/2025/02/09/aligning-change-initiatives-with-executive-leadership-priorities-the-bridge-between-strategy-and-continuous-transformation/.

Chertoff, M. (2023, April 13). Cyber risk is growing. Here's how companies can keep up. *Harvard Business Review.*. https://hbr.org/2023/04/cyber-risk-is-growing-heres-how-companies-can-keep-up

Conn, C., & McLean, R. (2020, September 15). Six problem-solving mindsets for very uncertain times. https://www.mckinsey.com/capabilities/strategy-and-corporate-finance/our-insights/six-problem-solving-mindsets-for-very-uncertain-times.

Davenport, T. H., & Tiwari, P. (2024, March 26). *Is your company's data ready for generative AI?* Harvard Business Review.

Day, R. (n.d.). What is the proper balance of formal and informal learning? Global Knowledge. https://www.globalknowledge.com/us-en/content/learning/the-702010-model-finding-the-right-balance-of-formal-and-informal-learning/#gref

Deming, D J, (2021). The growing importance of decision-making on the job.. https://www.nber.org/papers/w28733

Desai, V., Fountaine, T., & Rowshankish, K. (2022, August). A better way to put your data to work. *Harvard Business Review.*. https://hbr.org/2022/07/a-better-way-to-put-your-data-to-work

Eastwood, B. (2023, December 5). It's time for everyone in your company to understand generative AI. MIT Management Sloan School. https://mitsloan.mit.edu/ideas-made-to-matter/its-time-everyone-your-company-to-understand-generative-ai.

Edmondson, A. (1999). Psychological safety and learning behavior in work teams. *Administrative Science Quarterly, 44*(2), 350–383.

Galbraith, J. R. (1973). *Designing complex organizations.* Addison-Wesley.

Ghani, E. K., Ariffin, N., & Sukmadilaga, C. (2022). Factors influencing artificial intelligence adoption in publicly listed manufacturing companies: A technology, organisation, and environment approach. *International Journal of Applied Economics, Finance and Accounting, 14*(2), 108–117.

Hackman, J. R. (2002). *Leading teams: Setting the stage for great performances.* Harvard Business School Press.

Harvard Law School Forum (2025). Governance of AI: A critical imperative for today's boards. https://corpgov.law.harvard.edu/2025/05/27/governance-of-ai-a-critical-imperative-for-todays-boards2/#:~:text=Board%20Considerations%20for%20Opening%20Communication,strategic%20initiatives%20and%20how%20often?

Haslberger, M., Gingrich, J., & Bhatia, J. (2024). *No great equalizer: Experimental evidence on productivity effects of generative AI use in the UK labor market.* SSRN. https://papers.ssrn.com/sol3/papers.cfm?abstract_id=4594466

Horvat, D., Baumgartner, M., Kinkel, S., & Mikalef, P. (2023). Examining heterogeneous patterns of ai capabilities. *IFIP Advances in Information and Communication Technology*, 619–633.

Hossain, S., Fernando, M., & Akter, S. (2025). Digital leadership: Towards a dynamic managerial capability perspective of artificial intelligence-driven leader capabilities. *Journal of Leadership & Organizational Studies, 32*(2), 189–208.

Hossain, M. K., Srivastava, A., Oliver, G., Islam, M. E., Jahan, N., Karim, R., et al. (2024). Adoption of artificial intelligence and big data analytics: An organizational readiness perspective of the textile and garment industry in Bangladesh. *Business Process Management Journal, 30*(7), 2665–2683.

Hradecky, D., Kennell, J., Cai, W., & Davidson, R. (2022). Organizational readiness to adopt artificial intelligence in the exhibition sector in Western Europe. *International Journal of Information Management, 65*, 102497.

Huang, G., He, J., Chi, C. H., Zhou, W., & Zhang, Y. (2015, June). A data as a product model for future consumption of big stream data in clouds. In *In 2015 IEEE international conference on services computing* (pp. 256–263). IEEE.

Jain, S. C., & Ropple, L. M. (2018, December 14). Stopping data breaches will require help from governments. *Harvard Business Review.*. https://hbr.org/2018/12/stopping-data-breaches-will-require-help-from-governments

Kozlowski, S. W. J., & Ilgen, D. R. (2006). Enhancing the effectiveness of work groups and teams. *Psychological Science in the Public Interest, 7*(3), 77–124.

Krantz, T. & Jonker, A. (2025). What is a data product? *IBM.* https://www.ibm.com/think/topics/data-product#:~:text=A%20data%20product%20is%20a,built%20queries%20or%20data%20pipelines.

Lawrence, P. R., & Lorsch, J. W. (1967). *Organization and environment: Managing differentiation and integration.* Harvard Business School Press.

Marwell, G., & Hage, J. (1970). The organization of role-relationships: A systematic description. *American Sociological Review*, 884–900.

McKinsey. (2025, March). The state of AI: How organizations are rewiring to capture value. https://www.mckinsey.com/capabilities/quantumblack/our-insights/the-state-of-ai.

Mintzberg, H. (1979). *The structuring of organizations.* Prentice Hall.

MohanaSundaram, A., Yogesh, M., Atanasov, A. G., Kaneda, Y., Oami, T., Sathanantham, S. T.,... & Apostolopoulos, V. (2024). Prevalence and usage trends of large language models in health Science Publications: A cross-sectional analysis.

Mucci, T., & Stryker, C. (2024, February 23). What is data as a product (DaaP)? IBM. https://www.ibm.com/think/topics/data-as-a-product

Nasrollahi, M., & Ramezani, J. (2020). A model to evaluate the organizational readiness for big data adoption. *International Journal of Computers Communications & Control, 15*(3).

Nehme, M., Schneider, F., Amruthalingam, E., Schnarrenberger, E., Trëmeaud, R., & Guessous, I. (2024). Chatbots in medicine: Certification process and applied use case. *Swiss Medical Weekly, 154*(10), 3954–3954.

Noy, S., & Zhang, W. (2023, March 1). *Experimental evidence on the productivity effects of generative artificial intelligence.* SSRN. https://doi.org/10.2139/ssrn.4375283

Pahwa, S., Garg, A., & Colmenares, A. (2025). *The foundation of scalable enterprise AI: Building a robust framework for data and AI governance and security.* IBM Technology. https://www.ibm.com/think/insights/foundation-scalable-enterprise-ai

Pecan. (n.d.). *Better together: Generative AI and predictive analytics.* Pecan.ai. https://www.pecan.ai/blog/better-together-generative-ai-and-predictive-analytics/#:~:text=The%20Intersection%20of%20Generative%20AI,ensures%20consistent%20and%20accurate%20predictions.

Ransbotham, S., Kiron, D., Candelon, F., Khodabandeh, S., & Chu, M. (2022). Achieving individual—And organizational—Value with AI. *MIT Sloan Management Review.*. https://sloanreview.mit.edu/projects/achieving-individual-and-organizational-value-with-ai/?utm_source=chatgpt.com

Rudy, B. (2023). Evaluating ROI on your company's learning and development initiatives. *Harvard Business Review.*. https://hbr.org/2023/10/evaluating-roi-on-your-companys-learning-and-development-initiatives

Sambasivan, N., Kapania, S., Highfill, H., Akrong, D., Paritosh, P., & Aroyo, L. M. (2021). Everyone wants to do the model work, not the data work: Data cascades in high-stakes. In *Proceedings of the 2021 CHI conference on human factors in computing systems (CHI '21)* (Vol. 39, pp. 1–15).

Seamans, R. and Raj, M. (2018). AI, labor, productivity and the need for firm-level data.

Segalla, M., & Rouziès, D. (2023). The ethics of managing people's data. *Harvard Business Review.*. https://hbr.org/2023/07/the-ethics-of-managing-peoples-data

Singla, A., Sukharevsky, A., Yee, L., Chui, M., & Hall, B. (2024, May 30). The state of AI in early 2024: Gen AI adoption spikes and starts to generate value. McKinsey & Company. https://www.mckinsey.com/capabilities/quantumblack/our-insights/the-state-of-ai

Statista. (2025). Annual number of data compromises and individuals impacted in the United States from 2005 to 2024. https://www.statista.com/statistics/273550/data-breaches-recorded-in-the-united-states-by-number-of-breaches-and-records-exposed/#:~:text=In%202024%2C%20the%20number%20of,vulnerability%20before%20the%20cyber%20criminals.

Tushman, M. L., & Nadler, D. A. (1978). Information processing as an integrating concept in organizational design. *Academy of Management Review, 3*(3), 613–624.

Vaittinen, E., & Martinsuo, M. (2019). Industrial customers' organizational readiness for new advanced services. *Journal of Manufacturing Technology Management, 30*(7), 1073–1096.

Vishwanatha, J. K., Christian, A. H., Sambamoorthi, U., Thompson, E. L., Stinson, K., & Syed, T. A. (2023). Community perspectives on ai/ml and health equity: Aim-ahead nationwide stakeholder listening sessions. *PLOS Digital Health, 2*(6), e0000288.

Yukl, G. (2013). *Leadership in organizations* (8th ed.). Pearson.

Zhang, P., Dillard, N., & Cavallo, T. (2025). Navigating the limitations of algorithmic management: An integrative framework of sociotechnical systems theory and strategic HRD. *Human Resource Development Review.*

Leading Workforce Ecosystems in Industry 5.0: A Dynamic Capabilities Playbook

Walid El Mansour, Heesun Kim, Fredrick M. Nafukho, and Helen A. M. Muyia

Abstract Industry 5.0 reframes progress around human well-being, sustainability, and resilience while building on the tools of Industry 4.0 and redirecting them toward societal value. This chapter positions workforce ecosystems as the setting where that shift becomes real, since value creation now depends on coordination among employees, contractors, partner firms, platforms, and AI agents. We argue that dynamic capabilities serve as the leadership operating system for this setting by supplying repeatable routines for sensing, seizing, and reconfiguring. Sensing widens the aperture to markets, science, partners, and regulation, supported by external linkages, decentralized scanning, and analytics. Seizing converts options into boundary choices, viable business models, and joint commitments that align incentives across contributors. Reconfiguring then realigns assets, relationships, and structures through knowledge transfer and resources moves so architectures stay adaptive. We identify moderators that shape capability design and effectiveness, including technological change, platform dependency, institutional rules, and workforce voice. We define outcomes that match the aims of Industry 5.0, such as faster innovation, stronger resilience, higher engagement and well-being, improved sustainability, and lower cost to capability. The chapter contributes to a leadership model that turns cross boundary collaboration into a durable source of advantage.

Keywords Industry 5.0 · Workforce ecosystems · Dynamic capabilities · Leadership

W. El Mansour (✉) · H. Kim
Texas A&M University, College Station, TX, USA
e-mail: welmansour@tamu.edu

F. M. Nafukho · H. A. M. Muyia
University of Washington, Seattle, WA, USA

A. M. Waite et al. (eds.), *Leading in Industry 5.0*, Management and Industrial Engineering, https://doi.org/10.1007/978-3-032-19207-3_12

1 Leading Workforce Ecosystems in Industry 5.0: A Dynamic Capabilities Playbook

Industry 5.0 is the most recent stage in human development. It is the culmination of technologies, knowledge, and expertise that have been in the making for centuries. The first industrial revolution (IR) was the spark that triggered these series of revolutions leading to Industry 5.0. The first IR took humans from manual to machine-powered production, the second IR transformed manufacturing by harnessing the power of electricity, and the third IR was marked by automation through computers and the spread of communication technologies (Nafukho et al., 2024). The fourth IR, or Industry 4.0, is a technology-driven revolution characterized by the merging of information technologies and operational technologies (Leng et al., 2022). It employs Cyber-Physical Production Systems (CPPS) in making intelligent decisions (Xu et al., 2021). Big data, the Internet of Things, cloud computing, additive manufacturing, and autonomous robots are the key technologies that Industry 4.0 stands on.

Where Industry 4.0 emphasizes cyber-physical integration and automation, Industry 5.0 revolves around harnessing the power of industry to achieve societal goals that go beyond economic growth and jobs. Its ultimate goal is to provide prosperity to all humanity (Leng et al., 2022). Industry 5.0 stands on three core values: human-centricity, sustainability, and resilience. Putting humans at the center instead of replacing them is the most fundamental aspect of Industry 5.0. It seeks to improve collaboration and interaction between humans and machines. Moreover, Industry 5.0 puts sustainability and the protection of the planet as prerequisites for the introduction of technologies. With constant changes in the world, Industry 5.0 aspires to promote resilience not only in single enterprises but also across entire industries (Leng et al., 2022; Xu et al., 2021). The shift from Industry 4.0 to 5.0 commands a shift in leaders' focus from resource control to purposeful orchestration of diverse contributors.

As humanity goes from one industrial revolution to another, the workforce environment changes. Industry 5.0 is not immune to such change. In fact, the workforce ecosystem must adapt to Industry 5.0 and serve its core values. The workforce ecosystem is comprised of all actors (internal and external) that create value for an organization (Altman et al., 2023b). This constellation of actors is not owned by a single organization, requiring a shift from exploitation of resources to orchestration. Value creation, the core of workforce ecosystems' mission, hinges on architectural alignment (Adner, 2017). In the current workforce ecosystem, leaders win by coordinating between internal and external actors, going beyond the popular resource-based approach to a dynamic capabilities approach (Teece, 2007).

Despite the rapid adoption of the Industry 5.0 philosophy of a human-centric, sustainable, and resilient agenda, management and leadership theories and practices still assume single-employer control, firm-specific KPIs, and standardized jobs. In reality, today's workforce creates value through new workforce ecosystems that rely on effective coordination between interdependent actors inside and beyond the firm

(Altman et al., 2023a). Currently, leadership models offer little guidance on how to orchestrate and align interdependent roles across organizations (Adner, 2017). This gap becomes wider as organizations employ in-house algorithms to direct, evaluate, and reward work (Kellogg et al., 2020; Rani et al., 2021), as the complexity of psychological contracts increases with the diversity of external workers (Liu et al., 2020), and as heightened security risks emerge in safety, data protection, and fairness (Probst et al., 2018). Present leadership approaches, built on pre-Industry 5.0 employment, are ill-equipped for the Industry 5.0 workforce ecosystem. Leaders need frameworks that take into consideration interdependent actors, data, and algorithms, an approach that dynamic capabilities enable.

In this work, we argue that dynamic capabilities (DCs) provide an operating system for leading workforce ecosystems in Industry 5.0. DCs can be considered the strategic routines that utilize existing organizational resources, acquire new ones, shed resources, and combine resources into new ones to achieve competitive advantage (Eisenhardt & Martin, 2000). DCs comprise routinized ways of sensing changes, seizing opportunities through coordinated commitments, and reconfiguring assets and relationships to sustain performance under turbulence (Teece, 2007). Teece stresses the importance of a set of microfoundations necessary for the process of sensing, seizing, and reconfiguring.

2 Industry 5.0

The intervals between industrial revolutions have tended to shorten, and the transitions are increasingly blurred. Industry 4.0 was introduced and recognized in 2011. A few years later, Industry 5.0 was proposed in 2016 and officially defined in 2021 (Cotta et al., 2021). The transition from Industry 4.0 to 5.0 is unique compared to other transitions (e.g., from IR 1.0 to IR 2.0). Industry 5.0 complements 4.0 and stands on the technologies implemented in 4.0. Industry 4.0 focuses on improving efficiency and productivity while implementing Cyber-Physical Production Systems. It treats human outcomes as a consequence of technological change (Xu et al., 2021). Industry 5.0 retains the advantages of technological advancement while shifting the focus to humans. It explicitly targets human-centricity, sustainability, and resilience (Cotta et al., 2021). Industry 5.0 puts workers' well-being first, respects the planet, promotes human–machine synergy, and aims at achieving sustainable production (Leng et al., 2022; Xu et al., 2021).

Industry 5.0 stands on three core values: human-centricity, sustainability, and resilience.

- **Human-Centricity**: Human needs, rights, and well-being are at the center of any production chain. In Industry 5.0, technology adapts to humans and not the other way around, leading to work redesign where humans and machines collaborate, relying on human-centric AI and leveraging ergonomics to improve well-being and performance.

- **Sustainability:** In its mission to respect the planet, Industry 5.0 promotes resource-efficient production that reduces energy use, emissions, and waste. It employs advanced technologies to optimize resource efficiency.
- **Resilience:** Industry 5.0 aims at building robust and adaptive industrial systems that can withstand and recover from crises effectively and efficiently while securing key infrastructures and vital supply chains.

3 Workforce Ecosystems: The Setting for Leadership

Achieving the desired outcomes of industry 5.0 relies on leaders' ability to align and orchestrate contributors that are inside and outside the firm. Managing all the workforce ecosystems and not just one type of employee becomes crucial. According to Altman et al. (2023b, p. 15), the workforce ecosystem is defined as:

A workforce ecosystem is a structure that encompasses actors, from within the organization and beyond, working to create value for an organization. Within the ecosystem, actors work toward individual and collective goals with interdependencies and complementarities among the participants.

The definition highlights that the workforce ecosystem is a structure by which an organization organizes its workforce. This structure encompasses both internal employees and external contributors. Internal employees are the traditional employees hired by the organization. The external workforce includes contractors, consultants, other organizations, and leased or contracted technologies. All these actors work toward achieving a common goal—in this case, the core values of Industry 5.0. The interdependence of these actors is crucial to workforce ecosystems. The success of one member of the system depends on the success of other members. In addition, members of the workforce ecosystem are expected to complement one another to create value.

Leading a workforce ecosystem means orchestrating outcomes across players over whom leaders do not have full control. This requires a shift from merely supervising employees through direct guidelines to aligning interdependent contributors, including internal employees, contractors, suppliers, and platforms (Altman et al., 2023b). Altman et al. (2023b) provide comprehensive directions on how to approach leadership in the current workforce ecosystem. The first step is to relinquish direct control and establish indirect control mechanisms through clear outcomes, participation rules, and accountability. Leaders should expect a shift in power dynamics between internal and external actors. Leadership focus extends across organizational boundaries. Leaders must assume mixed-team realities, strive for alignment rather than uniformity, and promote an inclusive environment where everyone works toward the same outcome. Boards must also provide oversight of workforce ecosystem leadership. Finally, culture cannot be ignored. There is no one-size-fits-all approach. It is up to the leader to promote a culture that puts humans first and maximizes collaboration. The main challenge is less about who works for the organization and more about how to architect and lead the system that works alongside it.

4 Dynamic Capabilities: The Leadership "Operating System"

When the workforce ecosystem becomes complex, sustainability is a goal, and resilience in the face of change is crucial, innovative responses are needed to match the fluid environment of Industry 5.0. Such an environment requires a dynamic approach. Dynamic refers to the ability to renew competencies in order to attain harmony with the changing business environment. The word capabilities highlights the crucial function of strategic management in effectively integrating, reconfiguring, and adapting internal and external firm-specific resources, competencies, and skills to meet the demands of a changing environment. Human resources and organizational learning are among the areas that dynamic capabilities draw upon in their effort to develop and integrate different organizational functions and capabilities to attain competitive advantage (Teece et al., 1997). In the context of this chapter, dynamic capabilities are the routines of sensing, seizing, and reconfiguring, underpinned by specific microfoundations.

At the core of the dynamic capabilities framework reside three key components: sensing, seizing, and reconfiguring (Teece, 2007).

- *Sensing:* Involves environmental scanning, forming hypotheses about demands and technological solutions, and determining the right channels for relevant information to reach leaders.
- *Seizing:* Involves acting on the opportunities identified during the sensing stage. At this stage, decisions are made about resources and the avenues to which they are committed.
- *Reconfiguring:* Involves redesigning and realigning business interventions. The continuous alignment and realignment of intangible and tangible assets and resources sit at the core of this process.

Dynamic capabilities rest on microfoundations, defined as "distinct skills, processes, procedures, organizational structures, decision rules, and disciplines" (Teece, 2007, p. 1319). Effective sensing requires microfoundations such as leaders' cognitive capacity, their ability to make sense of information, and their access to external information (Kaur, 2023). Seizing opportunities requires the selection and design of the appropriate business model, a clear process for effective decision-making, and accurate resource allocation (Eisenhardt & Martin, 2000). Reconfiguring microfoundations include the transfer of knowledge, knowledge brokering, resource allocation routines, linking across units, and alliance, acquisition, and exit routines that add, combine, or shed resources as markets shift (Eisenhardt & Martin, 2000; Teece, 2007).

Viewed as a leadership operating system, dynamic capabilities supply the routines and decision logics to sense across organizational boundaries, seize with multi-party commitments, and transform resource configurations. This enables leaders to guide the enterprise and its workforce ecosystem toward sustainability and resilience in the face of change.

5 Conceptual Framework: Dynamic Capabilities for Workforce Ecosystems in Industry 5.0

In this section, we formally propose the mechanism by which dynamic capabilities operate in a workforce ecosystem spanning traditional employees, external contributors, partners, platforms, and communities. We treat the ecosystem interface (technical, organizational, legal, social) as the locus of leadership work and model three capability clusters, sensing, seizing, and reconfiguring, that produce outcomes aligned with the Industry 5.0 core values of human-centricity, sustainability, and resilience.

5.1 Sensing

According to O'Neill (2025), sensemaking is the act of pausing to reflect on a situation or challenge and creating shared understanding amid complexity and chaos. This is critical for leaders especially during this Industry 5.0 era. Effective sensing in Industry 5.0 serves its three core values and accounts for the complexities of current workforce ecosystems. Sensing must go beyond the traditional boundaries of firms. At this stage, leaders must pay close attention to labor markets and customer needs, promote internal R&D, learn from external research, and expect innovation in all units of the ecosystem (Teece, 2007). For effective sensing, leaders must follow a set of routines and structures.

We propose the following routines:

- External linkages: Build strong relationships with universities, key suppliers, and other complementors to access new knowledge (Eisenhardt & Martin, 2000).
- Decentralized network scanning: Nourish a well-connected intra-organizational network that extends beyond the firm so that information reaches leaders wherever it emerges (Teece, 2018).
- Analytics and scenario planning: Develop robust analytical tools to reduce noise, detect clear options, and reach specific solutions to lower uncertainty (Teece, 2007).

These routines require an act of collective sense making which is a social process that is most productive when leaders come together to engage in collaborative inquiry (O'Neill, 2025) with the goal of directly impacting Industry 5.0's mission. Changes in labor markets and customer needs require continuous skills and roles monitoring and capability development across the ecosystem. Leaders must also scan for external science and technological advancements to identify opportunities for human–machine collaboration. Because Industry 5.0 focuses on humans and their well-being, the sensing aperture must also include regulatory and institutional changes such as safety standards and data or AI governance.

5.2 *Seizing*

Seizing in Industry 5.0 involves leaders' decisions that capitalize on insights from the sensing process and translate them into coordinated commitments. Seizing turns sensed options into action. In Teece's (2007) formulation, this means selecting and redesigning the business model to capture value. In the fast-moving Industry 5.0 environment, getting the business model right is crucial. Without it, governance and leadership will fail to create value for all stakeholders.

Key areas of commitment include:

- Enterprise boundaries: Decisions about what takes place inside versus outside the firm, based on complementary assets, industry stage, and the firm's ability to protect unique gains. These boundary rules increase the likelihood that innovators, not imitators, benefit. Leaders must identify and combine specialized and cospecialized assets, a critical dynamic capability.
- Orchestration without owning: Decentralized decision-making allows faster responses since unit leaders have different information and can act on it. Yet decentralization alone is insufficient—it must be paired with leaders' integration and coordination skills. Leaders must also be able to select and adopt outside technologies that align with the ecosystem.

5.3 *Reconfiguring*

In the dynamic capabilities view, reconfiguring comprises the higher-order capacities that keep strategy, structure, processes, and assets continuously aligned as conditions change, often across organizational boundaries. Teece (2007) frames these as organizational-level capacities that realign, redeploy, and redesign assets, while also shaping the broader business ecosystem through collaboration.

For leaders in workforce ecosystems, reconfiguring is primarily a cross-boundary responsibility. Because production and development flow between units and firms, governance must allow technology and know-how to move while protecting intellectual property (Eisenhardt & Martin, 2000). Leaders should prepare for a constant flow of changes in corporate boundaries, organizational structures, business models, and products. Decentralized structures make these adjustments safer and faster (Teece, 2007).

In practice, reconfiguring appears as processes that can be orchestrated across internal and external contributors. These include knowledge transfer and brokering, efficient resource-allocation routines, coordination across the ecosystem, adding or shedding partnerships, and road maps for acquiring or releasing resources as conditions shift (Eisenhardt & Martin, 2000).

What makes reconfiguring leadership-ready for Industry 5.0 is the systemic bundle of microfoundations that account for mixed workforces. Teece's (2007) microfoundations emphasize incentive alignment and agency control, knowledge transfer and protection, cospecialization management, continuous realignment of tangible and intangible assets, loosely coupled structures, an open-innovation posture, and coordination skills (Teece, 2018).

Reconfiguring becomes the leadership aim when Industry 5.0 pushes for resilience, human-centered innovation, and ecosystem collaboration. Dynamic capabilities provide leaders with a systems lens, helping them identify which relationships matter at a given moment and placing external firms and institutions within the frame of action.

6 Boundary Conditions and Moderators

Dynamic capabilities are sensitive to context (Eisenhardt & Martin, 2000; Teece, 2007; Teece, 2018). They take on different forms depending on business environment characteristics. In moderately dynamic contexts, they are detailed, analytic, and robust. In high-velocity contexts, they become simple, experiential, and fragile, relying on a few rules and rapid iteration. In leading workforce ecosystems in Industry 5.0, four moderators are salient:

- Technological change: Increases the importance of sensing but can weaken seizing if commitment rules are poor (e.g., overcommitting to short-lived technologies).
- Platform dependency: Reliance on proprietary platforms raises the need for clear data and intellectual property rights and can slow reconfiguring.
- Institutional and regulatory regime: Safety, data protection, and labor rules may push toward tighter coupling and more closed participation, biasing DC design toward more codified and auditable routines.
- Workforce voice and collective representation: Strong voice requires inclusive seizing, with clear participation rules and transparency where algorithms allocate or evaluate work, to sustain cross-actor commitment.

These moderators act as levers leaders can adjust to fit dynamic capabilities to Industry 5.0 workforce ecosystems.

7 Outcomes

The DC lens explicitly concerns sustaining enterprise performance in open, innovative, globally dispersed environments by continuously renewing and aligning the asset base. Outcomes aligned with Industry 5.0 goals include:

- Innovation velocity: How quickly sensed options become scalable capabilities, moving from idea to pilot to scale across firms.
- Resilience: Improved recovery time and reduced performance dips during crises.

- Engagement and well-being: Stronger contributor engagement and energy across workforce types.
- Sustainability: Lower emissions and reduced energy per unit of value.
- Cost-to-capability: Lower total cost per unit of delivered capability.

8 Leadership Playbook for Orchestrating Workforce Ecosystems in Industry 5.0

Industry 5.0 shifts leadership from controlling employees inside a hierarchy to orchestrating value creation across workforce ecosystems that include employees, contractors, partner firms, platform contributors, and increasingly AI agents. The leader's job is to align these interdependent actors around outcomes and to install the higher-order abilities the dynamic capabilities literature calls sensing, seizing, and reconfiguring. Sensing brings weak signals and external knowledge into view. Seizing translates options into credible commitments and business-model choices. Reconfiguring continually realigns assets, structures, and relationships as conditions change. These are routines supported by concrete microfoundations such as decision rules, coordination mechanisms, knowledge transfer, and governance for data and intellectual property across firm boundaries (Altman et al., 2023b; Teece, 2007; Teece, 2018).

This section shifts from the earlier proposed framework to how leaders actually run it day to day. The task is no longer to argue for sensing, seizing, and reconfiguring, but to institutionalize them across a workforce ecosystem that blends employees, contractors, partner firms, platform contributors, and AI agents. Leadership becomes orchestration rather than control by setting direction, specifying interfaces, aligning incentives, and enabling distributed execution while protecting value capture and legitimacy (Altman et al., 2023a, b; Teece, 2007; Teece, 2018).

8.1 Sensing as an Operating Habit

Move from sporadic market scans to a standing "ecosystem radar" that feeds decisions on a fixed cadence. Each month, curate four inputs into a short decision brief: shifts in customers and segments interpreted as skills and roles foresight, internal R&D and technology trials that can be promoted or killed, exogenous science and technology with emphasis on AI and robotics for human-AI collaboration, and supplier and complementor signals drawn from partner signals. This is a faithful operationalization of Teece's sensing categories of scanning, learning, and interpretive work supported by analytical systems and decentralized attention, and it places external partners and platforms directly inside the aperture (Teece, 2007; Teece, 2018). At the behavioral level, treat psychological safety and digital fluency as sensing enablers through inviting dissent, normalizing upward voice, and coaching managers to detect weak signals in AI-augmented settings. Recent Industry 5.0 evidence links these behaviors to innovation and organizational learning outcomes (Alshaibani et al., 2025; Pratiwi et al., 2025; Santos et al., 2025). Use a compact dashboard, a skills and roles radar, a regulatory and AI tracker, and a partner telemetry view, and track sensing breadth and time-to-detection so the routine remains decision-relevant (Teece, 2007; Teece, 2018).

8.2 *Seizing without Owning*

Begin with a clear boundary for each initiative that tests what to do inside versus with partners, using appropriability and the location and nature of complementary assets as the decision logic. Then, lock a business model and an investment sequence that matches those boundary choices (Teece, 2007). Translate commitments into work through multi-party objectives, joint roadmaps, and cross-organization escalation routines so autonomy stays high but choices converge on time. Pratiwi et al. (2025)and Santos et al. (2025) both emphasize decentralization paired with authentic, inclusive leadership as the way to preserve engagement (Pratiwi et al., 2025; Santos et al., 2025). Formalize trust and value capture with a partnering playbook that standardizes data and intellectual property terms (access, improvement, reuse). Increase the effectiveness of algorithmic transparency, and cybersecurity. When engaging non-employees, make psychological contracts explicit by clearly outlining scope, feedback channels, attribution, opportunities for learning, and mechanisms for voice, rather than leaving these elements implicit (Altman et al., 2023b; Alshaibani et al., 2025).

8.3 *Reconfiguring as a Normal Muscle*

Treat reconfiguration as routine, not as a crisis response. Redesign one workflow at a time to establish human-AI collaboration patterns that move routine tasks to AI where safe and effective and elevate human roles toward creativity and complex problem solving. Embed reskilling so capability grows with each redesign (Alshaibani et al., 2025). Keep socio-technical architecture modular and auditable, version interfaces, codify data contracts, and instrument observability, so you can change one part of the system without breaking the rest (Teece, 2007). Manage your contributors as a portfolio that can be "patched," adding or swapping partners as opportunities and risks evolve. These are the identifiable processes the DC literature flags as effective in dynamic environments (Eisenhardt & Martin, 2000). Since Industry 5.0 elevates resilience to a high-level priority, leaders should run explicit recovery and rerouting exercises and track indicators such as reconfiguration lead time and change failure rate. Recent evidence in operations and supply suggests that repeated sensing, seizing, and reconfiguring cycles are what convert adaptation into robustness (Mohammed et al., 2025).

8.4 Governance, Ethics, and Voice as Design Constraints

Treat governance not as compliance after the fact but as part of the capability. Publish fairness, accountability, and transparency practices around algorithmic management and communicate them. Providing such information to actors reduces anxiety and accelerates learning. Evidence shows these outcomes mediate innovative performance and organizational learning in contexts of heavy AI use (Alshaibani et al., 2025). In settings with strong workforce voice, leaders should adopt inclusive commitment mechanisms. This means setting clear participation rules and ensuring contestability when algorithms allocate or evaluate work. These practices sustain cross-employer cooperation and align with the dynamic capabilities focus on incentives, complements, and co-specialized assets (Santos et al., 2025; Teece, 2007). Boards should also monitor ecosystem issues, including intellectual property leakage, dependence on proprietary platforms, rent loss from weak contracts, and the adequacy of sensing, seizing, and reconfiguring not just financial results (Altman et al., 2023b; Teece, 2018).

9 3-Month Plan for Leaders' Development

First Month

- Map contributors and interfaces.
- Launch the ecosystem radar.
- Adopt a lightweight charter for AI ethics and workforce voice while training managers in psychological safety.

Second Month

- Make boundary choices for one initiative.
- Publish a partnering playbook with data and intellectual property tiers and escalation rules.
- Agree on joint objectives and key results.
- Open a sandbox with a clear path to certification.

Third Month

- Run one human-AI redesign with embedded upskilling.
- Execute a reconfiguration drill that swaps a partner or splits a module.
- Measure lead time.
- Blast-radius containment.

- Change-failure rate.
- Second-source coverage to prove the system can adapt under load.

10 Conclusion

Industry 5.0 represents a turning point in how societies and organizations approach progress. Unlike earlier revolutions that placed efficiency and technology at the center, Industry 5.0 highlights human needs, planetary sustainability, and resilience. It builds on the foundations of Industry 4.0 but moves beyond the focus on automation and productivity. Its goal is to create prosperity that is inclusive, responsible, and future-oriented.

This chapter has shown that workforce ecosystems are central to the success of Industry 5.0. These ecosystems include employees, contractors, partner firms, platforms, and even artificial intelligence agents. No single actor can achieve the vision of Industry 5.0 alone. Value creation depends on coordination and alignment across boundaries. Leaders must shift from command and control to orchestration. They must design systems that allow different contributors to thrive together while advancing shared outcomes.

Dynamic capabilities provide the leadership operating system for this new reality. The routines of sensing, seizing, and reconfiguring equip leaders to navigate change, balance trade-offs, and guide ecosystems toward resilience. Sensing ensures that leaders recognize weak signals and opportunities across markets, technologies, and regulations. Seizing translates those insights into credible commitments and models for value creation. Reconfiguring allows leaders to continually realign resources, relationships, and structures as conditions evolve. These processes are not optional. They are necessary for organizations that must adapt quickly while protecting fairness, security, and legitimacy.

The chapter also underscored that leadership in Industry 5.0 must be deliberate in addressing ethics, transparency, and workforce voice. Governance should be proactive, not reactive. Boards must monitor ecosystem health beyond financial performance, asking how well leaders are building trust, securing intellectual property, and safeguarding contributors. Leaders must also invest in cultures that enable collaboration and continuous learning, since human–machine synergy depends on trust and adaptability.

In closing, Industry 5.0 challenges organizations to expand their vision of progress. It asks leaders to move from efficiency alone to human flourishing, from isolated firms to interconnected ecosystems, and from static strategies to dynamic capabilities. By adopting this mindset, leaders can orchestrate systems that are innovative, sustainable, and resilient. The path forward is not about controlling every detail. It is about building the conditions for collaboration and renewal. Those who succeed will shape industries that serve both people and the planet for generations to come.

References

Adner, R. (2017). Ecosystem as structure: An actionable construct for strategy. *Journal of Management, 43*(1), 39–58.

Alshaibani, E., Bakir, A., & Al-Atwi, A. (2025). The impact of leadership behaviors on organizational innovative performance and learning in AI-driven industry 5.0 environments. *Development and learning in organizations: An International Journal, 39*(3), 18–21.

Altman, E. J., Kiron, D., Schwartz, J., & Jones, R. (2023a). Manage your workforce ecosystem, not just your employees. *MIT Sloan Management Review, 64*(4), 80–84.

Altman, E. J., Kiron, D., Schwartz, J., & Jones, R. (2023b). *Workforce ecosystems: Reaching strategic goals with people, partners, and technologies*. MIT Press.

Cotta, J., Breque, M., De Nul, L., & Petridis, A. (2021). *Towards a sustainable, human-centric and resilient european industry*. Publications Office of the European Union.

Eisenhardt, K. M., & Martin, J. A. (2000). Dynamic capabilities: What are they?

Kaur, V. (2023). Knowledge-based dynamic capabilities: A scientometric analysis of marriage between knowledge management and dynamic capabilities. *Journal of Knowledge Management, 27*(4), 919–952.

Kellogg, K. C., Valentine, M. A., & Christin, A. (2020). Algorithms at work: The new contested terrain of control. *Academy of Management Annals, 14*(1), 366–410.

Leng, J., Sha, W., Wang, B., Zheng, P., Zhuang, C., Liu, Q., Wuest, T., Mourtzis, D., & Wang, L. (2022). Industry 5.0: Prospect and retrospect. *Journal of Manufacturing Systems, 65*, 279–295.

Liu, W., He, C., Jiang, Y., Ji, R., & Zhai, X. (2020). Effect of gig workers' psychological contract fulfillment on their task performance in a sharing economy—A perspective from the mediation of organizational identification and the moderation of length of service. *International Journal of Environmental Research and Public Health, 17*(7), 2208.

Mohammed, A., Al Balushi, F. A., Zubairu, N., & Govindan, K. (2025). Dynamics capabilities 5.0 toward inner business resiliency: A conceptual and evaluation panacea. *Computers & Industrial Engineering, 111223.*

Nafukho, F. M., El Mansour, W., & Van, H. T. M. (2024). The fourth and fifth industrial revolutions: HRD'S role in preparing graduates for future work. In T. S. Rocco, M. L. Morris, & R. F. Poell (Eds.), *The sage hand-book of human resource development* (pp. 248–263). Sage Publications Ltd.

O'Neill, K. (2025). *Collective sensemaking in an environment of constant disruption*. Center for Creative Leadership.

Pratiwi, E. C., Purwanto, F., & Irfan, M. (2025). Leadership 4.0: How transformational leadership shapes employee engagement in the era of industry 5.0. *Jurnal Informatika Ekonomi Bisnis*, 277–281.

Probst, T. M., Petitta, L., Barbaranelli, C., & Lavaysse, L. M. (2018). Moderating effects of contingent work on the relationship between job insecurity and employee safety. *Safety Science, 106*, 285–293.

Rani, U., Kumar Dhir, R., Furrer, M., Gőbel, N., Moraiti, A., & Cooney, S. (2021). *World employment and social outlook: The role of digital labour platforms in transforming the world of work*. International Labour Organisation.

Santos, C. F., Rodriguez, J. L. M., Fontes, C. H. D. O., & Santiago, S. B. (2025). Leadership perspectives on the transition from industry 4.0 to industry 5.0: A broad systematic literature review. *Journal of Scientometric Research, 14*(2), 467–478.

Teece, D. J. (2007). Explicating dynamic capabilities: The nature and microfoundations of (sustainable) enterprise performance. *Strategic Management Journal, 28*(13), 1319–1350.

Teece, D. J. (2018). Dynamic capabilities as (workable) management systems theory. *Journal of Management & Organization, 24*(3), 359–368.

Teece, D. J., Pisano, G., & Shuen, A. (1997). Dynamic capabilities and strategic management. *Strategic Management Journal, 18*(7), 509–533.

Xu, X., Lu, Y., Vogel-Heuser, B., & Wang, L. (2021). Industry 4.0 and industry 5.0—Inception, conception and perception. *Journal of Manufacturing Systems, 61*, 530–535.

Human-Centric Leadership in the Age of AI: Redefining HRD for Industry 5.0 Transformation

Robert M. Yawson and Daniel Woldeab

Abstract *Purpose* This chapter provides Human Resource Development (HRD) professionals with a conceptual framework for leading Industry 5.0 transformation, positioning them as architects of human-AI collaboration rather than victims of technological displacement. *Design/Methodology/Approach* The chapter employs a multi-theoretical framework integrating nested complexity theory, wicked problem constructs, and systems thinking approaches. The methodology combines a comprehensive literature review of recent empirical research with conceptual framework development, synthesizing insights from peer-reviewed sources. *Results* The framework addresses four critical HRD challenges: velocity of change outpacing traditional development cycles, epistemological shifts requiring non-linear thinking, authenticity gaps in diversity initiatives, and measurement paradoxes in quantifying qualitative outcomes. It produces actionable leadership competencies spanning human-centric, digital transformation, and antifragile capabilities, alongside practical implementation strategies. *Originality/Value* This work introduces the first comprehensive application of nested complexity theory to HRD contexts, reframes HRD professionals as "human translators," and provides novel frameworks for collaborative intelligence and antifragile organizational design. It advances both theoretical understanding and practical effectiveness for Industry 5.0 transformation. *Limits* The conceptual nature requires empirical validation across diverse contexts. Rapid AI advancement may necessitate continuous framework updates, and implementation complexity may challenge practitioners lacking systems thinking capabilities.

Keywords Industry 5.0 · Human resource development · Artificial intelligence · Human-centric leadership · Nested complexity · Wicked problems · Systems thinking · Organizational transformation · Collaborative intelligence · Antifragile organizations

---

R. M. Yawson (✉)
School of Business, Quinnipiac University, Hamden, CT, USA
e-mail: robert.yawson@qu.edu

D. Woldeab
College of Individualized Studies, Metropolitan State University, Saint Paul, MN, USA

A. M. Waite et al. (eds.), *Leading in Industry 5.0*, Management and Industrial Engineering, https://doi.org/10.1007/978-3-032-19207-3_13

1 Introduction

The industrial landscape is undergoing its most profound transformation since the dawn of the Industrial Revolution. As we transition from Industry 4.0's emphasis on automation, connectivity, and data-driven decision-making to Industry 5.0's human-centric paradigm, Human Resource Development (HRD) finds itself at the epicenter of this metamorphosis (Nahavandi, 2019; Xu et al., 2021). This transition represents more than a technological evolution; it embodies a fundamental reimagining of the relationship between humans and machines in organizational contexts (Breque et al., 2021).

Industry 5.0, characterized by its three foundational pillars: human-centricity, sustainability, and resilience, places people at the heart of industrial processes while leveraging advanced technologies such as cognitive artificial intelligence (AI), collaborative robots (cobots) , and smart wearables (Breque et al., 2021; Maddikunta et al., 2022). Unlike its predecessor, which often relegated humans to peripheral roles in increasingly automated systems, Industry 5.0 recognizes that the most significant organizational challenges are inherently human problems that require uniquely human solutions augmented by intelligent technologies (Demir et al., 2019; Nahavandi, 2019).

Recent empirical research validates this human-centric approach. Jia et al. (2024) demonstrate through field experiments that AI augments rather than replaces human creativity, particularly for higher-skilled workers, when tasks are sequentially divided between human and machine capabilities. These findings challenge traditional narratives of automation and support the emergence of collaborative intelligence models. Gilli et al. (2023) analyzed 239 job advertisements for digital transformation roles, revealing that primary competencies remain fundamentally human: collaboration, strategic thinking, leadership, customer orientation, and communication, capabilities that AI cannot easily replicate.

For HRD professionals, this paradigm shift presents both unprecedented opportunities and complex challenges. The traditional linear models of human capital development, rooted in predictable cause-and-effect relationships, prove inadequate for navigating the Volatile, Uncertain, Complex, and Ambiguous (VUCA) environments that define contemporary organizational reality (Yawson & Goryunova, 2025). The emergence of AI as a collaborative partner rather than a replacement tool fundamentally alters how we conceptualize learning, development, and human potential optimization (Raisch & Krakowski, 2021).

This chapter argues that HRD professionals are uniquely positioned to become the "human translators" in an increasingly algorithmic world, bridging the gap between technological capability and human flourishing. Rather than viewing AI as a threat to human employment, this perspective positions AI as an amplifier of human intelligence, creativity, and problem-solving capacity. This transformation requires HRD to evolve from a reactive, training-focused discipline to a proactive, systems-oriented practice that designs interventions capable of creating positive impact across multiple organizational levels simultaneously.

The chapter introduces an integrated conceptual framework that synthesizes three critical theoretical constructs: nested complexity theory for understanding AI implementation challenges (Goryunova & Yawson, 2025; Yawson & Goryunova, 2025), wicked problem constructs for addressing interconnected organizational challenges (Peterson et al., 2018; Rittel & Webber, 1973), and systems thinking approaches for designing holistic development interventions (Senge, 2006a; Senge, 2006b). This integrated framework provides HRD professionals with structured approaches for leading AI initiatives while prioritizing human development and organizational effectiveness.

2 Nested Complexity in Human-AI Collaboration

2.1 Understanding Nested Complexity Theory

The concept of nested complexity offers a lens for understanding the multifaceted challenges organizations face when implementing AI, while maintaining a focus on human resource development. Developed through extensive research on AI integration in organizational contexts, nested complexity theory recognizes that AI implementation challenges exist as interconnected layers spanning technological, ethical, regulatory, and environmental dimensions (Yawson & Goryunova, 2025). This framework builds upon earlier complexity science applications in organizational studies (Anderson, 1999; Burnes, 2005) while incorporating unique challenges posed by AI integration.

The framework identifies eight distinct types of complexity that HRD professionals must navigate: environmental complexity arising from external market forces and societal changes; regulatory complexity stemming from evolving legal frameworks and compliance requirements; integration complexity related to incorporating AI into existing organizational systems; collaboration complexity involving human-AI team dynamics; ethical complexity concerning fairness, bias, and transparency; data complexity addressing quality, privacy, and governance issues; algorithmic complexity related to AI system behavior and decision-making processes; and evolutionary complexity reflecting the rapid pace of AI advancement and organizational adaptation requirements (Goryunova & Yawson, 2025; Yawson & Goryunova, 2025). Gomez et al. (2025) analyzed 131 human-AI interaction sequences across 105 articles, revealing that current implementations fall short of realizing their genuine collaborative potential. Their systematic review found that 67% of interactions follow "AI-first" patterns and 21% follow "AI-follow" patterns, with limited sophisticated collaborative elements. This suggests that organizations are not yet fully leveraging the nested complexity framework to create truly synergistic human-AI partnerships.

These complexities are not isolated phenomena but exist in dynamic, reciprocal relationships where changes in one dimension cascade across others. For instance, the introduction of AI-powered performance evaluation systems (algorithmic complexity) immediately raises questions about fairness and bias (ethical complexity), requires new policies and procedures (regulatory complexity), demands integration with existing HR information systems (integration complexity), and necessitates training for both managers and employees (collaboration complexity) (Tambe et al., 2019).

2.2 *The Human-Centric Imperative in Industry 5.0*

Industry 5.0‘s human-centric approach represents a deliberate departure from Industry 4.0‘s technology-first mentality (Breque et al., 2021). This shift acknowledges that while technological capabilities continue to advance exponentially, the capacity to create meaningful and sustainable organizational change remains fundamentally dependent on human agency, creativity, and wisdom. Human-centricity in Industry 5.0 contexts involves three key dimensions: empowering employees through technology rather than replacing them (Sima et al., 2020), fostering collaboration between humans and intelligent systems (Demir et al., 2019), and creating organizational cultures that prioritize human well-being alongside operational efficiency (Xu et al., 2021). Abbu et al. (2025) validate this approach through the development of the Digital Leadership Scale, which measures seven dimensions of human-centric digital leadership: positive attitude toward change, ethical AI use, growth mindset, track record of success, transparent agenda, skills acquisition focus, and participative style. Organizations that implement these human-centric leadership approaches exhibit significantly higher digital transformation success rates.

Research demonstrates that organizations embracing human-centric approaches to Industry 5.0 transformation achieve superior performance outcomes compared to those pursuing purely technology-driven strategies (Castagnoli et al., 2024). A study of manufacturing organizations revealed that companies prioritizing human-machine collaboration experienced higher productivity gains and better employee retention rates compared to those implementing traditional automation approaches (Raiche, 2022). These findings, supported by similar results from Przegalinska et al. (2025) whose integrated Resource-Based View and Task Technology Fit theories, underscore the strategic importance of maintaining human agency within technologically advanced organizational systems.

The human-centric imperative extends beyond operational considerations to encompass broader questions of meaning, purpose, and values alignment (Kuchinke, 2010). Contemporary workforce research indicates that younger generations increasingly seek purpose-driven work experiences that connect individual contributions to societal impact (GALLUP, 2025). Organizations that successfully integrate AI technologies while maintaining clear connections between individual purpose and organizational mission demonstrate significantly higher levels of employee engagement, innovation, and adaptive capacity (Mayer et al., 2025).

2.3 Collaborative Intelligence: Redefining Human-AI Partnerships

The evolution from human-versus-machine to human-with-machine paradigms represents a fundamental shift in how organizations conceptualize work design and capability optimization. Collaborative intelligence emerges when human cognitive abilities, such as creativity, emotional intelligence, complex problem-solving, and ethical reasoning, combine synergistically with AI capabilities, such as pattern recognition, data processing, predictive analytics, and automation of routine tasks, to create outcomes that neither could achieve independently (Shenoi et al., 2024).

Raisch and Krakowski (2021) described hybrid problem-solving processes, identifying three distinct approaches: autonomous search, where AI generates distant solutions, sequential search, enabling local solutions, and interactive search, promoting recombinative solutions. This behavioral theory demonstrates that hybrid approaches broaden organizational search outcomes beyond traditional human-centric problem-solving limitations (Raisch & Krakowski, 2021).

Research on human-robot collaboration in manufacturing contexts provides compelling evidence for the potential of collaborative intelligence approaches. Studies of cobot implementation reveal that fluent human-robot interactions are positively associated with increased task performance, enhanced organizational citizenship behaviors, and improved creative performance among workers (Simões et al., 2020). These findings suggest that well-designed human-AI partnerships can simultaneously enhance operational efficiency and expand opportunities for meaningful human contribution.

The development of collaborative intelligence requires careful attention to interface design, skill complementarity, and trust-building processes (Glikson & Woolley, 2020). Organizations achieving successful human-AI collaboration invest heavily in training programs that help employees understand AI capabilities and limitations while developing competencies that complement rather than compete with AI functionalities (Choudhury et al., 2020). This approach transforms the traditional training paradigm from skills replacement to skills augmentation, positioning employees as active partners in AI-enabled work processes.

3 The Wicked Problem Construct: Addressing Interconnected Organizational Challenges

3.1 Conceptualizing Wicked Problems in HRD Practice

The wicked problem construct provides a powerful framework for understanding and addressing the complex, interconnected challenges that characterize contemporary organizational environments. Originally developed by Rittel and Webber (1973) to describe policy problems that resist traditional linear solutions, the concept has proven highly relevant for HRD practitioners navigating the complexities of Industry 5.0 transformation (Head, 2022; Peterson et al., 2018).

Wicked problems in HRD contexts exhibit several distinctive characteristics: they lack definitive formulations, have no clear stopping points, possess no true-or-false solutions, involve multiple interdependent factors, resist standardized approaches, and evolve continuously as intervention attempts modify the problem space (Yawson, 2015). Locklear (2025) examines AI and scenario planning applications for wicked problems, proposing integrated frameworks that combine artificial intelligence capabilities with scenario development methodologies. This approach is particularly relevant for addressing climate change and sustainability challenges that intersect with organizational transformation efforts.

The application of wicked problem constructs to HRD practice requires fundamental shifts in how practitioners approach organizational diagnosis, intervention design, and outcome evaluation. Traditional linear models that assume clear cause-and-effect relationships prove inadequate for addressing the emergent, adaptive nature of wicked problems. Instead, HRD professionals must develop comfort with ambiguity, iteration, and continuous sense-making processes that embrace rather than eliminate complexity.

3.2 Systems Thinking Approaches to Complex Problem-Solving

Systems thinking provides essential methodological approaches for engaging with wicked problems in organizational contexts. Unlike linear thinking, which seeks to identify root causes and implement targeted solutions, systems thinking recognizes that organizational phenomena emerge from the dynamic interactions among multiple interconnected elements (Senge, 2006a; Senge, 2006b). This perspective proves particularly valuable for HRD practitioners working in Industry 5.0 contexts where human, technological, and environmental factors interact in complex, non-predictable ways.

Effective systems thinking in HRD practice involves several key competencies: the ability to identify feedback loops and unintended consequences, recognition of delays between interventions and outcomes, understanding of how mental models shape organizational behavior, appreciation for the role of structure in creating behavioral patterns, and skill in designing leverage points where small changes can produce significant system-wide improvements (Senge et al., 2015; Yawson, 2013).

Research on HR transformation initiatives demonstrates that organizations applying systems thinking approaches achieve more sustainable change outcomes compared to those using traditional linear interventions (Senge, 2009). HR functions that utilize systems-based change strategies have a higher success rate in culture transformation initiatives and exhibit better long-term sustainability of development programs (HR Observer, 2024). These findings, corroborated by Evenseth et al. (2022) who systematically reviewed 59 articles on organizational resilience, highlight the practical value of systems thinking for addressing the complex challenges inherent in Industry 5.0 transformation.

3.3 Designing Interventions for Complex Adaptive Systems

The design of HRD interventions for complex adaptive systems requires approaches that embrace emergence, experimentation, and continuous learning. Traditional instructional design models, based on linear analysis-design-development-implementation-evaluation sequences, prove insufficient for environments characterized by rapid change, uncertainty, and emergent properties. Instead, HRD practitioners must develop capabilities for designing adaptive interventions that can evolve in response to changing conditions and emergent insights. Bender-Salazar (2023) provides a hybrid design thinking framework combining behavioral science and traditional design methods for addressing wicked problems. This three-phase approach (inspiration, ideation, implementation) demonstrates effectiveness through iterative problem-setting and double-loop learning processes, offering practical methodologies for HRD practitioners.

Effective intervention design for complex systems involves several key principles: starting with small-scale experiments rather than large-scale rollouts, building feedback mechanisms that enable real-time learning and adaptation, creating safe-to-fail environments that encourage innovation and risk-taking, focusing on capability building rather than problem-solving, and maintaining flexibility to modify approaches based on emergent understanding (Bansal & Birkinshaw, 2025).

The implementation of these principles requires new competencies for HRD professionals, including facilitation skills for navigating group complexity, analytical capabilities for interpreting emergent patterns, comfort with ambiguity and uncertainty, and leadership abilities for inspiring others through unclear and evolving situations (Heifetz et al., 2009). Developing these competencies represents a significant shift from traditional HRD practice but proves essential for effectiveness in Industry 5.0 contexts.

4 Challenges and Opportunities in Industry 5.0 HRD

4.1 The Velocity Challenge: Accelerating Change in Traditional Development Cycles

One of the most significant challenges facing HRD professionals in Industry 5.0 contexts is the acceleration of change cycles that threaten to outpace traditional human capital development approaches. While HRD has historically operated on monthly or yearly planning horizons, the digital economy increasingly demands real-time adaptation and continuous learning capabilities (Mayer et al., 2025). This velocity challenge forces fundamental reconsideration of how development interventions are conceptualized, designed, and delivered (Li & Herd, 2017).

The acceleration of change affects multiple dimensions of HRD practice. Skill requirements evolve rapidly as AI capabilities advance and job roles transform. Training content becomes obsolete more quickly, requiring continuous updating and refresh cycles. Employee expectations for just-in-time learning and personalized development experiences increase. Organizational strategies shift more frequently, necessitating agile approaches to building capabilities and driving culture change. Addressing the velocity challenge requires HRD professionals to develop new approaches that prioritize speed, flexibility, and continuous adaptation (Yawson, 2020). This includes implementing microlearning architectures that deliver targeted content in small, digestible units; creating modular development programs that can be rapidly reconfigured based on changing needs; establishing continuous feedback systems that enable real-time course corrections; and building organizational learning capabilities that support self-directed, peer-to-peer knowledge sharing.

4.2 Epistemological Shifts: New Ways of Knowing in Non-Linear Environments

The transition to Industry 5.0 requires fundamental epistemological shifts in how HRD professionals understand and approach human development in organizational contexts. Traditional cause-and-effect models, rooted in linear thinking and predictable relationships, prove inadequate for complex, interconnected environments where emergence and adaptation are primary characteristics (Yawson, 2013). This epistemological challenge demands new frameworks for understanding how learning, development, and behavior change occur in complex adaptive systems.

The epistemological shift involves moving from mechanistic to organic metaphors for understanding organizational functioning. Rather than viewing organizations as machines with predictable inputs and outputs, Industry 5.0 perspectives recognize organizations as living systems characterized by self-organization, adaptation, and emergence. This shift has profound implications for HRD practice, requiring approaches that work with rather than against natural organizational dynamics.

New epistemological frameworks for HRD practice emphasize pattern recognition over problem-solving, emergence over planning, relationship building over individual development, and continuous sense-making over predetermined outcomes. These approaches require HRD professionals to develop comfort with uncertainty, ambiguity, and multiple perspectives while maintaining focus on human flourishing and organizational effectiveness.

4.3 The Authenticity Gap in Diversity, Equity, and Inclusion Initiatives

Industry 5.0's emphasis on human-centricity brings increased scrutiny to organizational efforts around Diversity, Equity, and Inclusion (DEI). The authenticity gap, the disconnect between stated DEI commitments and actual organizational practices, represents a significant challenge for HRD professionals tasked with creating genuinely inclusive environments rather than performative diversity initiatives (Callaham, 2023; Dobbin & Kalev, 2022).

There are critical tensions between business cases and moral imperatives in diversity management. Qualitative studies of UK D&I consultants show that personal ethical values drive authentic DEI implementation, with external consultants positioned as valuable equality practitioners (Greene & Kirton, 2024). Furthermore, a critical examination of "whole self" discourse, using Foucault's concepts, reveals that inclusion becomes contingent upon conformity to corporate norms, with certain identities being legitimized while others are marginalized (Carr, 2025).

Creating authentic DEI requires embedding diversity and inclusion principles into every aspect of organizational operation rather than treating them as separate initiatives. This includes standardizing recruitment processes to minimize bias, implementing transparent decision-making procedures, creating accountability mechanisms for inclusive behavior, and measuring outcomes that reflect genuine culture change rather than compliance activities. The integration of AI technologies adds additional complexity to DEI efforts, requiring careful attention to algorithmic bias, equitable access to AI-enhanced opportunities, and fair representation in AI system design and implementation.

4.4 The Measurement Paradox: Quantifying Qualitative Development Outcomes

HRD professionals face the ongoing challenge of demonstrating value and impact in ways that satisfy organizational demands for return on investment while honoring the inherently qualitative nature of human development processes. This measurement paradox becomes more pronounced in Industry 5.0 contexts, where the most important outcomes: wisdom, creativity, resilience, and ethical reasoning, resist traditional quantification approaches.

The measurement paradox emerges from tensions between the desire for objective, quantifiable metrics, and the reality that meaningful human development involves subjective, contextual, and emergent qualities that cannot be easily measured. Traditional approaches often focus on easily quantifiable outputs, such as training hours completed, certification achievements, and performance ratings, rather than deeper outcomes related to human flourishing and organizational capability building. Addressing the measurement paradox requires HRD professionals to develop longitudinal tracking systems that reveal development patterns over time and utilize AI-powered analytics that can process both numerical data and unstructured text to provide comprehensive impact assessments (Kiron et al., 2014).

5 Leadership Frameworks for Industry 5.0 Transformation

5.1 Human-Centric Leadership Competencies

The successful navigation of Industry 5.0 transformation requires new leadership competencies that integrate technological sophistication with deep human understanding. Human-centric leadership in Industry 5.0 contexts involves the ability to create environments where humans and intelligent systems collaborate effectively while maintaining focus on human agency, creativity, and well-being (Santana & Díaz-Fernández, 2023).

There are several critical competencies for human-centric leadership in Industry 5.0: empathy and emotional intelligence for understanding and responding to human needs in technology-mediated environments; systems thinking capabilities for recognizing interconnections and unintended consequences; ethical reasoning skills for navigating complex moral questions raised by AI implementation; change leadership abilities for guiding organizations through continuous transformation processes; and collaborative intelligence for optimizing human-AI partnerships.

The development of these competencies requires experiential learning approaches that provide leaders with opportunities to practice complex decision-making in realistic scenarios. This includes simulation-based training programs that recreate the complexity of Industry 5.0 environments, mentoring relationships with experienced practitioners, cross-functional project assignments that develop systems thinking capabilities, and reflective practices that build self-awareness and ethical reasoning skills.

5.2 *Digital Transformation Leadership*

Digital transformation leadership represents a specialized form of leadership that combines traditional transformational leadership competencies with deep understanding of digital technologies and their organizational implications (Westerman et al., 2014). In Industry 5.0 contexts, digital transformation leaders must navigate the intersection of human development needs and technological capabilities while maintaining focus on sustainable organizational effectiveness.

Effective digital transformation leadership involves several key dimensions: technological fluency sufficient to make informed decisions about AI implementation; change management expertise for guiding organizations through complex transformation processes; stakeholder engagement skills for building coalitions and managing resistance; strategic thinking capabilities for aligning digital initiatives with organizational mission and values; and cultural intelligence for creating environments that support innovation and collaboration.

The development of digital transformation leadership capabilities requires integrated approaches that combine technical training with leadership development and organizational psychology education. Many organizations are creating specialized development programs that rotate high-potential leaders through various digital initiatives while providing coaching and mentoring support to build competency and confidence.

5.3 *Adaptive and Antifragile Leadership*

The concept of antifragile leadership extends beyond traditional resilience to encompass the ability to gain strength from volatility, uncertainty, and stress (Taleb, 2014). In Industry 5.0 contexts characterized by rapid change and continuous disruption, leaders must develop capabilities that enable their organizations not merely to survive challenges but to become stronger as a result of encountering them (Adobor & Kudonoo, 2025).

Antifragile leadership involves several key capabilities: embracing volatility as a source of learning and improvement rather than something to be avoided; decentralizing decision-making to increase organizational responsiveness and adaptability; building optionality into strategic plans to maintain flexibility in uncertain environments; fostering cultures of experimentation that treat failures as learning opportunities; and developing organizational capabilities that enable rapid pivoting and adaptation.

The cultivation of antifragile leadership requires deliberate exposure to challenging situations that build resilience and adaptive capacity. This includes stretch assignments that push leaders beyond their comfort zones, crisis simulation exercises that develop rapid decision-making skills, cross-cultural experiences that build cognitive flexibility, and reflective practices that help leaders extract learning from difficult experiences.

6 Organizational Design for Human-AI Collaboration

6.1 Creating Collaborative Work Environments

The design of work environments that optimize human-AI collaboration requires careful attention to both physical and virtual spaces, which support the seamless integration of human and artificial intelligence capabilities. Research on collaborative robotics demonstrates that successful human-robot teams require workplace designs that facilitate natural interaction, clear communication protocols, and shared situational awareness (Neura Robotics, 2022).

Effective collaborative work environments incorporate several design principles: flexible physical spaces that can be rapidly reconfigured to support different types of human-AI collaboration; intuitive user interfaces that minimize cognitive load and enable natural interaction with AI systems; transparent communication protocols that clarify roles, responsibilities, and decision-making authority between humans and AI; and feedback mechanisms that enable continuous improvement of collaborative processes.

The implementation of these design principles requires close collaboration between HRD professionals, information technology specialists, facilities managers, and operational leaders. This interdisciplinary approach ensures that workplace design decisions support both technological functionality and human well-being while advancing organizational objectives (Parker & Grote, 2022).

6.2 Developing AI Literacy and Human Capabilities

The successful integration of AI into organizational work processes requires comprehensive approaches to developing both AI literacy among employees and unique human capabilities that complement AI functionalities (Ng et al., 2021). AI literacy involves understanding AI capabilities and limitations, developing skills for effective human-AI collaboration, and building ethical frameworks for responsible AI use (Long & Magerko, 2020). Cetindamar et al. (2022) propose an AI competency framework encompassing four levels: AI awareness (understanding what AI is and is not), AI literacy (knowing how AI works), AI capability (using AI tools effectively), and AI fluency (creating with AI).

Effective AI literacy development programs address multiple competency areas: technical understanding sufficient to work effectively with AI systems without requiring programming expertise; collaborative skills for participating in human-AI teams; critical thinking capabilities for evaluating AI recommendations and outputs; ethical reasoning skills for navigating moral questions related to AI implementation; and continuous learning abilities for adapting to evolving AI capabilities (Littlejohn et al., 2016).

Simultaneously, organizations must invest in developing uniquely human capabilities that become more valuable as AI assumes responsibility for routine tasks. These include creative problem-solving abilities, emotional intelligence, interpersonal skills, complex ethical reasoning, strategic thinking, systems understanding, and cultural competency, as well as cross-functional collaboration skills.

6.3 Building Organizational Resilience and Antifragility

The development of organizational resilience and antifragility represents a critical capability for thriving in Industry 5.0 environments characterized by continuous change and uncertainty. Resilience involves the ability to maintain functionality during disruptions, while antifragility encompasses the capacity to become stronger as a result of stressors and challenges (Taleb, 2014).

Building organizational antifragility requires systematic approaches that embed adaptive capacity into organizational structures, processes, and cultures (Adobor & Kudonoo, 2025). This includes creating redundancy and backup systems that prevent single points of failure; developing sensing mechanisms that enable early detection of emerging challenges and opportunities; establishing rapid response capabilities that allow quick pivoting and adaptation; and fostering learning cultures that extract maximum value from both successes and failures.

HRD professionals play critical roles in building organizational antifragility through their influence on culture, capability development, and change management processes. This includes designing learning experiences that build adaptive capacity, creating feedback systems that enable continuous improvement, facilitating cross-functional collaboration that breaks down organizational silos, and developing leadership capabilities that enable effective navigation of uncertainty and complexity.

7 Implementation Strategies and Best Practices

7.1 Phased Approach to Industry 5.0 Transformation

The complexity of Industry 5.0 transformation requires carefully planned implementation approaches that balance the urgency of competitive pressures with the need for thoughtful change management (Gfrerer et al., 2021). Successful transformations typically follow phased approaches that begin with pilot initiatives and gradually scale successful interventions across broader organizational contexts.

Phase one involves assessment and preparation activities, including organizational readiness evaluation, stakeholder engagement and coalition building, infrastructure assessment and planning, and capability gap analysis. Phase two focuses on pilot implementation through small-scale experiments, rapid prototyping of new processes, continuous feedback and iteration, and documentation of lessons learned. Phase three encompasses scaling and integration, including broader rollout of successful initiatives, integration with existing organizational systems, culture change reinforcement, and sustainability planning.

Each phase requires different competencies and resources from HRD professionals. Early phases demand strong analytical and planning capabilities, while later phases require change management expertise and systems integration skills. Successful implementation requires sustained commitment from organizational leadership and adequate resource allocation throughout the transformation process.

7.2 Change Management for Complex Transformation

The Industry 5.0 transformation represents a complex form of organizational change that exceeds the scope of traditional change management approaches. The interconnected nature of technological, cultural, and structural changes requires change management strategies that address multiple dimensions simultaneously while maintaining focus on human well-being and organizational effectiveness (Yawson et al., 2025).

Effective change management for Industry 5.0 transformation incorporates several key elements: comprehensive stakeholder analysis and engagement strategies that address diverse interests and concerns; communication approaches that provide transparency about transformation goals, progress, and challenges; participation mechanisms that involve employees in designing and implementing changes; support systems that help individuals and teams navigate transition challenges; and measurement approaches that track both quantitative outcomes and qualitative experiences.

The implementation of these change management approaches requires HRD professionals to develop sophisticated facilitation skills, cultural intelligence, and emotional competency. This includes the ability to navigate resistance and conflict constructively, create safe spaces for vulnerable conversations, design participatory processes that generate authentic engagement, and maintain hope and momentum during difficult transition periods.

7.3 Measuring Impact and Continuous Improvement

The measurement of Industry 5.0 transformation impact requires sophisticated evaluation approaches that capture both quantitative performance improvements and qualitative changes in employee experience, organizational culture, and stakeholder relationships. Traditional metrics focused solely on financial outcomes prove insufficient for understanding the full impact of human-centric transformation initiatives (Bititci et al., 2012).

Comprehensive impact measurement incorporates multiple data sources and methodologies, including quantitative performance metrics (productivity, quality, efficiency), employee engagement and satisfaction surveys, cultural assessment tools, stakeholder feedback mechanisms, and longitudinal case studies that capture transformation stories and experiences. The integration of AI-powered analytics enables more sophisticated analysis of these diverse data sources while reducing the manual effort required for evaluation activities.

Continuous improvement processes use measurement results to inform ongoing refinement of transformation strategies and implementation approaches. This includes regular review cycles that assess progress against objectives, feedback sessions that gather insights from participants, adaptation processes that modify approaches based on learning, and documentation practices that capture and share lessons learned across the organization.

8 Future Directions and Implications

8.1 Emerging Trends in Human-AI Collaboration

The evolution of human-AI collaboration continues to accelerate, with several emerging trends that will significantly impact HRD practice in the coming decade. These include the development of more sophisticated conversational AI systems that enable natural language interaction, the emergence of AI agents capable of autonomous task completion, the integration of AI into creative and strategic work processes, and the evolution of AI systems that can learn and adapt based on individual user preferences and organizational contexts. Dwivedi et al. (2021), in a survey of leading AI researchers, identify five transformative capabilities emerging by 2030: explainable AI that provides transparent reasoning, emotional AI recognizing and responding to human emotions, collaborative AI that seamlessly integrates with human work processes, ethical AI with built-in moral reasoning, and general AI approaching human-level problem-solving.

These technological advances create both opportunities and challenges for HRD professionals. Enhanced AI capabilities enable more personalized and adaptive learning experiences while requiring new competencies for managing human-AI relationships and ensuring ethical implementation. The challenge for HRD will be maintaining focus on human development and organizational effectiveness while leveraging these expanding technological capabilities.

Future HRD practice will likely involve much closer integration with AI systems for tasks such as personalized learning path recommendations, real-time performance feedback, predictive analytics for identifying development needs, and automated administrative processes that free HRD professionals to focus on higher-value strategic activities. Success will require developing new competencies for human-AI collaboration while maintaining the essential human elements of empathy, creativity, and ethical reasoning that define effective HRD practice.

8.2 Implications for HRD Education and Professional Development

The transformation of HRD practice in Industry 5.0 contexts has significant implications for how HRD professionals are educated and developed. Traditional HRD education programs, which focused primarily on training design, organizational psychology, and change management, must expand to include technological literacy, systems thinking, and complex problem-solving capabilities (Ardichvili, 2022).

Future HRD education should incorporate several additional competency areas: understanding of AI capabilities and limitations sufficient to make informed implementation decisions; systems thinking skills for navigating complex, interconnected organizational challenges; design thinking approaches for creating innovative solutions to wicked problems; ethical reasoning capabilities for addressing moral questions raised by new technologies; and global and cultural competency for working in increasingly diverse and distributed organizational environments.

The development of these competencies requires experiential learning approaches that provide opportunities to practice complex decision-making in realistic scenarios. This includes simulation-based learning experiences, cross-functional project assignments, international exchange programs, and mentoring relationships with experienced practitioners who have successfully navigated Industry 5.0 transformations.

8.3 Research Priorities and Methodological Considerations

The emergence of Industry 5.0 contexts creates numerous opportunities for HRD research that can advance both theoretical understanding and practical effectiveness. Priority research areas include the effectiveness of different approaches to human-AI collaboration, the impact of Industry 5.0 transformation on employee well-being and job satisfaction, the role of organizational culture in successful technology integration, and the development of new evaluation methodologies appropriate for complex, adaptive systems.

Methodological considerations for Industry 5.0 HRD research include the need for mixed-methods approaches that capture both quantitative outcomes and qualitative experiences, longitudinal designs that track changes over extended time periods, participatory research methods that engage practitioners as co-researchers, and case study approaches that provide rich contextual understanding of transformation processes.

Future research should also address ethical considerations related to AI implementation in HRD contexts, including questions of privacy, bias, transparency, and human agency. This research will be essential for developing guidelines and best practices that ensure AI integration serves human flourishing rather than merely operational efficiency.

9 Conclusion

The transition to Industry 5.0 represents a fundamental transformation in how organizations conceptualize the relationship between humans and technology. For HRD professionals, this transition presents both unprecedented opportunities and significant challenges that require new competencies, frameworks, and approaches. The human-centric emphasis of Industry 5.0 positions HRD professionals as critical architects of organizational transformation rather than peripheral support functions.

This chapter has argued that successful Industry 5.0 transformation requires HRD professionals to adopt a role as "human translators," bridging the gap between technological capabilities and human flourishing. This involves developing competencies in nested complexity thinking, wicked problem solving, and systems approaches to organizational development. The integration of these theoretical frameworks provides practical guidance for navigating the interconnected challenges that characterize contemporary organizational environments.

The challenges facing HRD in Industry 5.0 contexts: the velocity of change, epistemological shifts, authenticity gaps, and measurement paradoxes, require fundamental reconceptualization of traditional practice approaches. Rather than viewing these challenges as obstacles to be overcome, they represent opportunities for HRD to evolve into a more strategic, impactful, and essential organizational capability.

The leadership frameworks and implementation strategies presented in this chapter provide roadmaps for HRD professionals seeking to lead successful Industry 5.0 transformations. The emphasis on human-centric leadership, collaborative intelligence, and antifragile organizational design offers concrete approaches for creating work environments where humans and AI can collaborate effectively while maintaining focus on human agency and well-being.

The future of HRD in Industry 5.0 contexts will be defined by the profession's ability to maintain its essential focus on human development while embracing the transformative potential of intelligent technologies. This requires ongoing commitment to learning, adaptation, and professional development coupled with unwavering dedication to the fundamental values of human dignity, equity, and flourishing that define effective HRD practice.

As we stand at the threshold of this new industrial age, HRD professionals have the opportunity and responsibility to shape a future where technological advancement serves human potential rather than replacing it. The frameworks, strategies, and insights presented in this chapter provide a foundation for this critical work, but success will ultimately depend on the creativity, wisdom, and commitment of HRD professionals who choose to embrace the transformative possibilities of Industry 5.0.

The journey toward human-centric Industry 5.0 transformation is neither simple nor straightforward, but it represents perhaps the most significant opportunity in our profession's history to demonstrate the essential value of investing in human potential. By embracing the complexity, maintaining focus on human flourishing, and developing the competencies necessary for success in this new environment, HRD professionals can ensure that the future of work is not only more efficient and productive, but also more meaningful, equitable, and sustainable for all.

References

Abbu, H., Khan, S., Mugge, P., & Gudergan, G. (2025). Building digital-ready leaders: Development and validation of the human-centric digital leadership scale. *Digital, 5*(1), 7. https://doi.org/10.3390/digital5010007

Adobor, H., & Kudonoo, E. C. (2025). Antifragility and organizations: An organizational design perspective. *Leadership & Organization Development Journal, 46*(2), 351–375. https://doi.org/10.1108/LODJ-03-2024-0185

Anderson, P. (1999). Perspective: Complexity theory and organization science. *Organization Science*, 10(3), 216–232. doi:https://doi.org/10.1287/orsc.10.3.216

Ardichvili, A. (2022). The impact of artificial intelligence on expertise development: Implications for HRD. *Advances in Developing Human Resources, 24*(2), 78–98. https://doi.org/10.1177/15234223221077304

Bansal, T., & Birkinshaw, J. (2025). Why you need systems thinking now. *Harvard Business Review, R2505M*, 1–16.

Bender-Salazar, R. (2023). Design thinking as an effective method for problem-setting and needfinding for entrepreneurial teams addressing wicked problems. *Journal of Innovation and Entrepreneurship, 12*(1), 24. https://doi.org/10.1186/s13731-023-00291-2

Bititci, U., Garengo, P., Dörfler, V., & Nudurupati, S. (2012). Performance measurement: Challenges for tomorrow. *International Journal of Management Reviews, 14*(3), 305–327. https://doi.org/10.1111/j.1468-2370.2011.00318.x

Breque, M., De Nul, L., & Petridis, A. (2021). *Industry 5.0—Towards a sustainable, human-centric and resilient European industry*. Publications Office of the European Union. https://doi.org/10.2777/308407

Burnes, B. (2005). Complexity theories and organizational change. *International Journal of Management Reviews, 7*(2), 73–90. https://doi.org/10.1111/j.1468-2370.2005.00107.x

Callaham, S. (2023). How authenticity can validate DEI and increase business success.. https://www.forbes.com/sites/sheilacallaham/2023/05/30/how-authenticity-can-validate-dei-and-increase-business-success/

Carr, M. (2025). Bring your whole self to work: Boundary conditions of subjectivity in diversity and inclusion discourse on investment Bank websites. *Journal of Business Ethics*. https://doi.org/10.1007/s10551-025-06055-0

Castagnoli, R., Cugno, M., Maroncelli, S., & Cugno, A. (2024). A new research agenda for human-centric manufacturing: A systematic literature review. *IEEE Transactions on Engineering Management, 71*, 15236–15253. https://doi.org/10.1109/TEM.2024.3479775

Choudhury, P., Starr, E., & Agarwal, R. (2020). Machine learning and human capital complementarities: Experimental evidence on bias mitigation. *Strategic Management Journal, 41*(8), 1381–1411. https://doi.org/10.1002/smj.3152

Demir, K. A., Döven, G., & Sezen, B. (2019). Industry 5.0 and human-robot co-working. *Procedia Computer Science, 158*, 688–695. https://doi.org/10.1016/j.procs.2019.09.104

Dobbin, F., & Kalev, A. (2022). *Getting to diversity: What works and what doesn't*. Harvard University Press. https://books.google.com/books?id=0OF-EAAAQBAJ

Dwivedi, Y. K., Hughes, L., Ismagilova, E., Aarts, G., Coombs, C., Crick, T., Duan, Y., Dwivedi, R., Edwards, J., Eirug, A., Galanos, V., Ilavarasan, P. V., Janssen, M., Jones, P., Kar, A. K., Kizgin, H., Kronemann, B., Lal, B., Lucini, B., et al. (2021). Artificial intelligence (AI): Multidisciplinary perspectives on emerging challenges, opportunities, and agenda for research, practice and policy. *International Journal of Information Management, 57*, 101994. https://doi.org/10.1016/j.ijinfomgt.2019.08.002

Evenseth, L. L., Sydnes, M., & Gausdal, A. H. (2022). Building organizational resilience through organizational learning: A systematic review. *Frontiers in Communication, 7*. https://doi.org/10.3389/fcomm.2022.837386

GALLUP. (2025). *State of the global workplace understanding employees*. Informing Leaders.

Gfrerer, A., Hutter, K., Füller, J., & Ströhle, T. (2021). Ready or not: Managers' and employees' different perceptions of digital readiness. *California Management Review, 63*(2), 23–48. https://doi.org/10.1177/0008125620977487

Gilli, K., Nippa, M., & Knappstein, M. (2023). Leadership competencies for digital transformation: An exploratory content analysis of job advertisements. *German Journal of Human Resource Management: Zeitschrift Für Personalforschung, 37*(1), 50–75. https://doi.org/10.1177/23970022221087252

Glikson, E., & Woolley, A. W. (2020). Human Trust in Artificial Intelligence: Review of empirical research. *Academy of Management Annals, 14*(2), 627–660. https://doi.org/10.5465/annals.2018.0057

Gomez, C., Cho, S. M., Ke, S., Huang, C.-M., & Unberath, M. (2025). Human-AI collaboration is not very collaborative yet: A taxonomy of interaction patterns in AI-assisted decision making from a systematic review. *Frontiers in Computer Science, 6*. https://doi.org/10.3389/fcomp.2024.1521066

Goryunova, E., & Yawson, R. M. (2025). "Nested complexity" framework for human-Centered augmented leadership. *Journal of Leadership Studies, 19*(2). https://doi.org/10.1002/jls.70016

Greene, A., & Kirton, G. (2024). "Doing the right thing" and "making a difference": The role of personal ethical values in diversity and inclusion consulting. *Journal of Business Ethics, 193*(1), 179–191. https://doi.org/10.1007/s10551-023-05514-w

Head, B. W. (2022). Wicked problems in public policy: Understanding and responding to complex challenges. *Springer International Publishing*. https://doi.org/10.1007/978-3-030-94580-0

Heifetz, R. A., Grashow, A., & Linsky, M. (2009). *The practice of adaptive leadership: Tools and tactics for changing your organization and the world*. Harvard Business Press.

Jia, N., Luo, X., Fang, Z., & Liao, C. (2024). When and how artificial intelligence augments employee creativity. *Academy of Management Journal, 67*(1), 5–32. https://doi.org/10.5465/amj.2022.0426

Kiron, D., Prentice, P. K., & Ferguson, R. B. (2014). The analytics mandate. In *2014 Data & Analytics Global Executive Study and research project*. MIT Sloan Management Review.

Kuchinke, K. P. (2010). Human development as a central goal for human resource development. *Human Resource Development International, 13*(5), 575–585. https://doi.org/10.1080/13678868.2010.520482

Li, J., & Herd, A. M. (2017). Shifting practices in digital workplace learning: An integrated approach to learning, knowledge management, and knowledge sharing. *Human Resource Development International, 20*(3), 185–193. https://doi.org/10.1080/13678868.2017.1308460

Locklear, K. (2025). Wicked problems: A novel approach using artificial intelligence and scenarios. *Journal of Leadership & Organizational Studies, 32*(3), 230–248. https://doi.org/10.1177/15480518251330728

Long, D., & Magerko, B. (2020). What is AI literacy? Competencies and design considerations. In *Proceedings of the 2020 CHI conference on human factors in computing systems* (pp. 1–16). https://doi.org/10.1145/3313831.3376727

Maddikunta, P. K. R., Pham, Q.-V., Deepa, N., Dev, K., Gadekallu, T. R., Ruby, R., & Liyanage, M. (2022). Industry 5.0: A survey on enabling technologies and potential applications. *Journal of industrial information Integration, 26*, 100257. https://doi.org/10.1016/j.jii.2021.100257

Mayer, H., Yee, L., Chui, M., & Roberts, R. (2025). *Superagency in the workplace. Empowering people to unlock AI'S full potential*. McKinsey & Company.

Nahavandi, S. (2019). Industry 5.0—A human-centric solution. *Sustainability, 11*(16), 4371. https://doi.org/10.3390/su11164371

Robotics, N. (2022). *September*. Cognitive robots—the new era of true collaboration between human and machine. https://neura-robotics.com/cognitive-robots-the-new-era-of-true-collaboration/

Ng, D. T. K., Leung, J. K. L., Chu, S. K. W., & Qiao, M. S. (2021). Conceptualizing AI literacy: An exploratory review. *Computers and Education: Artificial Intelligence, 2*, 100041. https://doi.org/10.1016/j.caeai.2021.100041

Parker, S. K., & Grote, G. (2022). Automation, algorithms, and beyond: Why work design matters more than ever in a digital world. *Applied Psychology, 71*(4), 1171–1204. https://doi.org/10.1111/apps.12241

Peterson, G., Yawson, R. M., Sherman, J., & Johnson Kanda, I. (2018). A systems model of using the deliberate leadership® framework for addressing wicked problems. *International Journal of Business and Systems Research, 12*(3), 262–289. https://doi.org/10.1504/IJBSR.2018.091155

Przegalinska, A., Triantoro, T., Kovbasiuk, A., Ciechanowski, L., Freeman, R. B., & Sowa, K. (2025). Collaborative AI in the workplace: Enhancing organizational performance through resource-based and task-technology fit perspectives. *International Journal of Information Management, 81*, 102853. https://doi.org/10.1016/j.ijinfomgt.2024.102853

Raiche, J.-P. (2022). Industry 5.0: Revolutionizing work by putting people first. *ProAction International*. https://blog.proactioninternational.com/en/industry-5.0-the-next-industrial-revolution-is-people-centric

Raisch, S., & Krakowski, S. (2021). Artificial intelligence and management: The automation–augmentation paradox. *Academy of Management Review, 46*(1), 192–210. https://doi.org/10.5465/amr.2018.0072

Rittel, H. W. J., & Webber, M. M. (1973). Dilemmas in a general theory of planning. *Policy Sciences, 4*(2), 155–169. https://doi.org/10.1007/BF01405730

Santana, M., & Díaz-Fernández, M. (2023). Competencies for the artificial intelligence age: Visualisation of the state of the art and future perspectives. *Review of Managerial Science, 17*(6), 1971–2004. https://doi.org/10.1007/s11846-022-00613-w

Senge, P. (2006a). The fifth discipline: The art & practice of the learning organization. *New York Currency, 1*, 445.

Senge, P. (2006b). The leader's new work: Building learning organizations. In J. V. Gallos (Ed.), *Organization development: A Jossey-bass reader* (pp. 765–792). Jossey-Bass.

Senge, P. (2009). The necessary revolution. *Leader to Leader, 2009*(51), 24–28. https://doi.org/10.1002/ltl.314

Senge, P., Hamilton, H., & Kania, J. (2015). The dawn of system leadership. *Stanford Social Innovation Review, 13*(1), 26–33.

Shenoi, V. V., Sreeram, P., Sai Varma, C. L., Goud, K. R., & Afroz, S. N. (2024). Analysing the role of human-AI collaboration in workforce transformation. *International Conference on Advances in Data Engineering and Intelligent Computing Systems (ADICS), 2024*, 1–7. https://doi.org/10.1109/ADICS58448.2024.10533640

Sima, V., Gheorghe, I. G., Subić, J., & Nancu, D. (2020). Influences of the industry 4.0 revolution on the human capital development and consumer behavior: A systematic review. *Sustainability, 12*(10), 4035. https://doi.org/10.3390/su12104035

Simões, A. C., Lucas Soares, A., & Barros, A. C. (2020). Factors influencing the intention of managers to adopt collaborative robots (cobots) in manufacturing organizations. *Journal of Engineering and Technology Management, 57*, 101574. https://doi.org/10.1016/j.jengtecman.2020.101574

Taleb, N. N. (2014). *Antifragile things that gain from disorder*. Random House Publishing Group.

Tambe, P., Cappelli, P., & Yakubovich, V. (2019). Artificial intelligence in human resources management: Challenges and a path forward. *California Management Review, 61*(4), 15–42. https://doi.org/10.1177/0008125619867910

Westerman, G., Bonnet, D., & McAfee, A. (2014). Leading digital: Turning technology into business transformation. *Harvard Business Review Press.*. https://books.google.com/books?id=Fh9eBAAAQBAJ

Xu, X., Lu, Y., Vogel-Heuser, B., & Wang, L. (2021). Industry 4.0 and industry 5.0—Inception, conception and perception. *Journal of Manufacturing Systems, 61*, 530–535. https://doi.org/10.1016/j.jmsy.2021.10.006

Yawson, R. (2020). Strategic flexibility analysis of HRD research and practice post COVID-19 pandemic. *Human Resource Development International, 23*(4), 406–417. https://doi.org/10.1080/13678868.2020.1779169

Yawson, R. M. (2013). Systems theory and thinking as a foundational theory in human resource development—A myth or reality? *Human Resource Development Review, 12*(1), 53–85. https://doi.org/10.1177/1534484312461634

Yawson, R. M. (2015). The "wicked problem construct" for organisational leadership and development. *International Journal of Business and Systems Research, 9*(1), 67–85. https://doi.org/10.1504/IJBSR.2015.066812

Yawson, R. M., & Goryunova, E. (2025). Nested complexity: A conceptual framework for leveraging AI for sustainable organizations and human resource development. *Advances in Developing Human Resources, 27*(2/3), 91–123. https://doi.org/10.1177/15234223251335908

Yawson, R. M., Yarney, L., & Paros, A. K. B. (2025). Nested hierarchies and multi-level interventions in organizations: A comprehensive approach to organizational change. In *Envisioning HRD'S legacy in 2025—advocacy, leadership, and sustained change. Proceedings of the 32nd AHRD international research conference in the Americas.*

Index

A. M. Waite et al. (eds.), *Leading in Industry 5.0*, Management and Industrial Engineering, https://doi.org/10.1007/978-3-032-19207-3

E

F

G

P

Q

R

T

U

V

MIX
Papier aus verantwortungsvollen Quellen
Paper from responsible sources
FSC® C105338

If you have any concerns about our products,
you can contact us on
ProductSafety@springernature.com

In case Publisher is established outside the EU,
the EU authorized representative is:
Springer Nature Customer Service Center GmbH
Europaplatz 3, 69115 Heidelberg, Germany

Printed by Libri Plureos GmbH
in Hamburg, Germany